T0309085

Handbook of Mobile Robotics

Handbook of Mobile Robotics

Edited by Andy Evans

CLANRYE INTERNATIONAL
www.clanryeinternational.com

Clanrye International,
750 Third Avenue, 9ᵗʰ Floor,
New York, NY 10017, USA

Copyright © 2023 Clanrye International

This book contains information obtained from authentic and highly regarded sources. Copyright for all individual chapters remain with the respective authors as indicated. All chapters are published with permission under the Creative Commons Attribution License or equivalent. A wide variety of references are listed. Permission and sources are indicated; for detailed attributions, please refer to the permissions page and list of contributors. Reasonable efforts have been made to publish reliable data and information, but the authors, editors and publisher cannot assume any responsibility for the validity of all materials or the consequences of their use.

Trademark Notice: Registered trademark of products or corporate names are used only for explanation and identification without intent to infringe.

ISBN: 978-1-64726-675-2

Cataloging-in-Publication Data

Handbook of mobile robotics / edited by Andy Evans.
    p. cm.
Includes bibliographical references and index.
ISBN 978-1-64726-675-2
1. Mobile robots. 2. Robots. 3. Robotics. I. Evans, Andy.
TJ211.415 .H36 2023
629.892--dc23

For information on all Clanrye International publications
visit our website at www.clanryeinternational.com

Contents

Permissions

List of Contributors

Index

Preface

A mobile robot is an automatic machine that is capable of moving around in a physical environment. Mobile robotics is a subfield of robotics and information engineering concerned with the research and development of mobile robots. This field integrates the technological advancements in machine learning with physical environment, which enables the mobile robots to navigate their surroundings. Mobile robots can be classified into autonomous mobile robots and non-autonomous mobile robots. Autonomous robots do not require any external guidance for locomotion, while non-autonomous mobile robots move with the assistance of a guidance system. Mobile robots have applications in hospitals, industries and military. This book is a compilation of chapters that discuss the most vital concepts and emerging trends in the field of mobile robotics. A number of latest researches have been included to keep the readers up-to-date with the global concepts in this area of study. The book aims to serve as a resource guide for students and experts alike and contribute to the growth of the discipline.

This book is a result of research of several months to collate the most relevant data in the field.

When I was approached with the idea of this book and the proposal to edit it, I was overwhelmed. It gave me an opportunity to reach out to all those who share a common interest with me in this field. I had 3 main parameters for editing this text:

1. Accuracy – The data and information provided in this book should be up-to-date and valuable to the readers.

2. Structure – The data must be presented in a structured format for easy understanding and better grasping of the readers.

3. Universal Approach – This book not only targets students but also experts and innovators in the field, thus my aim was to present topics which are of use to all.

Thus, it took me a couple of months to finish the editing of this book.

I would like to make a special mention of my publisher who considered me worthy of this opportunity and also supported me throughout the editing process. I would also like to thank the editing team at the back-end who extended their help whenever required.

Editor

Real-Time Dynamic Path Planning of Mobile Robots: A Novel Hybrid Heuristic Optimization Algorithm

Qing Wu [1], Zeyu Chen [1], Lei Wang [1], Hao Lin [1], Zijing Jiang [1], Shuai Li [2] and Dechao Chen [1,*]

[1] School of Computer Science and Technology, Hangzhou Dianzi University, Hangzhou 310018, China; wuqing@hdu.edu.cn (Q.W.); zeno_chen@163.com (Z.C.); 181050059@hdu.edu.cn (L.W.); 171050026@hdu.edu.cn (H.L.); jzj@hdu.edu.cn (Z.J.)

[2] Department of Computing, The Hong Kong Polytechnic University, Hung Hom, Kowloon, Hong Kong 999077, China; shuaili@polyu.edu.hk

* Correspondence: chdchao@hdu.edu.cn

Abstract: Mobile robots are becoming more and more widely used in industry and life, so the navigation of robots in dynamic environments has become an urgent problem to be solved. Dynamic path planning has, therefore, received more attention. This paper proposes a real-time dynamic path planning method for mobile robots that can avoid both static and dynamic obstacles. The proposed intelligent optimization method can not only get a better path but also has outstanding advantages in planning time. The algorithm used in the proposed method is a hybrid algorithm based on the beetle antennae search (BAS) algorithm and the artificial potential field (APF) algorithm, termed the BAS-APF method. By establishing a potential field, the convergence speed is accelerated, and the defect that the APF is easily trapped in the local minimum value is also avoided. At the same time, by setting a security scope to make the path closer to the available path in the real environment, the effectiveness and superiority of the proposed method are verified through simulative results.

Keywords: hybrid optimization algorithm; mobile robot; real-time path planning; dynamic obstacle avoidance; beetle antennae search algorithm (BAS)

1. Introduction

In recent decades, path planning has had important applications in many areas, such as mobile robots [1–6], unmanned aerial vehicle (UVA) [7,8], game artificial intelligence (AI) automatic pathfinding [9], etc. [10–13]. In particular, mobile robots have been put into practical use in many industries. For example, [14] introduced the practical application of sweeping robots for daily household cleaning, [15] proposed a robotic automatic battery sorting system for improving battery sorting efficiency, and [16] introduced many applications of mobile robots in agriculture: tilling, seeding, harvesting, etc.

For mobile robots, path planning consists of finding a feasible path to the target point in the workspace [17–23]. There are many related research results for this research direction [24–28]. For example, some traditional methods are: artificial potential field (APF), probabilistic roadmap method (PRM), and rapidly-exploring random trees (RRT) [29–31]. The paper presented in [32] used a hybrid algorithm of A* and RRT for indoor navigation of unmanned aerial vehicle. In [33], they proposed a bidirectional RRT based on potentially guidance to quickly plan paths in a messy environment. Most of these methods have certain advantages in terms of speed, but the path obtained lacks optimization. Some heuristic algorithms, such as A*, D*, are widely used in industry [34–36]. Xin et al. [37] proposed an improved A* algorithm, which extends the neighborhood propagation of the standard A* algorithm. In [38], the modified A* algorithm is used to plan the path of the mobile robot. The connection between the path points of the A* algorithm is mainly modified. But these methods are mainly used in globally-known environments. There are also some intelligent optimization algorithms, such as ACO algorithm, genetic algorithm, and PSO algorithm, which also has corresponding results [39–42]. In [43], by combining artificial bee colony algorithm and evolutionary programming algorithm, they proposed a new path planning algorithm applied to path planning in two-dimensional static environment. In [44], they designed a non-dominated sorting genetic algorithm for multi-objective path planning in static environments. In [45], a heuristic PSO algorithm is proposed, which improves the PSO planning deficiency to a certain extent but only verifies the effectiveness of the algorithm in static environment. However, due to the characteristics of the group intelligence algorithm, the heuristic PSO algorithm is not particularly ideal in terms of time.

As can be seen from the above introduction, it is still necessary to study real-time dynamic path planning. It can also be seen that the research trend of path planning in recent years is also realized by a mixture of various algorithms to obtain better results.

The structure of this paper is organized as follows. Section 2 includes the definitions and formula descriptions for real-time path planning problems. In Section 3, we introduce the design ideas and implementation steps of the proposed beetle antennae search (BAS)-APF method in stages. Demonstration of the effectiveness of the algorithm through some numerical simulations for maps with dynamic obstacles is shown in Section 4. Finally, the conclusions are presented in Section 5. The main contributions of this paper are listed below.

- A novel hybrid intelligent optimization method, named BAS-APF, is proposed and applied to mobile robot dynamic path planning. The proposed method is divided into two phases. First, a feasible path from the current point to the end point is initialized based on the proposed algorithm. Then, path tracking is performed, and the path is optimized during path tracking without affecting the real-time performance of the algorithm.
- The proposed method has a good performance in both the planning time and the path length. And the advantages in planning speed are outstanding.
- The proposed method is simulative in our pre-set simulation environment maps and the real map collected by sensors, and its effectiveness and superiority are verified by comparison with other algorithms.

2. Problem Formulation

This section describes the definition of the path planning problem discussed in this article, some of the symbols used, and a brief introduction to the design basis of the proposed method. The method we proposed is called the BAS-APF. Therefore, as a background, as well as a brief introduction to the BAS algorithm and the APF algorithm, given below.

BAS is a relatively novel intelligent optimization algorithm. The design idea is inspired by the foraging behavior of beetles in nature. By optimizing the beetle's foraging process and the concept of pheromone, an optimization algorithm with faster optimization is proposed [46].

APF is a classic obstacle avoidance method. The basic idea is to abstract the motion of the robot in the surrounding environment into the motion in the artificial potential field and guide the robot movement through the force of the potential field. But this method easily falls into local extremum.

The above is a brief introduction to the basic algorithm of the proposed method. The following is a brief introduction and definition of the problem to be solved in this paper. The specific discussion in this paper is the real-time path planning problem in dynamic environment. For this problem, we need to plan a safe path from the start point to the end point as quickly as possible. The next points are taken into consideration for the hybrid BAS-APF path planning method.

- x_{sta} and x_{tar} represents the start and target points of the plan, respectively.
- Express the configuration space in this paper as a set of $Z \subset \mathbb{R}^n, n \in \mathbb{N}$ and $n \geq 2$, where n is the dimension of the configuration space. $Z_{obs} \subset Z$ is a set of obstacle areas in our configuration

space. $Z_{\text{fre}} \subset Z$ is a set of passable areas in our configuration space that can be obtained according to Formula (1). Another thing to note is that the dynamic obstacles used in this paper, the value of Z will change in real time:

$$Z_{\text{fre}} = Z \setminus Z_{\text{obs}}. \tag{1}$$

- The path obtained by the planning defined as **P** is a set of points $\{p_1, p_2, p_3 \dots p_{\max}\}$, where p_i is the i-th path point.
- Path planning can be defined as: setting the function to be optimized $f(\cdot)$ of the sampling point, and let $f(\cdot) \to 0^+$ during the planning process.
- The path optimization process can be defined as: setting the cost function $g(\cdot)$ of the path and obtaining $\min.g(\cdot)$ during each optimization process.

In general, we evaluate the path based on some criteria, such as path length, planning time, etc. Therefore, we only need to minimize $g(\cdot)$, while ensuring the planned time t. In addition, this article is mainly focused on two-dimensional environment, but this method is applicable to the case of higher dimensions.

3. Methodology

In this paper, we proposed a novel hybrid BAS-APF method for real-time path planning. This method can be applied to mobile robot path planning in both static and dynamic environments. The entire method flow is shown in Figure 1.

The proposed method is used to generate a safe path from a preset starting point to a target point. Our approach aims to solve the real-time path planning problem of mobile robots in dynamic environments. The map information of the environment needs to be processed after being collected from the sensor. Some of the map files used in this paper are in our own preset simulation map, and the other part is obtained after the real environment is collected and processed by the sensor. Below, we have a detailed introduction to the proposed method.

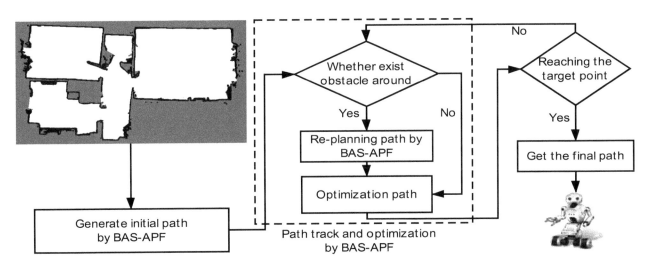

Figure 1. The schematic diagram of the proposed beetle antennae search (BAS)-artificial potential field (APF) method applied to mobile robot path planning in both static and dynamic environments.

3.1. Proposed BAS-APF Method

Path planning in a dynamic environment places high demands on the real-time and security of the planning algorithm. For these two requirements, we first avoid collisions by setting up a layered detector and a well-designed cost function. At the same time, in order to meet the requirements of real-time, the proposed method first gives the initial path, and then gradually optimizes and updates the path during the tracking process of the mobile robot.

The proposed method can be divided into two phases as a whole. The first phase is to generate the initial path, and the second phase is to trace the trajectory of the initial path and optimize the iterative update.

3.1.1. Initial Path Generation

First, according to the pre-setting starting point x_{sta}, the detection range τ, the direction vector $\mathbf{d}$ and the Formulas (2) and (3), select two candidate points x_l, x_r within the perimeter detection range. x_{sta} is the initial position of the longicorn in BAS algorithm. $\mathbf{d}$ is the direction that beetle will expore.

$$\mathbf{d} = \frac{\text{rands}\,(k,1)}{\|\text{rands}\,(k,1)\|_2}, \tag{2}$$

where $\text{rands}(\cdot)$ is a random function, each time a k-dimensional direction vector $\mathbf{d}$ is randomly selected, and k is the dimension of the space to be planned.

$$\begin{aligned} x_l &= x + \mathbf{d}\tau\eta^{i-1}, \\ x_r &= x - \mathbf{d}\tau\eta^{i-1}, \end{aligned} \tag{3}$$

where x_l and x_r are the coordinates of the candidate points. τ is the size of the detection range, and η is the attenuation rate of the detection range. As the number of iterations increases, the detection range of the agent will gradually shrink, which can also reduce the possibility of falling into local extremum to some extent.

Then, according to Formula (12), the cost function of current point, we calculate $f(x_l)$ and $f(x_r)$ to select a better candidate point, respectively. And according to the Formula (4) and (5), the mobile robot moves one step in the direction of the better point. x_{nex} is the position that the robot will move next, and its formula is as follows:

$$x_{nex} = x + s\chi^n \text{sign}(f(x_l) - f(x_r))\mathbf{d}, \tag{4}$$

where s represents the current step size, χ represents the step decay rate, and $\text{sign}(\cdot)$ is a symbolic function that extracts the sign of a real number.

$$\text{sign}\,(x) = \begin{cases} 1, & x > 0, \\ 0, & x = 0, \\ -1, & x < 0. \end{cases} \tag{5}$$

Add the coordinates to the alternate path after getting x_{nex}. The candidate path is evaluated according to the evaluation Formula (6). If the evaluation value is better, the current path is replaced; otherwise, the candidate point is re-selected. Iterate through the above process until the target point is reached:

$$g(p) = \sum_{i=1}^{N} (p_i - p_{i-1})^2, \tag{6}$$

where N is the number of path points, and i is the i-th path point. The above steps can already achieve path planning from the start point to the end point, but it may fall into local extremes, resulting in slow planning. Therefore, we speed up planning by designing appropriate cost functions and introducing artificial potential fields. A spline curve is also introduced to fit the path to achieve a smooth path effect.

3.1.2. Add Artificial Potential Field Function

Artificial potential field (APF) is a trajectory planning method based on artificial space field. The basic idea is as follows: First, the robot is regarded as a point in the space to be planned, and the artificial potential field fills the entire space to be planned. The construction method of the artificial potential field can cause the robot to be attracted to the target point away from the obstacle space. If the potential field is constructed reasonably, the artificial potential field will reach a global minimum at the end point. However, it is difficult to construct such a potential field, and even if such a potential field is constructed, it is not common to all environments. The implementation of the entire artificial potential field can be seen more intuitively through Figure 2.

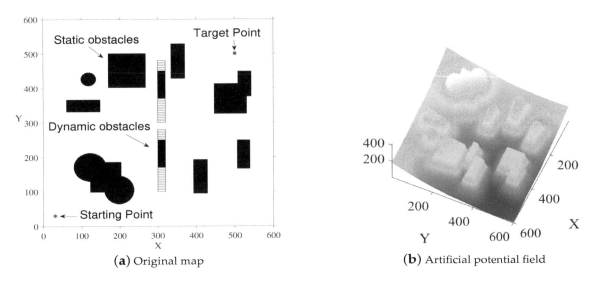

(**a**) Original map   (**b**) Artificial potential field

Figure 2. Original map and artificial potential field (APF) visualization.

Next, we will explain the construction methods of the gravitational field and the repulsive field in the artificial potential field. The gravitational field needs to satisfy the increase in the distance between the current point and the target point, so the easiest way is to set a function that increases linearly with distance. However, this is a defect in the potential field; that is, when the current point is too far from the target point, the gravity is too large. Therefore, an additional quadratic form function is set, and the threshold is set to counteract the effect of the linearly increasing gravitational pull. The specific gravitational field Formula (7) is as follows.

$$U_{\text{att}} = \begin{cases} \dfrac{1}{2}\epsilon d_g^2, & \text{if} \quad d_g < d_{\text{gra}}, \\ \epsilon d_{\text{gra}} d_g - \dfrac{1}{2}\epsilon d_{\text{gra}}^2, & \text{if} \quad d_g > d_{\text{gra}}, \end{cases} \tag{7}$$

where ϵ is the gravitational factor that determines the gravitational pull of the current point, and d_g is the distance from the current point to the target point. d_{gra} is the distance threshold of the gravitational field. The calculation of gravitation is to obtain the gradient information of the gravitational field and take the negative gradient. See Formula (8) for specific gravity settings,

$$F_{\text{att}} = \begin{cases} -\epsilon d_g, & \text{if} \quad d_g < d_{\text{gra}}, \\ -\epsilon d_{\text{gra}} + \dfrac{\epsilon d_{\text{gra}}}{d_g}, & \text{if} \quad d_g > d_{\text{gra}}. \end{cases} \tag{8}$$

Then, we need to build a repulsive field. The repulsive field needs to ensure that the farther away from the obstacle, the smaller the repulsive force, and the distance threshold d_{rep} is set in order to prevent repelling interference from distant obstacles. When the current point distance obstacle is greater than the threshold, the obstacle does not generate a repulsive force to the current point. The specific repulsive field structure is shown in Formula (9).

$$U_{\text{rep}} = \begin{cases} \dfrac{1}{2}\mu(\dfrac{1}{d_r} - \dfrac{1}{d_{\text{rep}}})^2, & \text{if} \quad d_r < d_{\text{rep}}, \\ \\ 0, & \text{if} \quad d_r > d_{\text{rep}}, \end{cases} \tag{9}$$

where μ is the repulsion factor, d_r is the distance from the current point to the obstacle, and d_{rep} is the distance threshold for the repulsion. Similarly, we can get a specific repulsion value by grading the repulsion field. The specific form is as in Formula (10),

$$F_{\text{rep}} = \begin{cases} \mu(\dfrac{1}{d_r} - \dfrac{1}{d_{\text{rep}}})^2\dfrac{1}{d_r^2}, & \text{if} \quad d_r < d_{\text{rep}}, \\ \\ 0, & \text{if} \quad d_r > d_{\text{rep}}. \end{cases} \tag{10}$$

Finally, according to the Formula (11), we combined it with attraction and repulsive force to obtain a complete force to guide the robot to the target point.

$$F = F_{\text{att}} + F_{\text{rep}}. \tag{11}$$

3.1.3. Design Cost Function

In the path planning process, due to the scope of the sensor, there may be cases where the agent is trapped at a local minimum and cannot reach the target point. This situation is called a local extremum problem. As shown in Figure 3a, in this example, the robot is stuck at a local minimum and cannot reach the target point. We avoid this problem by designing a suitable cost function. By layering the scope of the robot's detection, we add the corresponding penalty to the cost function according to the location of the obstacle, which helps to escape the local minimum. See Figure 3 for specific effects.

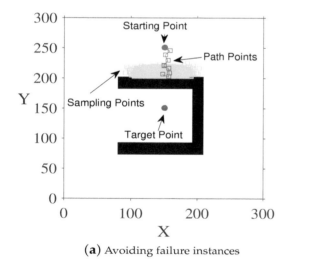

(**a**) Avoiding failure instances

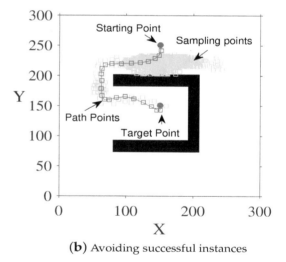

(**b**) Avoiding successful instances

Figure 3. Avoid instances of local extremum.

The green squares in Figure 3 represent sampling points, and the blue squares represent path points. Figure 3a is an example of local extremum avoidance when the detection range of the agent is not layered, and the agent is more likely to fall into local extremum. In Figure 3b, we layer the detection range of the agent. Comparing the two sub-figures, it is found that layering the detection range is beneficial to improve the local extremum avoidance ability of the proposed method.

The specific cost function is shown in Formula (12). We divide the detection range of the agent into two layers. When the agent detects obstacles $\mathbb{O}$ in the internal detection range, it adds a penalty factor α to the cost function. When the outer layer of the detection range has an obstacle, the reward coefficient β is added to the cost function. When there is no obstacle within the detection range, the resultant force of the potential field is directly used as the cost function value.

$$f(\mathbf{x}) = \begin{cases} \alpha(\mathbf{x} - \mathbf{x}_{\text{tar}})^2 + F(\mathbf{x}), & \text{if} \quad \mathbb{O} \in D_{\text{in}}, \\ \beta(\mathbf{x} - \mathbf{x}_{\text{tar}})^2 + F(\mathbf{x}), & \text{if} \quad \mathbb{O} \in D, \\ F(\mathbf{x}), & \text{if} \quad \mathbb{O} \notin D, \end{cases} \tag{12}$$

where α, β is the reward coefficient and penalty factor, $F(\mathbf{x})$ is the potential field force of the current point, D_{in} is the inner detection range, and D is the detection range.

3.1.4. Spline Curve

The initial path we get may be too tortuous. Therefore, we have adopted a more common curve fitting method: B-spline curve [47]. The B-spline curve can achieve the effect of modifying the local path without changing the shape of the entire path [48–50]. $N_{i,p}$ is a normalized B-spline basis function defined by the following Cox-deBoor recursive Formula (13),(14).

$$N_{i,0}(u) = \begin{cases} 1, & \text{if} \quad u_{i+1} \le u < u_{i+1}, \\ 0, & \text{if} \quad ; otherwise, \end{cases} \tag{13}$$

where $N_{i,0}(\cdot)$ is piecewise constant 1 or zero.

$$N_{i,p}(u) = \frac{u - u_i}{u_{i+p} - u_i} N_{i,p-1}(u) + \frac{u_{i+p+1} - u}{u_{i+p+1} - u_i} N_{i+1,p-1}(u), \tag{14}$$

where $N_{i,p}(\cdot)$ is the i-th B-spline basis function of degree p. The u_i represents the i-th item in a set of non-decreasing numbers $[u_0, u_1, ..., u_{\max}]$.

In summary, the entire path planning process is now complete, and the pseudo code that generates the initial path can be viewed in Algorithm 1. It is worth noting that we input the map environment information to simulate the overall simulation environment, but the global environment information is not used in the planning process. That is to say, in every step of path planning, the whole environment information is not used; only the environment information around the robot is used. The content of the next subsection is the tracking optimization process for the initial path.

Algorithm 1: BAS-APF design process

Input: x_{sta}, x_{tar}, Map;

Output: Path P;

Set some related parameters: step size s, detection range τ, iteration number n, gravitation factor ϵ, repulsion factor μ, convergence threshold d_{thr}, current best objective function value g_{bes};

$x_n = x_{sta}$;

for $i = 1{:}n$ **do**

> Randomly generated **d** according to Formula (2);
>
> Calculate candidate point coordinates x_l and x_r according to Formula (3);
>
> Calculate cost function value f_l and f_r of x_l and x_r according to Formula (12);
>
> Compare f_l and f_r to select a better point to join Candidate Path P_{can};
>
> Calculate the evaluation value of the P_{can};
>
> **if** $g(P_{can}) < g_{bes}$ **then**
>
> > Add x_n to P;
> >
> > Continue;
>
> **else**
>
> > Continue;
>
> **if** *Distance between* x_n *and* $x_{tar} < d_{thr}$ **then**
>
> > x_n join P;
> >
> > break;

return P;

3.2. Real-Time Path Tracking

The whole process of path tracking is roughly as shown in Algorithm 2. First, in order to speed up planning and reduce the amount of computation, the proposed method does not re-plan the path at each step of the tracking but, instead, constructs the detector during the tracking process. The detector has a radius that is twice the step size. Algorithm 1 is called to re-plan the subsequent path only if there is an obstacle in the probe's detection range. At the same time, we also set limits on the number of re-planned algorithm iterations, so although the obtained path may not be globally optimal, it guarantees the real-time performance of the algorithm.

Algorithm 2: Real-time path tracking

Input: Initial path, Map;

Output: Final path;

Set the detection range, get the number of path points in the initial path pn;

for $pn > 2$ **do**

> **if** *There are obstacles in the* x_{now} *detection range* **then**
>
> > P = Call Algorithm 1;
> >
> > $x_{now} = P_{nex}(x, y)$;
>
> **else**
>
> > $x_{now} = P_{nex}(x, y)$;
>
> $pn = pn{-}1$;

return Final path;

The above is the complete process of the proposed method.

4. Simulations

In order to verify the effectiveness of the proposed method, we performed a series of simulations in MATLAB. In this section, we selected a few representative maps to visualize the results: one is our pre-set simulation map. The other two were obtained by modeling the real room environment. The length of the following maps is in centimeters. At the same time, comparison simulations with several other algorithms were also carried out. By comparing the simulative results, the superiority of the algorithm is verified. In addition, we also made a visual process diagram of the pathfinding process of the BAS-APF method, which can visually see the entire path optimization process.

4.1. Simulation Map Results Using Virtual Map

This section mainly shows the simulation results of the proposed method on the simulated map and the visualization of the planning process. In addition, a comparison of the selection criteria of the comparison algorithm and the results of the single path planning is also introduced. The preset simulation map (Map 1) resolution is 600×600, and the map contains multiple static obstacles and two regularly moving dynamic obstacles. Taking Figure 4 as an example, in an environment of 600 cm $\times$ 600 cm, dynamic obstacles move 5 cm in each step of the robot, with 130 steps in total, i.e., 650 cm in total, with a total planning time of 0.287 s. The obstacles move with the velocity being 22.65 m/s, and the white hollow square is the trajectory of the obstacles.

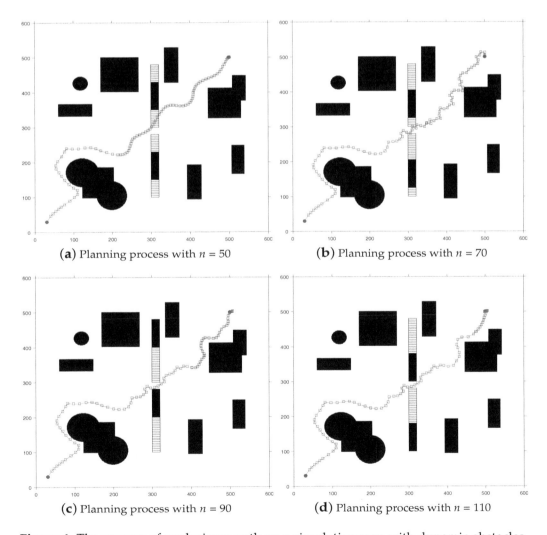

(**a**) Planning process with $n = 50$

(**b**) Planning process with $n = 70$

(**c**) Planning process with $n = 90$

(**d**) Planning process with $n = 110$

Figure 4. The process of exploring a path on a simulative map with dynamic obstacles.

A visual representation of the proposed method on Map 1 is shown in Figure 4, where the blue open squares indicate the determined path points, and the red open squares indicate the path points to be determined. The black area represents the obstacle, and the white hollow square represents the movement trajectory of the dynamic obstacle. Each square represents one step of the robot's movement, and n represents the number of iteration steps.

Regarding the choice of comparison algorithm, we make a decision based on the following considerations. Firstly, because our proposed method refers to the idea of APF algorithm, in order to highlight the superiority of hybrid algorithm, we choose APF as one of the comparison algorithms. In addition, since the BAS algorithm is an intelligent optimization algorithm, we chose a classic intelligent optimization algorithm: ACO algorithm as a comparison algorithm. Based on the comprehensive consideration of the comparison algorithm, we also selected a sampling-based algorithm: RRT. The superiority of the BAS-APF method is verified by comparison with multiple different types of comparison algorithms.

In Figure 5, we show the results of four different contrast algorithms on a simulated map. It is worth mentioning that the number of iterations of ACO is 1000, and the number of ants is 50. The blue dot represents the starting point and the red dot represents the ending point. By comparing the four subgraphs, we find that the final path obtained by our proposed algorithm is smoother and shorter, and the final path is not too close to the obstacle, and the safety factor is higher.

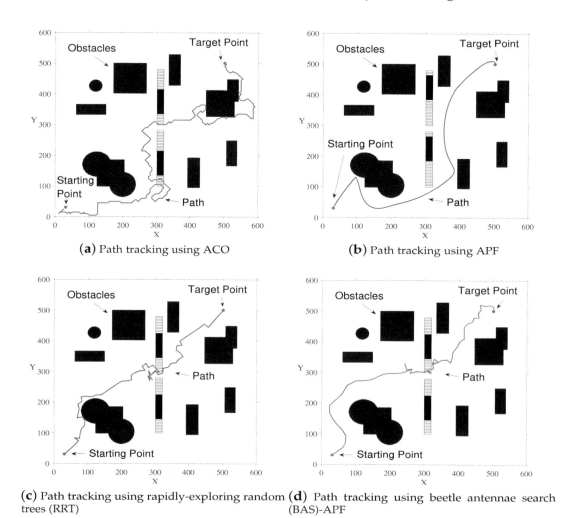

(**a**) Path tracking using ACO

(**b**) Path tracking using APF

(**c**) Path tracking using rapidly-exploring random trees (RRT)

(**d**) Path tracking using beetle antennae search (BAS)-APF

Figure 5. Simulation results of simulation map.

4.2. Simulated Results Using Real Map

Considering the practical applicability of the algorithm, in this section, we have selected two processed actual maps for simulation. Since this paper mainly verifies the obstacle avoidance ability of the proposed method in the dynamic environment, we added two dynamic obstacles that move regularly along the horizontal direction in the first real map (Map 2). And the size of Map 2 is 637×355. Similarly, the second real map (Map 3) adds two dynamic obstacles that move regularly in the vertical direction. The specification of Map 3 is 597×375.

Figure 6 shows the planning process of our algorithm on Map 2. At the same time, in order to better reflect the entire tracking process, we select the process screenshot of the re-planning. It can be seen that, since in Figure 6a,b, there are no obstacles in the surrounding area. In order to improve the tracking speed, no re-planning is performed. In Figure 6a,d, there are dynamic obstacles in the peripheral area where collisions may occur. The algorithm re-plans and successfully avoids obstacles in reaching the end point.

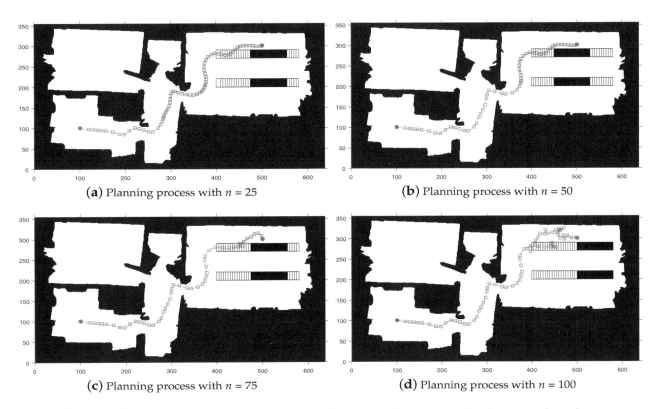

(a) Planning process with $n = 25$

(b) Planning process with $n = 50$

(c) Planning process with $n = 75$

(d) Planning process with $n = 100$

Figure 6. The process of exploring a path on the map with horizontally dynamic obstacles.

Similarly, Figure 7 shows the tracking process on Map 3 in its entirety. It can be seen from the two figures that the algorithm has good adaptability in different environments and can plan a safe path, while dynamically avoiding obstacles.

In order to more intuitively reflect the superiority of BAS-APF algorithm, we also conducted a comparison simulation on Map 2 and Map 3. Figure 8 shows the tracking results of the BAS-APF and comparison algorithm in Map 2. As can be seen from the figure, the path obtained by ACO algorithm is too close to the obstacle, and the path obtained by APF is smooth, but there are too many useless turns. The path obtained by RRT algorithm is too tortuous, and there is also a case where a useless return path is taken. The path obtained by BAS-APF takes into account the balance between path length and path smoothness.

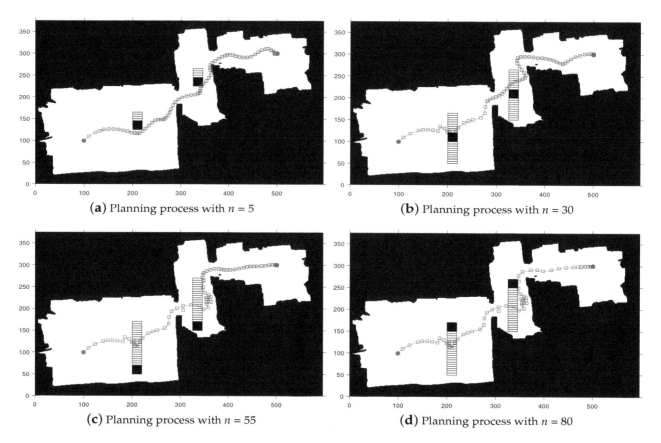

(a) Planning process with $n = 5$

(b) Planning process with $n = 30$

(c) Planning process with $n = 55$

(d) Planning process with $n = 80$

Figure 7. The process of exploring a path on a map with vertical dynamic obstacles.

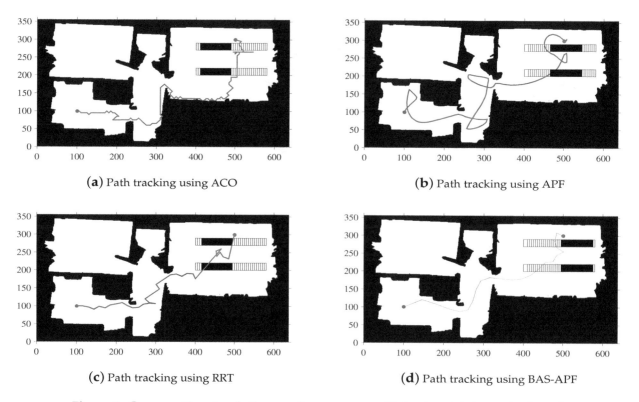

(a) Path tracking using ACO

(b) Path tracking using APF

(c) Path tracking using RRT

(d) Path tracking using BAS-APF

Figure 8. Comparative simulation results on a map with horizontal dynamic obstacles.

Similarly, in Figure 9, we show the results of planning for multiple planning algorithms under a map with vertically moving dynamic obstacles. In Figure 9a, the path planned by ACO is almost attached to the obstacle, which lacks practical application value. In Figure 9c, it can be seen that the path obtained by APF is cluttered. The path obtained by the RRT shown in sub-figure rrtMap 3 is found to have a large angle occasionally, which will affect the tracking speed of the robot during the actual tracking process. The final path obtained by our proposed algorithm, shown in Figure 9d, maintains a certain distance from the obstacle, while maintaining a certain degree of smoothness.

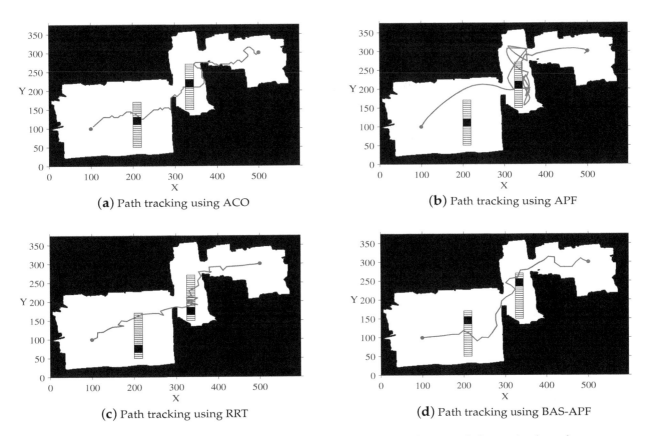

(**a**) Path tracking using ACO

(**b**) Path tracking using APF

(**c**) Path tracking using RRT

(**d**) Path tracking using BAS-APF

Figure 9. Comparative simulation results on a map with vertical dynamic obstacles.

4.3. Comparisons with Other Algorithms

This section mainly shows the comparison of the results of our method and comparison algorithms in batch simulations. The batch simulations in this paper refers to the average of 200 simulative results. We have shown the single-planning results of the comparison algorithm in Sections 4.1 and 4.2. In order to make a more rigorous and scientific comparison of the proposed method and other comparison algorithms, we conducted batch simulations on each of the three maps and summarized the numerical results. Like most researchers, we chose path length and planning time as a measure of the superiority of the algorithm. Table 1 expresses the comparison results of the path lengths of the various algorithms. By comparison, it is found that, although the advantage is not particularly obvious, the proposed method has the shortest path on all maps. Table 2 exhibits the comparison results of planning time for all algorithms. And the BAS-APF method has obvious advantages in planning time.

Table 1. Summary of comparison results of various algorithm path length.

Algorithm / Map	BAS-APF	APF	RRT	ACO
Map 1	**989.8**	1060.4	1028.6	1457
Map 2	**762.2**	1104.2	938.8	842.9
Map 3	**666.5**	1203	670.9	801.3

Table 2. Summary of comparison results of various algorithm planning time.

Algorithm Map	BAS-APF	APF	RRT	ACO
Map 1	**0.287**	1.75	1.9	8.27
Map 2	**0.276**	0.871	3.31	11.24
Map 3	**0.244**	0.804	1.79	5.14

From the above two tables, we can see that BAS-APF method has certain advantages in terms of path length and planning time, and the advantage in planning time is significant. In order to more intuitively demonstrate the advantages of the proposed method, we visualized the results.

The results of the comparison are shown in detail in Figures 10 and 11. Figure 10 is a comparison result of path lengths on respective maps. It can be seen that the performance of other algorithms is not stable, and the proposed method is more stable and optimal. Figure 11 displays the comparison of the planning time of each algorithm on the simulative map. The comparison of the three sub-figures in Figure 11 illustrates that our proposed method has significant advantages in planning time.

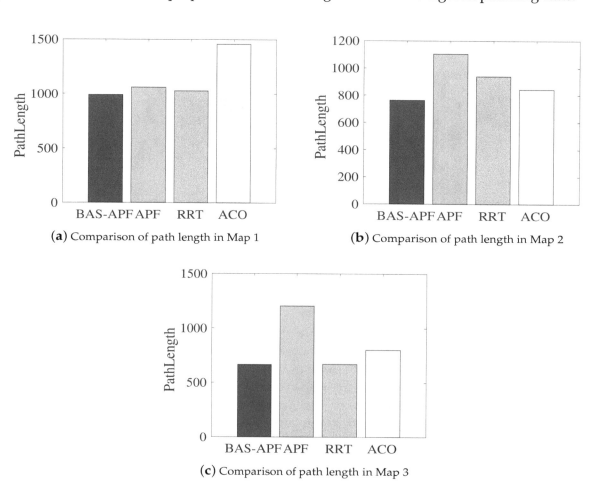

(**a**) Comparison of path length in Map 1

(**b**) Comparison of path length in Map 2

(**c**) Comparison of path length in Map 3

Figure 10. Comparison of path lengths on simulative maps.

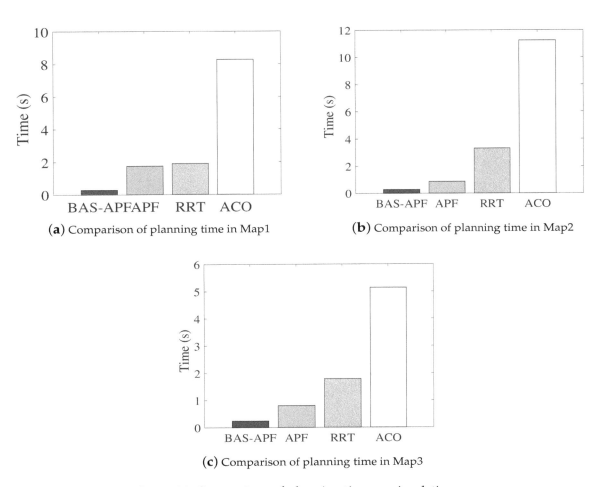

(**a**) Comparison of planning time in Map1

(**b**) Comparison of planning time in Map2

(**c**) Comparison of planning time in Map3

Figure 11. Comparison of planning time on simulative maps.

5. Conclusion and Future Work

This paper is concerned with the path planning problem of mobile robots in dynamic environments. We proposed a new two-stage hybrid method called BAS-APF to solve this problem. It could find a safe and feasible path, while ensuring real-time performance in a dynamic environment. Based on the real-time requirements of dynamic programming, the algorithm weighs path optimization and planning time, and it achieved good results in both aspects. In addition, we verified the effectiveness and robustness of the BAS-APF method through simulations. Moreover, the superiority of the proposed method was further verified by comparing it with several different types of classical path planning algorithms. The experiment test with real robots in real environments is not considered at the current state of this work. On the one hand, this is because the conditions of our laboratory are really limited, and the price of mobile robots is expensive, which we cannot afford, for the time being. On the other hand, the focus of this paper was mainly the research of algorithm, not the construction of the physical platform of mobile robot.

In the future, we intend to continue work on the following two aspects: One is that we would like to use real environment and real mobile robots for experiment tests in future work and realize the optimization on the accuracy of obstacle avoidance, so as to improve our algorithm and reflect the advantages of the proposed algorithm in the real environment. The other is to extend it to multi-objective planning so that it can simultaneously plan multiple mobile robots.

Author Contributions: Conceptualization, Q.W., Z.C., L.W. and H.L.; methodology, Q.W., Z.C., L.W. and H.L.; software, Q.W. and D.C.; validation, Q.W. and S.L.; formal analysis, Q.W., Z.C., L.W. and H.L.; investigation, Q.W., Z.C., L.W. and H.L.; resources, Q.W., Z.C., L.W. and H.L.; data curation, Q.W. and S.L.; writing—original draft preparation, Q.W.; writing—review and editing, Q.W., Z.C., L.W., H.L. and Z.J.; visualization, Q.W., Z.C., L.W. and

H.L.; supervision, Q.W., Z.C., L.W. and H.L.; project administration, Q.W., Z.C., L.W. and H.L.; funding acquisition, Z.C., L.W. and H.L. and D.C. All authors have read and agreed to the published version of the manuscript.

Acknowledgments: In this section you can acknowledge any support given which is not covered by the author contribution or funding sections. This may include administrative and technical support, or donations in kind (e.g., materials used for experiments).

References

1. Chen, D.; Li, S.; Lin, F.; Wu, Q. New Super-Twisting Zeroing Neural-Dynamics Model for Tracking Control of Parallel Robots: A Finite-Time and Robust Solution. *IEEE Trans. Cybern.* **2019**. [CrossRef]
2. Jin, L.; Li, S. Distributed task allocation of multiple robots: A control perspective. *IEEE Trans. Syst. Man Cybern. Syst.* **2018**, *48*, 693–701. [CrossRef]
3. Chen, D.; Zhang, Y.; Li, S. Tracking Control of Robot Manipulators with Unknown Models: A Jacobian-Matrix-Adaption Method. *IEEE Trans. Ind. Inform.* **2018**, *14*, 3044–3053. [CrossRef]
4. Sariff, N.; Buniyamin, N. An Overview of Autonomous Mobile Robot Path Planning Algorithms. In Proceedings of the 2006 4th Student Conference on Research and Development, Selangor, Malaysia, 27–28 June 2006; pp. 183–188.
5. Montiel, O.; Orozco-Rosas, U.; Sepúlveda, R. Path planning for mobile robots using Bacterial Potential Field for avoiding static and dynamic obstacles. *Expert Syst. Appl.* **2015**, *42*, 5177–5191. [CrossRef]
6. Chen, D.; Li, S.; Li, W.; Wu, Q. A Multi-Level Simultaneous Minimization Scheme Applied to Jerk Bounded Redundant Robot Manipulators. *IEEE Trans. Autom. Sci. Eng.* **2019**. [CrossRef]
7. Wu, Q.; Shen, X.; Jin, Y. Intelligent beetle antennae search for UAV sensing and avoidance of obstacles. *Sensors* **2019**, *19*, 1758. [CrossRef] [PubMed]
8. Roberge, V.; Tarbouchi, M.; Labonté, G. Comparison of parallel genetic algorithm and particle swarm optimization for real-time UAV path planning. *IEEE Trans. Ind. Inform.* **2013**, *9*, 132–141. [CrossRef]
9. Chen, Y.; Luo, G.; Mei, Y. UAV path planning using artificial potential field method updated by optimal control theory. *Int. J. Syst. Sci.* **2016**, *47*, 1407–1420. [CrossRef]
10. Chen, D.; Li, S.; Wu, Q. Rejecting Chaotic Disturbances Using a Super-Exponential-Zeroing Neurodynamic Approach for Synchronization of Chaotic Sensor Systems. *Sensors* **2019**, *19*, 74. [CrossRef]
11. Li, S.; He, J.; Li, Y.; Rafique, M.U. Distributed recurrent neural networks for cooperative control of manipulators: A game-theoretic perspective. *IEEE Trans. Neural Netw. Learn. Syst.* **2017**, *28*, 415–426. [CrossRef]
12. Wu, Q.; Lin, H.; Jin, Y. A new fallback beetle antennae search algorithm for path planning of mobile robots with collision-free capability. *Soft Comput.* **2019**. [CrossRef]
13. Chen, D.; Li, S.; Wu, Q.; Luo, X. Super-twisting ZNN for coordinated motion control of multiple robot manipulators with external disturbances suppression. *Neurocomputing* **2019**. [CrossRef]
14. Vourchteang, S.; Sugawara, T. Area partitioning method with learning of dirty areas and obstacles in environments for cooperative sweeping robots. In Proceedings of the 2015 IIAI 4th International Congress on Advanced Applied Informatics, Okayama, Japan, 12–16 July 2015; pp. 523–529.
15. Zhang, G.Q.; Li, L.M.; Choi, S. Robotic automated battery sorting system. In Proceedings of the 2012 IEEE International Conference on Technologies for Practical Robot Applications (TePRA), Woburn, MA, USA, 23–24 April 2012; pp. 117–120.
16. Aravind, K.R.; Raja, P.; Pérez, R.M. Task-based agricultural mobile robots in arable farming: A review. *Span. J. Agric. Res.* **2017**, *15*, 1–16. [CrossRef]
17. Tsardoulias, E.G.; Iliakopoulou, A. A review of global path planning methods for occupancy grid maps regardless of obstacle density. *J. Intell. Robot. Syst.* **2016**, *84*, 829–858. [CrossRef]
18. Jin, L.; Zhang, Y.; Li, S.; Zhang, Y. Modified ZNN for time-varying quadratic programming with inherent tolerance to noises and its application to kinematic redundancy resolution of robot manipulators. *IEEE Trans. Ind. Electron.* **2016**, *63*, 6978–6988. [CrossRef]

19. Chen, D.; Li, S.; Wu, Q.; Luo, X. New Disturbance Rejection Constraint for Redundant Robot Manipulators: An Optimization Perspective. *IEEE Trans. Ind. Inform.* **2019**. [CrossRef]
20. Chen, D.; Li, S. A recurrent neural network applied to optimal motion control of mobile robots with physical constraints. *Appl. Soft Comput.* **2019**. [CrossRef]
21. Jauwairia, N.; Malik, F.I.U.; Yasar, A.; Osman, H.; Mushtaq, K.; Muhammad, M.S. RRT*-SMART: A Rapid Convergence Implementation of RRT*. *Int. J. Adv. Robot. Syst.* **2013**. [CrossRef]
22. Noreen, I.; Khan, A.; Asghar, K.; Habib, Z. A Path-Planning Performance Comparison of RRT*-AB with MEA* in a 2-Dimensional Environment. *Symmetry* **2019**, *11*, 945. [CrossRef]
23. Iram, N.; Amna, K.; Zulfiqar, H. A Comparison of RRT, RRT* and RRT*-Smart Path Planning Algorithms. *IJCSNS Int. J. Comput. Sci. Netw. Secur.* **2016**, *16*, 20.
24. Raja, P.; Pugazhenthi, S. Optimal path planning of mobile robots: A review. *Int. J. Phys. Sci.* **2012**, *7*, 1314–1320. [CrossRef]
25. Li, S.; Zhang, Y.; Jin, L. Kinematic control of redundant manipulators using neural networks. *IEEE Trans. Neural Netw. Learn. Syst.* **2017**, *28*, 2243–2254. [CrossRef]
26. Mac, T.T.; Copot, C.; Tran, D.T.; De Keyser, R. Heuristic approaches in robot path planning: A survey. *Robot. Auton. Syst.* **2016**, *86*, 13–28. [CrossRef]
27. Chen, D.; Zhang, Y. A hybrid multi-objective scheme applied to redundant robot manipulators. *IEEE Trans. Autom. Sci. Eng.* **2017**, *14*, 1337–1350. [CrossRef]
28. Guo, D.; Xu, F.; Yan, L. New Pseudoinverse-Based Path-Planning Scheme With PID Characteristic for Redundant Robot Manipulators in the Presence of Noise. *IEEE Trans. Control Syst.* **2017**, *26*, 2008–2019. [CrossRef]
29. LaValle, S.M. Rapidly-Exploring Random Trees: A New Tool for Path Planning, 1998. Available online: http://janowiec.cs.iastate.edu/papers/rrt.ps (accessed on 11 November 2019).
30. Warren, C.W. Global path planning using artificial potential fields. In Proceedings of the 1989 International Conference on Robotics and Automation, Scottsdale, AZ, USA, 14–19 May 1989; pp. 316–321.
31. Wang, Z.; Cai, J. Probabilistic roadmap method for path-planning in radioactive environment of nuclear facilities. *Prog. Nucl. Energy* **2018**, *109*, 113–120. [CrossRef]
32. Zammit, C.; Van Kampen, E.J. Comparison between A* and RRT algorithms for UAV path planning. In Proceedings of the 2018 AIAA Guidance, Navigation, and Control Conference, Kissimmee, FL, USA, 8–12 January 2018; p. 1846.
33. Tahir, Z.; Qureshi, A.H.; Ayaz, Y. Potentially guided bidirectionalized RRT* for fast optimal path planning in cluttered environments. *Robot. Auton. Syst.* **2018**, *108*, 13–27. [CrossRef]
34. Ferguson, D.; Stentz, A. Using interpolation to improve path planning: The Field D* algorithm. *J. Field Robot.* **2006**, *23*, 79–101. [CrossRef]
35. Sudhakara, P.; Ganapathy, V. Trajectory planning of a mobile robot using enhanced A-star algorithm. *Indian J. Sci. Technol.* **2016**, *9*, 1–10. [CrossRef]
36. Loong, W.Y.; Long, L.Z.; Hun, L.C. A star path following mobile robot. In Proceedings of the 2011 4th International Conference on Mechatronics (ICOM), Kuala Lumpur, Malaysia, 17–19 May 2011; pp. 1–7.
37. Xin, Y.; Liang, H.; Du, M. An improved A* algorithm for searching infinite neighbourhoods. *Robot* **2014**, *36*, 627–633.
38. Duchoň, F.; Babinec, A.; Kajan, M. Path planning with modified a star algorithm for a mobile robot. *Procedia Eng.* **2014**, *96*, 59–69. [CrossRef]
39. Châari, I.; Koubaa, A.; Bennaceur, H.; Trigui, S.; Al-Shalfan, K. SmartPATH: A hybrid ACO-GA algorithm for robot path planning. In Proceedings of the 2012 IEEE congress on evolutionary computation, Brisbane, QLD, Australia, 10–15 June 2012; pp. 1–8.
40. Chen, D.; Li, S.; Wu, Q.; Liao, L. Simultaneous identification, tracking control and disturbance rejection of uncertain nonlinear dynamics systems: A unified neural approach. *Neurocomputing* **2019**. [CrossRef]
41. Brand, M.; Masuda, M. Ant colony optimization algorithm for robot path planning. In Proceedings of the 2010 International Conference On Computer Design and Applications, Qinhuangdao, China, 25–27 June 2010; Volume 3, p. V3-436.
42. Chen, D.; Zhang, Y. Robust Zeroing Neural-Dynamics and Its Time-Varying Disturbances Suppression Model Applied to Mobile Robot Manipulators. *IEEE Trans. Neural Netw. Learn. Syst.* **2018**, *29*, 4385–4397. [CrossRef] [PubMed]

43. Contreras-Cruz, M.A.; Ayala-Ramirez, V.; Hernandez-Belmonte, U.H. Mobile robot path planning using artificial bee colony and evolutionary programming. *Appl. Soft. Comput.* **2015**, *30*, 319–328. [CrossRef]

44. Xue, Y. Mobile Robot Path Planning with a Non-Dominated Sorting Genetic Algorithm. *Appl. Sci.* **2018**, *8*, 2253. [CrossRef]

45. Wang, H.; Zhou, Z. A Heuristic Elastic Particle Swarm Optimization Algorithm for Robot Path Planning. *Information* **2019**, *10*, 99. [CrossRef]

46. Jiang, X.; Li, S. BAS: Beetle Antennae Search Algorithm for Optimization Problems. *Int. J. Robot. Control* **2018**, *1*, 1–2. [CrossRef]

47. Catmull, E.; Clark, J. Recursively generated B-spline surfaces on arbitrary topological meshes. *Comput.-Aided Des.* **1978**, *10*, 350–355. [CrossRef]

48. Berglund, T.; Brodnik, A.; Jonsson, H.; Staffanson, M.; Soderkvist, I. Planning smooth and obstacle-avoiding B-spline paths for autonomous mining vehicles. *IEEE Trans. Autom. Sci. Eng.* **2010**, *7*, 167–172. [CrossRef]

49. Foo, J.L.; Knutzon, J.; Kalivarapu, V.; Oliver, J.; Winer, E. Path planning of unmanned aerial vehicles using B-splines and particle swarm optimization. *J. Aerosp. Inf. Syst.* **2009**, *6*, 271–290. [CrossRef]

50. Tsai, C.C.; Huang, H.C.; Chan, C.K. Parallel elite genetic algorithm and its application to global path planning for autonomous robot navigation. *IEEE Trans. Ind. Electron.* **2011**, *58*, 4813–4821. [CrossRef]

2

Integrate Point-Cloud Segmentation with 3D LiDAR Scan-Matching for Mobile Robot Localization and Mapping

Xuyou Li, Shitong Du *, Guangchun Li and Haoyu Li

College of Automation, Harbin Engineering University, Harbin 150001, China; lixuyou@hrbeu.edu.cn (X.L.); lgc_67@hrbeu.edu.cn (G.L.); 2012071507@hrbeu.edu.cn (H.L.)

* Correspondence: dushitong@hrbeu.edu.cn

Abstract: Localization and mapping are key requirements for autonomous mobile systems to perform navigation and interaction tasks. Iterative Closest Point (ICP) is widely applied for LiDAR scan-matching in the robotic community. In addition, the standard ICP algorithm only considers geometric information when iteratively searching for the nearest point. However, ICP individually cannot achieve accurate point-cloud registration performance in challenging environments such as dynamic environments and highways. Moreover, the computation of searching for the closest points is an expensive step in the ICP algorithm, which is limited to meet real-time requirements, especially when dealing with large-scale point-cloud data. In this paper, we propose a segment-based scan-matching framework for six degree-of-freedom pose estimation and mapping. The LiDAR generates a large number of ground points when scanning, but many of these points are useless and increase the burden of subsequent processing. To address this problem, we first apply an image-based ground-point extraction method to filter out noise and ground points. The point cloud after removing the ground points is then segmented into disjoint sets. After this step, a standard point-to-point ICP is applied into to calculate the six degree-of-freedom transformation between consecutive scans. Furthermore, once closed loops are detected in the environment, a 6D graph-optimization algorithm for global relaxation (6D simultaneous localization and mapping (SLAM)) is employed. Experiments based on publicly available KITTI datasets show that our method requires less runtime while at the same time achieves higher pose estimation accuracy compared with the standard ICP method and its variants.

Keywords: ICP; ground point; dynamic environments; segmentation; closed loops; 6D SLAM

1. Introduction

Localization and mapping are crucial tasks for autonomous mobile robot navigation in unknown environments. GPS is one of the widely used solutions for localization, while it suffers from some drawbacks, such as multi-path effect, latency, which limit its application in the city areas and indoor environments [1]. Pose estimation based on inertial navigation systems (INS) and visual sensors has been widely studied over recent decades. INS estimates pose information through integrating acceleration and angular velocity, which are subject to unbounded accumulation errors due to bias and noise from inertial sensors [2]. Vision-based methods can obtain robust and accurate motion estimation; however, they are vulnerable to ambient lighting conditions [3]. As an active sensor, the LiDAR is invariant to light. On the other hand, a typical 3D LiDAR, such as Velodyne VLP-16, can acquire environmental information at around 10 Hz scanning rate with a horizontal field of view (FOV) of 360 degrees and 30($\pm$15) degrees in the vertical direction. High resolution allows the LiDAR to capture a large amount of detailed information in an environment with long ranges. These advantages make LiDAR widely used in robot systems [4].

Point-cloud registration is the basis of the LiDAR-based robot system for localization and mapping. Given two adjacent point-cloud scans with different poses, the goal is to find the transformation that best aligns these two scans [5]. LiDAR-based point-cloud registration methods, also called scan-matching, are generally divided into three categories: point-based methods, feature-based methods and distributions-based methods [6]. The typical point-based method is the iterative closest point (ICP) [7], which iteratively calculates the point correspondences. In each iteration, ICP minimizes a distance function to calculate the transformation between two points clouds according to the selected closest points. Point-to-point ICP uses the point-to-point distance for calculating the closest points, which is the most popular method in the ICP family due to good performance in practice. Many variants of ICP have been proposed (point-to-plane ICP, Generalized-ICP (GICP) [8] for example) to improve the precision, efficiency and robustness of the algorithm.

In addition to point-based methods, scan-matching can also be performed by extracting some low-level attributes in a point cloud. Low-level attributes are geometric features that do not contain semantic information such as normal, intensity, planar surface, edge and some custom descriptor. These methods first automatic extract feature by geometric attributes. Then, feature points are used to find the point correspondences between scans [9]. Lidar Odometry and Mapping (LOAM) achieves a high pose estimation accuracy by extracting edge and plane features [10]. Feature points-based methods find corresponding points by extracting feature points. However, many feature descriptors are designed for applying to specific environmental conditions. Moreover, this method achieves poor estimation accuracy in environments with low geometric information, such as highways [11]. Another category is distribution-based methods. The Normal Distribution Transform(NDT) represents points as a set of Gaussian probability distribution. Instead of working directly on points, this method iteratively calculates point-to-distribution or distribution-to-distribution correspondences and minimizes a distance function in each iteration step [12].

Although there are many excellent point-cloud registration algorithms, the pose estimation suffers from error accumulation in long-term or large-scale scene [13]. A solution is to combine feature-based mapping(e.g., edge-based [14]) with point-based scan-matching algorithm which can limit this accumulation error. Simultaneous localization and mapping (SLAM) method has been shown great success over the past decade [15]. It uses the scan or image data to create a globally consistent representation. Commonly, simultaneous localization and mapping (SLAM) consists of two parts, the frontend and the backend. The frontend involves data association and sensor pose initialization. In the backend either filtering methods or pose-graph-optimization methods are used. This process aims to obtain a globally consistent mapping. Currently, graph-based optimization is the most popular technology in the SLAM field. In a graph-based network, nodes represent robot poses at different locations, and edges correspond to neighbor relations between them [16].

In this work, we propose a segment-based scan-matching framework for six degree-of-freedom pose estimation and mapping. There are four contributions in this paper. First, a 2D image-based ground points extraction method is introduced as a preprocessing step for ICP matching. LiDAR acquires a large number of 3D points while scanning the surrounding scene, which contains many ground points. The computation of the closest points is an expensive step in the standard ICP algorithm, which does not meet real-time requirements. Ground points on flat roads contain little geometric information while the standard point-to-point ICP algorithm only considers the point-to-point Euclidean distance for searching for closest points. Hence, these ground points can cause large corresponding point errors. Furthermore, Ground point extraction is also a key step in point-cloud segmentation. Secondly, point cloud after removing the ground points is then grouped into many clusters. By clustering, some outliers that do not have common attributes are removed. After this step, these different clusters are merged into a new point cloud. Compared to the original point cloud, the ground points of the new point cloud are removed and some false ground points and noise points are also filtered out. This will greatly increase the efficiency and accuracy of ICP matching. Thirdly, we extended the work of the 6D SLAM by combining the segmentation algorithm which has improved the pose estimation accuracy and efficiency with respect to the standard 6D SLAM. On this basis, a systematic evaluation based on urban, country and even highways with both absolute and relative error metrics is presented. The results validate that removing ground points can indeed improve the pose estimation accuracy of ICP and 6D SLAM. It also demonstrates that 6D SLAM performs better in pose optimization for point clouds without ground points with respect to raw point cloud. Furthermore, we also analyzed the possible error sources in different scenarios in detail. In addition, the effective evaluation of standard ICP variants and 6D SLAM in KITTI benchmark enriches the application research of these algorithms which can be considered to be a supplement to the performance of these methods in highly dynamic and complex scenarios.

The remainder of the paper is organized as follows. In Section 2, we summarize related works in ground points extraction, ICP, SLAM and segment-based localization and mapping methods. In Section 3, the proposed algorithm is described in detail. Experimental results are presented in Section 4. The paper ends with discussion in Section 5 and conclusion in Section 6.

2. Related Work

There is an increasing body of scholarly work regarding localization and mapping with LiDAR-based method. In this section, we present a brief literature review that is related to our current work.

The point cloud obtained by LiDAR contains many ground points, which poses a challenge to the classification, registration and tracking of subsequent point-cloud processing. Therefore, ground points removal is important in the point-cloud preprocessing step. The typical approach is *Bounding Box Filter* [17]. Points can be excluded from a rectangular bounding region through using this filter. The volume of the box is specified by defining the maximum and minimum coordinate values in the x,y,z directions. For example, in a coordinate system with z-axis up, ground points can be filtered out by setting the appropriate minimum coordinate value of the z-axis. This method is simple and easy to understand but parameters need to be adjusted according to different scenes and where the lidar is installed. Na et al. [18] computed local features with normal and gradient, then ground points were extracted by performing region growing. However, this method increases the computational burden which cannot meet real-time requirements. In [19], a probability occupancy grid-based ground segmentation method is proposed which can run online in different traffic scenarios. Shan et al. [20] projected point cloud onto a range image then extracted ground points by calculating the neighborhood relationship between adjacent scan lines. It is obvious that the neighborhood relationship on the 2D image is easier to calculate. At the same time, operating on 2D images enables a fast segmentation for each scan.

Point-cloud segmentation based on machine learning is also a mature research area. Pomares et al. [21] compared 23 state-of-the-art machine learning-based ground point extraction methods (e.g., SVM and KNN) through the MATLAB Classification Learner App which shows a promising ground extraction accuracy. Hackel et al. [22] developed a supervised learning framework for point-wise semantic classification. Feature descriptors considering neighborhood relationships are input into a random forest classifier, which can accurately and efficiently segment the semantic attributes of the scene, such as ground, cars, and traffic lights. However, traditional machine learning methods rely heavily on hand-crafted feature descriptors. In recent years, deep learning technologies have been applied to the field of 3D point-cloud processing. Velaset et al. [23] segmented the ground and non-ground points by employing a convolutional neural network (CNN) framework. Qi et al. [24] proposed the first deep learning network (PointNet) which directly consumes raw point clouds. PointNet differs from other frameworks in that it only uses fully connected layers to extract features instead of CNNs. Although traditional machine learning or currently popular deep learning frameworks achieves excellent segmentation performance, these supervised learning methods require pre-labeled data sets to train the model. In addition, the GPU must be used to speed up the training process. All these limit the application of learning-based methods.

Iterative closest point (ICP) is the most popular method in point-cloud matching. The most mature and widely used method is the point-to-point ICP method, which uses the point-to-point distance for calculating the closest points [7]. There are also many variants of ICP, such as point-to-plane ICP and GICP [8]. The former uses the point-to-plane distance to search for the closest points, while the latter unifies the point-to-point and point-to-plane iterative closest point algorithms into a probability framework. These two methods need to calculate the tangent plane of each point, while the point-to-point ICP algorithm performs directly on the raw points. Obviously, the point-to-point ICP algorithm is simple and more efficient. Non-geometric information has been also integrated into scan matching to improve the accuracy and efficiency of point-cloud registration. Huhle et al. [25] took color information as an additional dimension on the Normal Distributions. Although this method improves the accuracy of point-cloud registration, the color information is not often included in the raw point-cloud data. Algorithms that only deal with 3D point-cloud coordinates are obviously more general and practical.

In [26], the authors first segmented a single scan into three different semantic categories, i.e., floor, object and ceiling points. After this step, ICP-based transformation was estimated for each individual semantic segment. Since the introduction of semantic information, the corresponding points are only searched within the same semantic category, which greatly improves the possibility of searching for the correct corresponding point while at the same time accelerates the convergence of the ICP. However, the algorithm only uses the gradient relationship between adjacent points to segment the scene, which cannot satisfy complex scenes. In addition, the hand-crafted classifier cannot be extended to outdoor scenes. Inspired by [26], Zaganidis et al. [11,27] integrated semantic information into Normal Distributions Transform (NDT) instead of ICP for point-cloud registration. The method differs from [26] in the semantic segmentation. The method in [27] partitioned the point cloud into sharp edges and planar surface patches according to smoothness while deep learning framework is applied to semantic segmentation in [11]. However, deep learning requires large-scale training data sets, which limits its application in the field of point-cloud registration.

SLAM technology has been widely applied to the robot community in recent years. In the backend, either filter-based methods or pose-graph-optimization methods are used. This process aims to obtain a globally consistent mapping. There are many popular techniques in filter-based methods, such as the Extended Kalman filter [28] and Particle Filters [29]. The differences between these methods mainly focus on sensors, dynamic modes and state-estimation algorithms [30]. However, the main drawback is that the filtering strategy updates probability distributions through time without the convergence guarantee, and suffers from computational complexity or large amounts of particles [31]. In cases where it is difficult to obtain uncertainties and sensor models, these values are often guessed by researchers.

Pose-graph-optimization methods currently have greater advantages in the SLAM over filtering-based methods. Borrmann et al. [32] proposed a 6D SLAM framework that uses ICP to register all scans until convergence. Once closed loops are detected, a GraphSLAM for global relaxation is employed. This algorithm does not require additional point features such as normal, nor does it require high-level features. In [20], a lightweight and real-time six degree-of-freedom pose estimation framework called LeGO-LOAM, is presented. LeGO-LOAM first projects the point cloud into a 2D image. Then, the point cloud is further segmented into the ground and non-ground points. Feature point extraction and matching and error functions are used to estimate six degree-of-freedom pose. In addition, a pose-graph SLAM is also integrated into to obtain more accurate results. LOAM does achieve high pose estimation accuracy at the same time meeting real-time operations. However, feature points-based methods may lead to inaccurate registration and large drift in environments with low geometric information, such as highways.

3. Methodology

3.1. System Overview

The architecture of the system is shown in Figure 1, which can be divided into six main modules: point reduction, point-cloud projection, ground points removal, segmentation, ICP and pose-graph optimization (6D SLAM). We first apply an octree-based data structure to reduce the 3D point cloud. An image-based ground point removal method is then introduced. The point cloud after removing the ground point is further segmented into disjoint sets. After this step, a standard point-to-point ICP is applied to calculate the six degree-of-freedom transformation between consecutive scans. In addition, once closed loops are detected in the environment, a 6D graph-optimization algorithm for global relaxation is employed. Our system features a right-handed coordinate system with the z-axis pointing upwards and the x-axis in forward direction. The detailed algorithm principle of each modules will be introduced in the following sections.

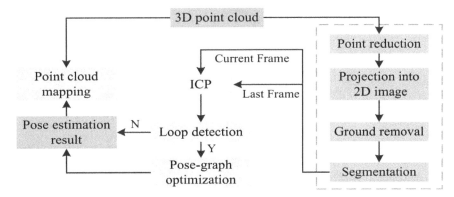

Figure 1. Overview of the proposed LiDAR localization and mapping system architecture.

3.2. Point Reduction

The high resolution of the LiDAR acquires large-scale data when scanning. For example, Velodyne HDL-64E can generate 1.8 million range measurements per second. Therefore, to process a huge amount of 3D data points efficiently, point-cloud storage and reduction are crucial steps. Octree is a spatial data structure used to describe three-dimensional space which enables efficient storage, compression and search of 3D point cloud. As shown in Figure 2, 3D space is assumed to be a cube and the root node represents a cubic bounding box that stores all points of a point cloud, i.e., 3D coordinates and additional attributes such as reflectance. The octree divides the space into 8 parts, and each node is a part. The sum of the volumes represented by the eight child nodes is equal to the volume of the parent node. In this work, we use an octree-based point-cloud reduction method which is described in detail in [33].

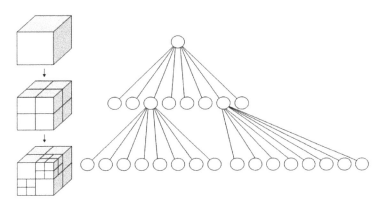

Figure 2. The sparse datastructure of octree.

3.3. Projection into 2D Range Image

Since the subsequent ground points removal and segmentation algorithms are based on 2D range images, we first need to obtain the cylindrical range image. The widely used LiDAR such as the Velodyne family acquires the environmental information by horizontal and vertical scanning. For example, the 16-channel VLP-16 has a horizontal field of view of 360 degrees and 30($\pm$15) degrees for the vertical field of view. If the horizontal azimuth angle θ_h is set to 0.2°and we know from the datasheet that the vertical resolution θ_v is 2°, the corresponding resolution of 2D range image is 1800 by 16. Given a point $P = (x, y, z)$, the corresponding 2D range image is calculated by:

$$h = \arctan(\frac{x}{y}) * 57.3$$
$$v = \arctan(\frac{z}{\sqrt{x^2 + y^2}}) * 57.3$$
$$r = \frac{v + v_b}{\theta_v} \tag{1}$$
$$c = \lfloor \frac{h - 90}{\theta_h} \rfloor + \frac{h_s}{2}$$
$$d = I(r, c) = \sqrt{x^2 + y^2 + z^2}$$

where h and v are the horizontal and vertical angles of P in the LiDAR coordinate system, cf. Figure 3. v_b represents the maximum vertical field of view of the LiDAR. For the VLP-16, $v_b = 15$. In addition, $h_s = 1800$ refers to the horizontal resolution while $\lfloor \rfloor$ indicates the corresponding number is rounded down. By projecting, each 3D coordinate point P is represented by a unique pixel $I(r, c)$of the image. For the 2D range image, each row of the range image has the same vertical angle θ_v, i.e., the same scan lines. Each column indicates points with the same horizontal angle and different scan lines.

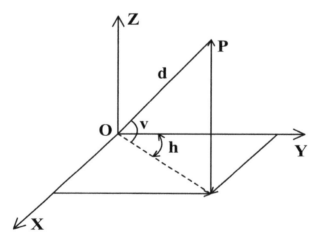

Figure 3. Velodyne LiDAR coordinate system.

3.4. Ground Removal

Ground point extraction is a key step in point-cloud processing. In this part, we adopt an image-based ground point extraction method which is similar to [34]. Liu et al. [13] used Equation (2) to extract ground points which is based on an intuitive understanding that the differences in the z-direction between two adjacent points from the same column is much smaller than x and y directions, When the LiDAR scans the ground. However, this assumption is applicable only for ground vehicles. For 3D mobile robots, such as drones, the sensor attitude with respect to the ground must be considered. Moreover, the algorithm traverses points of m rows from the bottom of the image. If α_i is smaller than a threshold θ, This corresponding point is considered to be the ground point. However, the user must set different m values and threshold θ according to the installation height of the LiDAR.

$$\alpha_i = \arctan \left(\frac{\delta^c_{z,i}}{\sqrt{\delta^c_{x,i} * \delta^c_{x,i} + \delta^c_{y,i} * \delta^c_{y,i}}} \right) \qquad (2)$$

where $\delta^c_{x,i}$, $\delta^c_{y,i}$, $\delta^c_{z,i}$ indicate the differences in x-, y-, and z-direction between two adjacent points from the cth column.

Therefore, in this work, we introduce a more robust and efficient approach. Algorithm 1 depicts the algorithm that we use to extract ground points. First, the 2D range image is converted to an angle image based on Equation (2) (line 2). After conversion, each pixel of the angle image is represented by the corresponding α_i. Next, a Savitzky–Golay filtering algorithm [35] is applied to the angle image (line 3). This aims to smooth the data and remove noise. After this step, we traverse each pixel from the bottom left of the filtered image. Whenever a non-labeled pixel is encountered, a breadth-first search (BFS) based on the pixel is carried out (line 7–15). The basic idea is BFS starts from the pixel, and find 4 neighborhood from the up, down, left, and right pixels. If the difference between the pixel and its 4 neighborhoods falls into the threshold γ, the pixel is added to the queue, i.e., it is assigned to the ground point (line 12–15). Please note that Label=1 refers to the ground point class. This procedure stops utile the whole connected component receives the same label. Intuitively, this algorithm starts from the bottom left of the image which is generally considered to be a ground point. We assign a label to this point (line 11). BFS is then employed to continuously expand the search until all points belonging to the same label (Label=1) are found. This algorithm traverses all points of the entire image, hence we do not have to manually select m for different hardware platforms.

Algorithm 1 Ground points extraction

Input: Range image R, ground angle threshold γ, Label=1, windowsize
Output: L
1: **function** GROUNDPOINTEXTRACTION(R,Label,L,γ,windowsize)
2:     A=CreateAngleImage(R)
3:     S=ApplySavitskyGolaySmoothing(A,windowsize)
4:     L=zeros($R_{rows} \times R_{cols}$)
5:     **for** r=S_{rows}; r $\geq$ 1; $r --$ **do**
6:         **for** $c = 1; c \leq S_{cols}; c ++$ **do**
7:             **if** $L(r,c) = 0$ **then**
8:                 queue.push(r,c)
9:                 **while** queue is not empty **do**
10:                     r,c=queue.top()
11:                     L(r,c)=Label
12:                     **for** $(r_n, c_n) \subset Neighborhood(r,c)$ **do**
13:                         g=$S(r,c) - S(r_n, c_n)$
14:                         **if** $fabs(g) < \gamma$ **then**
15:                             queue.push(r_n, c_n)
16:                         **end if**
17:                     **end for**
18:                     queue.pop()
19:                 **end while**
20:             **end if**
21:         **end for**
22:     **end for**
23: **end function**

3.5. Segmentation

To further remove noise points and outliers, we use the algorithm in [34] to segment the range image after removing the ground point. The idea of this algorithm is similar to the ground points extraction. The method of deciding whether points belong to the same label is shown in Figure 4. As right figure in Figure 4 depicts, β can be used to segment the point cloud if the appropriate threshold ϵ is set. we assume the one with a relatively long distance between OA and OB is d_1 ($\|OA\|$) and the other is d_2 ($\|OB\|$), then, β is calculated:

$$\beta = \arctan \frac{\|BH\|}{\|HA\|} = \arctan \frac{d_2 \sin \theta}{d_1 - d_2 \cos \theta} \tag{3}$$

where θ is the horizontal azimuth angle or vertical resolution which is described in Section 3.3.

The pseudocode of the algorithm is presented in Algorithm 2. The algorithm differs from Algorithm 1 in input images, the criteria for classification, and the number of labels. R^{ng} represents the image which is directly projected by the point cloud but does not include the ground points. Since the ground point is a category, Algorithm 1 has only one label. However, the segmentation includes many categories. Therefore, the label is automatically incremented by 1 when a cluster is completed.

Algorithm 2 Segmentation

Input: Range image R^{ng}, segmentation threshold ϵ, Label=1

Output: L

1: **function** LABELRANGEIMAGE(R^{ng},Label,L,ϵ)

2:      L=zeros($R^{ng}_{rows} \times R^{ng}_{cols}$)

3:      **for** $r = 1; r \le R^{ng}_{rows}; r + +$ **do**

4:          **for** $c = 1; c \le R^{ng}_{cols}; c + +$ **do**

5:              **if** $L(r,c) = 0$ **then**

6:                  queue.push(r,c)

7:                  **while** queue is not empty **do**

8:                      (r,c)=queue.top()

9:                      L(r,c)=Label

10:                     **for** $(r_n, c_n) \subset Neighborhood(r,c)$ **do**

11:                         $d_1 = \max(R^{ng}(r,c), R^{ng}(r_n,c_n))$

12:                         $d_2 = \min(R^{ng}(r,c), R^{ng}(r_n,c_n))$

13:                         **if** $\arctan \frac{d_2 \sin \alpha}{d_1 - d_2 \cos \alpha} > \epsilon$ **then**

14:                             queue.push(r_n, c_n)

15:                         **end if**

16:                     **end for**

17:                     queue.pop()

18:                 **end while**

19:                 Label=Label + 1

20:             **end if**

21:         **end for**

22:     **end for**

23: **end function**

Please note that after the segmentation algorithm is implemented, the 2D image grouped into many sub-images can be easily converted into sub-segments which are represented by 3D coordinate points. We aim to use the segmentation algorithm to remove noise and outliers. Therefore, these different clusters are then merged into a new point cloud. Compared to the original point cloud, the ground point of the new point cloud is removed and some noise and outlier points have also been filtered out. Finally, a standard point-to-point ICP algorithm is then applied to calculate the six degree-of-freedom transformation between consecutive scans. The specific calculation process will be described in the next section.

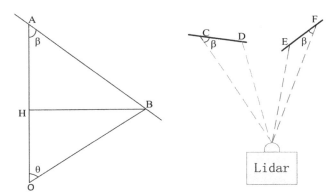

Figure 4. Left: O represents the center of the LiDAR while OA and OB are two laser beams that also represent the distance between the obstacle and the laser sensor. If $\beta > \epsilon$, where ϵ is a threshold, the two points are considered to be the same cluster. Right: An intuitive example which illustrates the relationship between the β and whether the two points belong to the same object. The blue dotted line is an example that shows C and D belong to the same object and β is larger than the angle in the red dotted line where E and F are from two different objects.

3.6. ICP and 6D SLAM

In this part, point-to-point ICP and a globally consistent scan-matching algorithm are used to calculate six degree-of-freedom pose. In addition, we also compared our result with the standard point-to-planar ICP method and *Bounding Box Filter*-based point-to-point ICP that first removes the ground point by *Bounding Box Filter* and then performs ICP algorithm. The concept of ICP is simple: given an initial guess, it calculates the point correspondences iteratively. Please note that an initial guess is not strictly needed when performing ICP-based scan-matching for LiDAR-based odometry. In fact, the ICP algorithm can be run assuming that the initial rotation and translation are zero as soon as the sensor dynamics is not too fast with respect to the frame rate. In each iteration, ICP minimizes a distance function to calculate the transformation between two points clouds according to the selected closest points. The distance function of point-to-point ICP is defined as:

$$E(\mathbf{R}, \mathbf{t}) = \sum_{i=1}^{N_m} \sum_{j=1}^{N_d} \|\mathbf{s}_i - (\mathbf{R}\mathbf{d}_j + \mathbf{t})\|^2 \tag{4}$$

where N_m and N_d are the number of points in the source point cloud **S** and target point cloud **D**.

Point-to-plane ICP minimizes the sum of the squares of the distances between the source points and the tangent plane of the target points. This specific formula is as follows:

$$\mathbf{T}_{opt} = \arg \min_{\mathbf{T}} \sum_{i=1}^{N} ((\mathbf{T}\mathbf{s}_i - \mathbf{d}_i)\mathbf{n}_i)^2 \tag{5}$$

where N is the number of points, and n_i is the normal vector corresponding to the target point. T is the rigid transformation between the source and the target points. Compared with the point-to-point ICP, point-to-plane ICP calculates the tangent plane of the point. Therefore, it can achieve better results in environments with low geometric information. However, it needs to calculate the normal vector, which will reduce the efficiency. Hence, point-to-point ICP is used in this work.

ICP obtains a trajectory by calculating the pose between two adjacent scans and then constantly updating it. However, the pose estimation suffers from error accumulation in the long-term or large-scale scene. To address this issue, the pose estimation result of the ICP is input into the 6D SLAM framework, i.e., globally consistent scan-matching [32], once closed loops are detected. It is available in *3DTK-The 3D Toolkit* [36]. 6D SLAM is similar to the point-to-point ICP method but taking into account all scans instead of only two adjacent scans. It solves for all poses at the same time and iterates like in the original ICP. It is actually a pose-graph-optimization method and uses the Mahalanobis distance to represent the global error of all poses. The specific formula is:

$$
\begin{aligned}
W &= \sum_{j\to k} (\bar{\mathbf{E}}_{j,k} - \mathbf{E}_{j,k})^T \mathbf{C}_{j,k}^{-1} (\bar{\mathbf{E}}_{j,k} - \mathbf{E}_{j,k}) \\
&= \sum_{j\to k} (\bar{\mathbf{E}}_{j,k} - (\mathbf{X}_j - \mathbf{X}_k))^T \mathbf{C}_{j,k}^{-1} (\bar{\mathbf{E}}_{j,k} - (\mathbf{X}_j - \mathbf{X}_k))
\end{aligned}
\tag{6}
$$

where j and k represent scans of the SLAM graph, $\mathbf{E}_{j,k}$ is the linearized error metric and $(\bar{\mathbf{E}}_{j,k}, \mathbf{C}_{j,k})$ is the Gaussian distribution. $\mathbf{X}_j$ and $\mathbf{X}_k$ are two connected nodes in the graph which represent the corresponding poses. we give only a brief overview here and a detailed description is given in [32].

4. Experimental Results

4.1. Experimental Platform and Evaluation Method

To evaluate the performance of the proposed algorithm, we test our method in the KITTI benchmark [37]. The datasets are acquired with a vehicle equipped with a Velodyne HDL-64E laser scanner, stereo color video cameras and a high accuracy GPS/INS for ground truth. It contains 11 sequences training data sets, which provide ground truth and 11 test data sets without ground truth. These sequences include three types of environments: urban with buildings around, the country on small roads with vegetations in the scene, and the highway where roads are wide, and the vehicle speed is fast. The HDL-64E has a horizontal FOV of 360°and 26.9°Vertical FOV with 64 Channels whose range reaches 120 m. All data in our experiments are processed on a desktop computer with an i7-7700 3.60 GHz CPU and both algorithms are implemented in C++ and executed in Ubuntu Linux.

The proposed method is evaluated using the absolute metric proposed in [38] and KITTI metric [37], respectively. The absolute metric computes absolute root-mean-square error (RMSE) of translation rotation errors according to Equation (7) to (11)

$$
\Delta \mathbf{T}_{abs,i} = \begin{pmatrix} \Delta \mathbf{R}_{abs,i} & \Delta \mathbf{t}_{abs,i} \\ \mathbf{0} & 1 \end{pmatrix} = \mathbf{T}_{r,i} \mathbf{T}_{e,i}^{-1},
\tag{7}
$$

where $\mathbf{T}_{r,i}$ and $\mathbf{T}_{e,i}$ represent the pose matrices of ground truth and estimated pose, respectively in ith frame. Furthermore, the absolute translation error $e_{abs,i}$ and rotation error $\Delta\theta_{abs,i}$ are computed by Equation (8) and Equation (9), respectively.

$$
e_{abs,i} = \|\Delta \mathbf{t}_{abs,i}\|
\tag{8}
$$

$$
\Delta\theta_{abs,i} = f_\theta(\Delta \mathbf{R}_{abs,i}),
\tag{9}
$$

where $\| \cdot \|$ indicates Euclidean metric. Then the root-mean-square error(RMSE) of absolute translation errors and absolute rotation errors are calculated by

$$\sigma_t = \sqrt{\frac{1}{n+1} \sum_{i=0}^{n} e_{abs,i}^2} \qquad (10)$$

and

$$\sigma_r = \sqrt{\frac{1}{n+1} \sum_{i=0}^{n} \Delta\theta_{abs,i}^2} \qquad (11)$$

4.2. Results

In this section, we analyze the results of four modules including ground point removal, segmentation, ICP and 6D SLAM. To test the robustness and accuracy of the proposed method to different scenarios, the results of four typical data sequences including urban with buildings around, the country on small roads with vegetations in the scene and a highway where roads are wide, and the vehicle speed is fast are presented.

4.2.1. Ground Points Removal

We compared *Bounding Box Filter* with the ground point extraction method used in this paper, i.e., Algorithm 1. For *Bounding Box Filter*, points can be excluded by designing a rectangular bounding region. The box is specified by defining the maximum and minimum coordinate values in the x,y,z directions. Ground points can be filtered out by setting the appropriate minimum coordinate value of z-axis. According to the installation height and range of the Velodyne HDL-64E laser scanner, the box is set as:

$$-120 < x < 120$$
$$-120 < y < 120 \qquad (12)$$
$$-1.1 < z < 120$$

where x, y, z refer to 3D point coordinates and the unit is the meter.

As for Algorithm 1, ground angle threshold γ and windowsize are set to 5 degrees and 7, respectively. Here, we only qualitatively compare the accuracy of ground point extraction. Two scenarios, including the urban and the highway, are selected to test our algorithm. Please note that Figure 5a,b are the visual inspection from *Bounding Box Filter*, where only non-ground points are presented. For our method, i.e., Figure 5c,d, ground points and non-ground points are displayed in different colors, where the yellow part indicates the ground point and the pink is non-ground points. As shown in Figure 5, two methods have achieved similar accuracy. However, when the same parameters of *Bounding Box Filter* are applied in sequence 01, a large number of ground points are not removed cf. Figure 6a,b.

To help identify ground points, the corresponding real scene is shown in Figure 7. If we want to use the box filtering method to remove all the ground points of Figure 6, the parameters must be changed. Instead, our method achieves the desired results with the same threshold, although some ground points have not been completely removed (blue arrows in Figures 5c,d and 6c,d). The next section will show that these outliers will be removed after using segmentation.

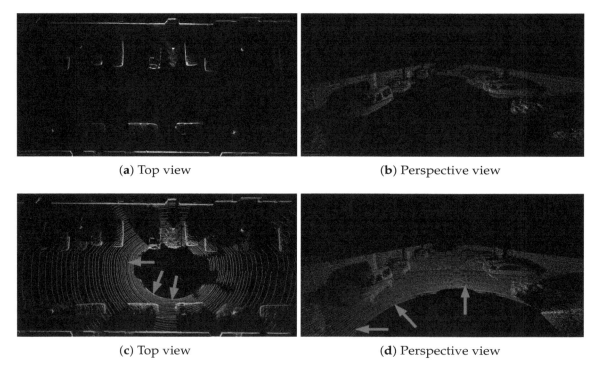

(a) Top view   (b) Perspective view

(c) Top view   (d) Perspective view

Figure 5. Comparison between *Bounding Box Filter*-based method and the algorithm used in this paper. The above images are a certain frame point cloud of sequence 07 which is collected on the urban. (a) ground removal results using the box filtering method from a bird's eye view. (b) the corresponding perspective views. Please note that only non-ground points are displayed in (a,b). (c,d) are the results from our method. The yellow part indicates the ground point and the pink color are non-ground points.

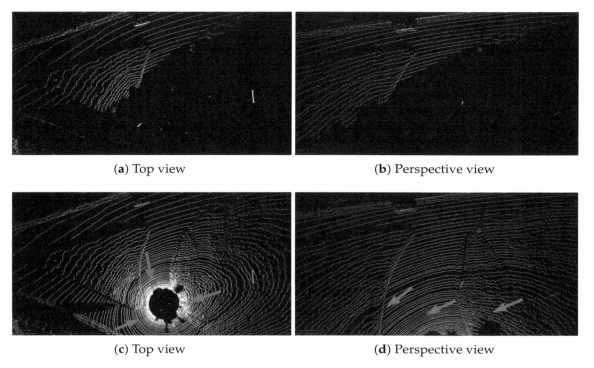

(a) Top view   (b) Perspective view

(c) Top view   (d) Perspective view

Figure 6. The same as Figure 5, but these images are from sequence 01 which is acquired on a highway.

Figure 7. A photo showing the scene corresponding to the point clouds seen in Figure 6.

4.2.2. Segmentation

To further remove noise points and outliers, we use the method in [34] to segment the range image after removing the ground point. Please note that after the segmentation algorithm is implemented, the 2D image grouped into many sub-images can be easily converted into sub-segments which are represented by 3D coordinate points. By using the segmentation algorithm, those points with the same attributes are assigned to the same labels and the entire point cloud is divided into many sub-segments. We aim to use the segmentation algorithm to remove noise and outliers. Therefore, these different clusters are then merged into a new point cloud. The clusters with fewer than 30 points will be discarded which are most likely to be noise and outliers. Figure 8 shows visual results after running segmentation algorithm. Compared to Figure 5c,d and Figure 6c,d, the ground points of the new point cloud are removed and some false ground points (blue arrows) have also been filtered out.

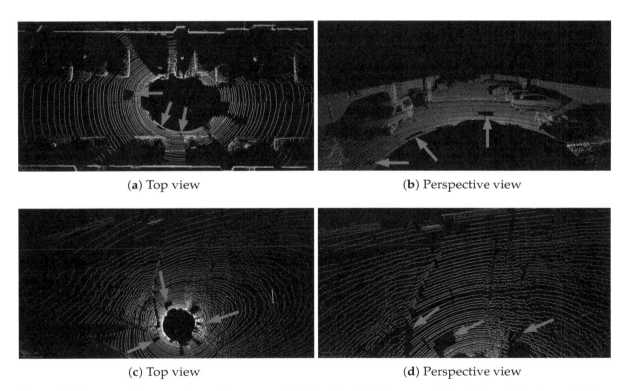

(**a**) Top view   (**b**) Perspective view

(**c**) Top view   (**d**) Perspective view

Figure 8. Some results come from the new point cloud, which is merged by different clusters. (**a**) The visual inspection of sequence 07 from a bird's eye view. (**b**) The visual inspection of sequence 07 from perspective views. (**c,d**) are the corresponding results from se01.

4.2.3. Comparison of Trajectory Results

In this part, four different scenarios from the KITTI dataset are selected to test the robustness, accuracy and efficiency of the proposed method. We compare the proposed method (SE+PTP) with the standard point-to-point ICP algorithm (PTP), the *Bounding Box Filter*-based ICP method (BBF+PTP), and the point-to-surface ICP method (PTS). Here, BBF+PTP-based method refers to a method that first uses Bounding Box Filter to remove ground points which is then input a standard point-to-point ICP framework. Furthermore, once closed loops are detected, 6D SLAM is used to improve pose estimation accuracy.

Figure 9 compares the 2D trajectory and 3D absolute translation and rotation error of the sequence 01 which is collected on the motorway. As Figure 9a(1) shows, SE+PTP achieves similar performance to BBF+PTP on the first part of the sequence and is slightly better than PTP and PTS. This shows that ICP can find the correct corresponding points with higher probability by removing ground points. On the second part, i.e., Figure 9a(2), SE+ICP is inferior to others but keep similar performance to PTP and PTS. Figure 10a shows the visual inspection corresponding to the Figure 9a(2). PTP and PTS exhibits low-precision in Figure 9a(3) while SE+PTP still maintained within a certain accuracy which can also be seen from Figure 9b(1). Figure 10b shows an example of a point cloud corresponding to Figure 9a(3). Figure 10b contains less geometric and semantic information relative to Figure 10a. This causes PTP and PTS to fail here. Although BBF+PTP does not suffer from large errors here, it finally failed to estimate the pose due to the lack of geometric and semantic information which caused the BBF+PTP-based algorithm to think that the vehicle stayed in place without moving. In contrast, SE+PTP is more robust, which is mainly due to the introduction of the segmentation algorithm. However, our method still cannot accurately estimate the pose of se01. Because there are too many moving vehicles running with high speed.

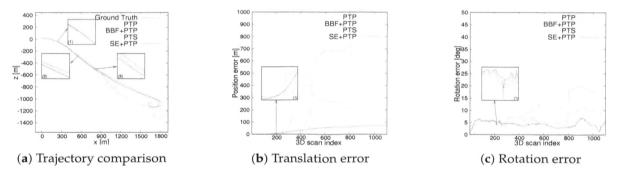

(a) Trajectory comparison   (b) Translation error   (c) Rotation error

Figure 9. Trajectory and translation as well as rotational error comparison of seq01. (**a**) the trajectory comparison between different ICP. (**b**) the translation error. (**c**) the rotational error.

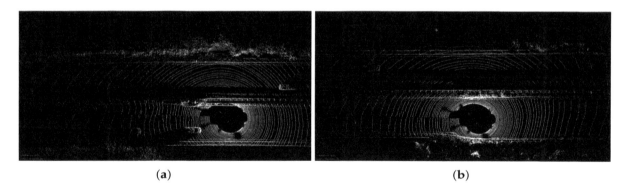

(a)   (b)

Figure 10. (**a**) The visual inspection corresponding to the Figure 9a(2). (**b**) The visual inspection corresponding to the Figure 9a(3).

The absolute translation and rotation error of corresponding sequences to ground truth are given in Table 1, which shows SE+PTP is superior to other methods. An intuitive conclusion is drawn from Table 1 is both BBF+ICP and SE+ICP have improved the accuracy of pose estimation relative to the standard ICP method. This is the result that the segmentation algorithm removes those ground points and noise points. Table 1 also demonstrates the performance of 6D SLAM in different scenarios. 6D SLAM does improve the accuracy of point-to-point ICP alone, cf. PTP and PTP+6DSLAM of se09 in Table 1. The reason is 6D SLAM taking into account all scans instead of only two adjacent scans which limits this accumulation error. Although the position accuracy of PTP+6DSLAM in se14 is similar to PTP, the rotation error has been eliminated. However, PTP+6DSLAM shows worse results than PTP in the urban scene (se07). This is because se07 contains a lot of dynamic vehicles which can cause larger error. The performance of standard 6D SLAM may degrade in a high dynamic environment. In contrast, since SE+PTP+6DSLAM includes a segmentation algorithm, which removes the noise points caused by dynamic objects to a certain extent. Consequently, SE + 6DSLAM achieves excellent results. Another issue that must be noted is the performance of PTS+6DSLAM degrades compared to PTS. This problem is caused by the 6DSLAM algorithm itself. Since 6D SLAM is similar to the point-to-point ICP method but taking into account all scans instead of only two adjacent scans. It solves for all poses at the same time and iterates as in the original ICP. Hence, 6D SLAM is more suitable for point-to-point ICP.

In addition, we also compared the execution time of the programs in Table 2. Compared with PTP (point-to-point ICP), point-to-plane ICP (PTS) needs to calculate the normal vector, which increases the computational. In addition, SE+PTP largely reduces the calculation time compared to standard point-to-point ICP (PTP) due to the ground point removal. For se01, although SE+PTP takes more time than BBF+ICP, cf. se01, the accuracy is much higher. In summary, this experiment of se01 shows that the proposed method can assist ICP to estimate the pose more accurately and efficiently in an environment with low geometric information.

Table 1. Results of the proposed method (SE+PTP) compared with point-to-point ICP (PTP), BBF+ICP and point-to-plane ICP on the KITTI dataset using absolute metric. *+6D SLAM refers to 6D SLAM is used in the corresponding method. Since sequences 01 does not contain closed loops, 6D SLAM is not employed on sequences. n.a. in this table indicates the corresponding method is not available. t_{err} represents RMSE(root-mean-square error) of absolute translation error, while r_{err} represents RMSE of absolute rotation errors.

Method	se01(Highway)		se07(Urban)		se09(Urban+Country)		se14(Country)	
	t_{abs} [m]	r_{abs} [deg]	t_{abs} [m]	r_{abs} [deg]	t_{abs} [m]	r_{abs} [deg]	t_{abs} [m]	r_{abs} [deg]
PTP	183.8600	6.7351	7.0136	3.4204	61.6887	8.8175	12.3578	3.1265
BBF+PTP	543.1686	12.0947	6.7612	3.7937	30.5351	4.8372	4.3729	3.0239
PTS	261.8181	22.4791	10.9247	3.6953	**27.0114**	5.0372	**2.3129**	**2.0502**
SE+PTP	**49.0841**	**4.4752**	**3.8886**	**2.7021**	27.4586	**4.7811**	5.5984	4.4430
PTP+6DSLAM	n.a.	n.a.	30.5522	15.9400	39.4745	5.4160	12.6070	1.6986
BBF+PTP+6DSLAM	n.a.	n.a.	8.6707	3.2422	24.7700	4.5936	3.5380	1.6498
PTS+6DSLAM	n.a.	n.a.	30.5522	15.9400	39.4745	5.4160	12.6070	1.6986
SE+PTP+6DSLAM	n.a.	n.a.	**3.0454**	**2.6856**	**18.5825**	**4.1382**	**1.0114**	**0.8563**

Table 2. Total program running time of every sequence with different methods. Please note that the time of point-cloud projection, ground point removal and segmentation have been included in SE+PTP. Only the time before using 6D SLAM is given here.

Sequences	Number of Scans	PTP(s)	BBF+PTP(s)	PTS(s)	SE+PTP(s)
se01	1101	534.9202	**235.6870**	1518.7555	351.6523
se07	1101	256.9504	199.3139	962.3334	**191.0013**
se09	1591	592.2633	368.2708	2000.6465	**311.0511**
se14	631	242.4361	183.1523	774.3427	**132.9431**

Figure 11 compares the trajectory error from an urban scene. The first row is some results without using 6D SLAM. Overall, the start and end positions of the trajectory from SE-PTP are perfectly coincident, while other methods suffer from significant accumulative errors, cf. Figure 11a. As Figure 11b,c depicts, from the starting point to scan325, PTS presents smaller translation and rotation error than other methods. However, at scan325 (Figure 11a(1)), which is a crossroad, the accuracy of PTS drops rapidly. Starting from scan560, which corresponds to arrow 2 in Figure 11a, the error of BBF+PTP and PTP increases rapidly. In contrast, the error produced by SE+PTP has not changed significantly. Figure 12 are the visual inspections corresponding to the Figure 11a(1,2) which shows that the big error at the corner is caused by the lack of geometric information and the existence of many dynamic objects. As Table 1 shows, SE+PTP achieves better performance compared with PTP, BBF+PTP and PTS, while PTS has larger error with respect to other methods. This shows that point-to-point ICP is more suitable for urban environments, and removing ground points can indeed improve estimation accuracy and efficiency (se07 in Table 2).

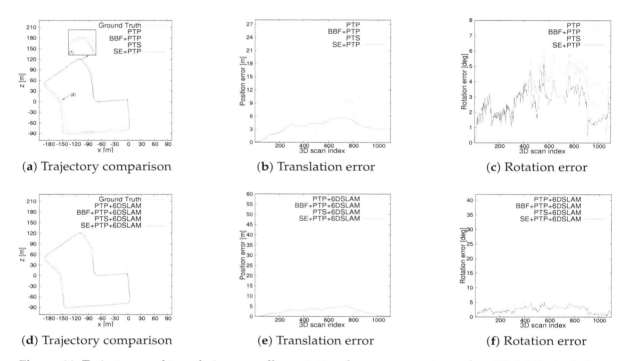

Figure 11. Trajectory and translation as well as rotational error comparison of seq07. (**a**) the trajectory comparison between different ICP. (**b**,**c**) are the translation error and rotational error from ICP. (**d**–**f**) are the corresponding results after applying 6D SLAM.

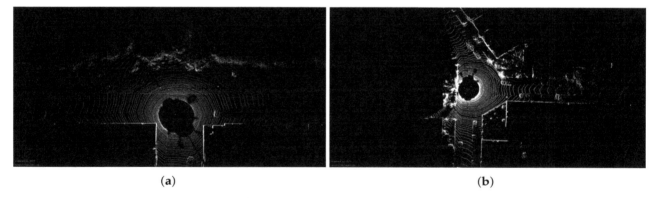

Figure 12. (**a**) The visual inspection of (a)(1) in Figure 11. (**b**) The visual inspection of (a)(2) in Figure 11.

After using the 6D SLAM, the trajectory has changed significantly, cf Table 1. First, PTP+6DSLAM and PTS+6DSLAM fail to estimate pose. This is because se07 contains a lot of dynamic vehicles which eventually leads to the performance degradation of the standard 6D SLAM. In contrast, since SE+PTP+6DSLAM includes a segmentation algorithm, which removes the noise points caused by dynamic objects to a certain extent. As a consequence, SE+PTP+6DSLAM achieves excellent results. The result of BBF+PTP+6DSLAM is slightly worse than before but better than PTP+6DSLAM and

PTS+6DSLAM, which shows removing ground points helps the convergence of 6D SLAM. Another issue that must be noted is PTS+6DSLAM obtains the same result as PTP+6DSLAM, cf. Figure 11d and Table 1. This shows that 6D SLAM is designed for point-to-point ICP. Overall, compared with other methods, our method requires the less time and achieves higher accuracy.

We also compared these methods in a complex scene mixing urban area and the country. As Figure 13a,b show, the translation accuracy of SE+PTP is inferior to PTS before using 6D SLAM (scan 500 to scan 1200 in Figure 13b). However, it has less rotation error (Table 1) and takes much less time to run than PTS (Table 2). Table 1 shows SE+PTP achieves similar performance to PTS, while PTP suffers from large errors, which Demonstrates PTS performs better in unstructured environments, such as roads and rural areas. In addition, the proposed method can achieve similar performance to PTS after combining segmentation, but it requires less calculation time. In addition, SE+PTP can better close the loop than other methods, cf. Figure 13a.

After 6D SLAM, SE+PTP+6DSLAM is superior to other methods in trajectory error and rotation error, cf. Figure 13e,f. We also find 6DSLAM does improve the accuracy of ICP alone. cf. se09 in Table 1. The reason is 6D SLAM taking into account all scans instead of only two adjacent scans which limits this accumulation error. Although the translation error was reduced from 27.0114 to 18.5825, this error is still rather large, which is caused by the complexity of the scene. Large changes between urban and villages have led to large errors in the middle of this trajectory (scan 300 to scan 800 in Figure 13e). Despite this, our algorithm can still close the loop well, cf. Figure 13d.

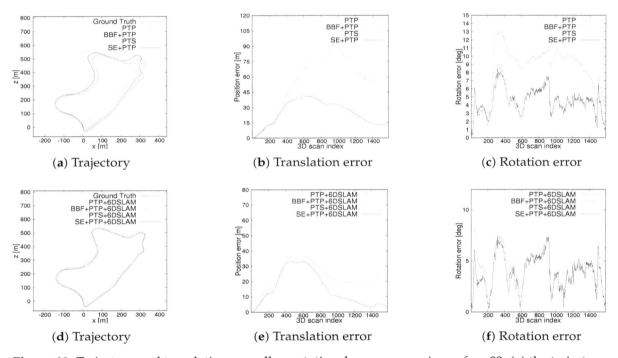

Figure 13. Trajectory and translation as well as rotational error comparison of seq09. (**a**) the trajectory comparison between different ICP. (**b,c**) are the translation error and rotational error from ICP. (**d–f**) are the corresponding results after applying 6D SLAM.

The last experiment was conducted in a rural environment, which is a vegetated road and contains little structural information. Please note that this data set is different from the above three groups, because it is a test data set in the KITTI benchmark which only provides the original LiDAR data but does not provide ground truth. To quantitatively analyze the trajectory error, we use the trajectory calculated by the SOFT2 [39] algorithm as the ground truth. SOFT2 is a state-of-the-art stereo visual odometry based on feature selection and tracking. This replacement is reasonable because the accuracy of SOFT2 algorithm is ranked fifth on the KITTI benchmark.

Figure 14a–c show the performance of SE+PTP is worse than both PTS and BBF+PTP and the gap between the initial position and the end position is larger, cf. Figure 14a(1). However, compared with PTP, SE+PTP reduces the translation error from 12.3578 to 5.5984 (Table 1), and the execution time of the algorithm decreased from 242.4361s to 132.9432 (Table 2). These improvements of performance are mainly due to the introduction of ground point removal and segmentation algorithms. Although PTS achieves higher accuracy before 6D SLAM, it consumes nearly 6 times more time than SE+PTP. The performance of our method has been greatly optimized after 6D SLAM. As shown in Figure 14d(1), the gap between the starting point and the ending point has been largely reduced. Table 1 reports, after 6D SLAM, the translation error was reduced from 5.5984 to 1.0114 while the rotation error is decreased to 0.8563. This shows that our method is superior to similar methods in terms of efficiency and accuracy.

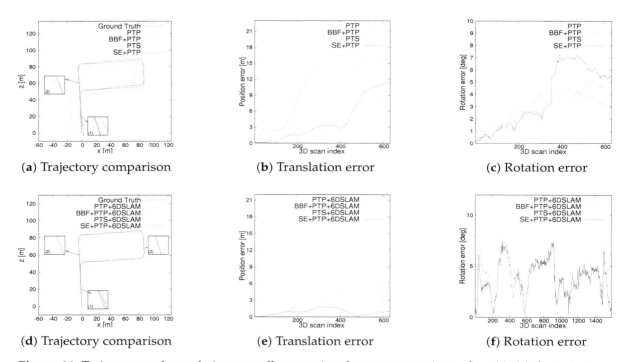

(a) Trajectory comparison   (b) Translation error   (c) Rotation error

(d) Trajectory comparison   (e) Translation error   (f) Rotation error

Figure 14. Trajectory and translation as well as rotational error comparison of seq14. (**a**) the trajectory comparison between different ICP. (**b**,**c**) are the translation error and rotational error from ICP. (**d**–**f**) are the corresponding results after applying 6D SLAM.

Figure 15 shows the point-cloud map of four experiments, which is calculated by SE+PTP+6DSLAM. To further test the effectiveness of the proposed algorithm, we evaluate the algorithm using the KITTI metric which calculated the accuracy by averaging relative position and rotation errors using segmented trajectory lengths. The average relative error of all four experiments based on the KITTI metrics is given in Table 3. Please note that only the methods with relatively high accuracy are given here according to Table 1. As shown Table 3, our method achieves higher accuracy. In addition, PTS is slightly inferior to our method in sequences 09 and 14, which demonstrates that point-to-plane ICP performs well in rural areas. This can be attributed to the tangent plane calculated by point-to-plane ICP, which is more robust to unstructured environments. However this also poses a challenge to computing efficiency. As Table 2 shows, PTS consumes nearly 6 times more time than SE+PTP. In sum, the proposed algorithm is superior to the ICP method in both accuracy and efficiency. In addition, our method is more suitable for 6D SLAM.

(**a**) Highway from se01 (**b**) Urban from se07

(**c**) Urban+Country from se09 (**d**) Country from se14

Figure 15. The point-cloud map of four experiments, which is calculated by SE+PTP+6DSLAM. The red line is the trajectory of the vehicle. (**a**) se01. (**b**) se07. (**c**) se09. (**d**) se14.

Table 3. Results of the proposed method(SE+PTP) compared with PTS and BBF+PTP+6DSLAM using relative (KITTI metric) metric, where t represents translation error, while r represents rotation error. Since sequences 01 do not contain closed loops, 6D SLAM is not employed on this sequence, hence, for se01, the results before and after 6D SLAM have the same error values.

Method	se01(Highway)		se07(Urban)		se09(Urban+Country)		se14(Country)	
	t_{rel} [%]	r_{rel} [deg]	t_{rel} [%]	r_{rel} [deg]	t_{rel} [%]	r_{rel} [deg]	t_{rel} [%]	r_{rel} [deg]
PTS	12.0361	0.0245	3.1390	0.0151	4.2922	0.0198	1.4128	0.0101
SE+PTP	3.9835	0.0080	1.5691	0.0136	4.1093	0.0189	3.0411	0.0301
BBF+PTP+6DSLAM	34.0821	0.0127	2.8382	0.0182	4.4884	0.0201	2.1407	0.0122
SE+PTP+6DSLAM	3.9835	0.0080	1.4075	0.0131	3.9607	0.0183	0.9815	0.0143

5. Discussion

The core idea of the proposed algorithm is to develop a highly accurate localization and mapping module in unknown environments. We have integrated ground point removal and segmentation modules with the standard point-to-point ICP method. Four experimental results show that both BBF+PTP and SE+PTP greatly improve efficiency and accuracy when compared with the standard ICP method(Table 1). As previously discussed, the LiDAR data contains a large number of ground points, which increase the computational burden as well as the possibility of ICP mismatch. Hence removing ground points is a necessary step. Compared with BBF + PTP, the introduction of the segmentation algorithm leads to higher accuracy of SE+PTP. This is the result that the segmentation algorithm removes those false ground points and noise points. It is worth emphasizing that our method often closes the loop well. After applying 6D SLAM, we also concluded that 6DSLAM is more suitable for optimizing point-to-point ICP, especially for the proposed method.

Our experiments also demonstrated some characteristics about ICP and 6D SLAM. First, the standard point-to-point ICP performs better in urban scene, cf. se07 in Table 1. This is because the environment contains more structured information, such as buildings. However, it has a large error in the country, cf. se09 and se14 in Table 1, while the point-to-plane ICP is more robust to these environments due to the introduction of the tangent plane. Moreover, 6DSLAM does improve the accuracy of point-to-point ICP alone, cf. PTP and PTP+6DSLAM of se09 in Table 1. The reason is 6D SLAM taking into account all scans instead of only two adjacent scans which limits this accumulation error. Although the position accuracy of PTP+6DSLAM in se14 is similar to PTP, the rotation error has been eliminated. However, PTP+6DSLAM shows worse results than PTP in the urban scene (se07 in Table 1). This is because se07 contains a lot of dynamic vehicles which can cause larger error. The performance of standard 6D SLAM degrade in a high dynamic environment. Another issue that must be noted is the performance of PTS+6DSLAM degrades compared to PTS. This problem is caused by the 6DSLAM algorithm itself. Since 6D SLAM is similar to the point-to-point ICP method but taking into account all scans instead of only two adjacent scans. It solves for all poses at the same time and iterates like in the original ICP. Hence, 6D SLAM is more suitable for point-to-point ICP. Furthermore, it must be noted that the point-to-plane ICP method always produces the same result as point-to-point ICP after they are input into 6D SLAM, which is because the 6D SLAM framework is specifically designed for the point-to-point ICP method.

In terms of application scenarios, all methods perform poorly on the highway, which is mainly due to the lack of rich geometric and semantic information on the highway, cf. se01 in Table 1 and Table 3. Due to the lack of semantic information, BBF+PTP finally failed to estimate the pose. This leads to the BBF+PTP-based algorithm to think that the vehicle stayed in place without moving. Hence the scale of this trajectory is reduced by a certain proportion, cf. Figure 9a. In contrast, SE+PTP is more robust, which is mainly due to the introduction of the segmentation algorithm. However, our method still cannot accurately estimate the pose of se01. Because there are too many moving vehicles running with high speed. Although the proposed algorithm perform better than the other methods in se09, it still suffers from large errors due to the complexity of the environment, which is a combination of rural and urban scenes. All methods perform better in the rural environment, i.e., se14, especially the proposed method greatly improves pose accuracy, which is the reason that se14 contains much structural information, e.g., this road is surrounded by trees on both sides and few dynamic objects are contained in this environment. As se14 of Table 1 shows, PTS achieves higher accuracy before 6D SLAM, which is due to it calculates the tangent plane of the point. However, it consumes nearly 6 times more time than SE+PTP, cf. se14 in Table 2. Moreover, the result of SE+PTP+6DSLAM is better than PTS.

Dynamic objects such as high-speed vehicles, are the main error sources affecting pose accuracy. By comparing the locations of errors, we also find that large errors often occur at intersections. As Figure 12 shows, intersections either lack sufficient geometry or contain a large number of dynamic vehicles which are the main cause of errors. In future work, we will carry out research based on dynamic objects removing to further improve the pose estimation accuracy.

6. Conclusions

This paper presented a method for enhancing pose estimation accuracy of 3D point clouds by properly processing ground point and point-cloud segmentation. Since the ground points are removed, the proposed method is mainly applied to estimate the pose of ground vehicles. First, a 2D image-based ground point extraction method is introduced as a preprocessing step for ICP matching. Secondly, the point cloud after removing the ground points is then grouped into many clusters. By clustering, some outliers that do not have common attributes are removed. After this step, these different clusters are merged into a new point cloud. Compared to the original point cloud, the ground points of the new point cloud are removed and those false ground points and noise points have also been filtered out, which will greatly increase the efficiency and accuracy of ICP matching. Thirdly, A standard point-to-point ICP is then applied to calculate the six degree-of-freedom transformation between consecutive scans. Once closed loops are detected in the environment, a 6D graph optimization algorithm for global relaxation is employed, which aims to obtain a globally consistent trajectory and mapping.

In addition, we validated the proposed algorithm in four different scenarios including the city, the country and a highway. To test the proposed algorithm, the accuracy and runtime between our method and point-to-point ICP, point-to-plane ICP and Bounding Box Filter-based ICP are presented. Four experimental results show that both BBF+ICP and SE+ICP have improved the accuracy and speed of pose estimation relative to the standard ICP method, demonstrating that removing ground points improve the accuracy, efficiency and robustness of pose estimation based on ground vehicles. Compared with BBF + ICP, the introduction of the segmentation algorithm leads to higher accuracy of SE+ICP. This is the result that the segmentation algorithm removes those false ground points and noise points. Furthermore, we also concluded that 6DSLAM is more suitable for optimizing point-to-point ICP, especially for the proposed method.

In future work, removing dynamic targets of the scene will be fused into this proposed algorithm. Moreover, since our algorithm does not perform well in environments with less geometric information, such as highways, future work will integrate semantic information into our method, which is expected to inevitably improve the efficiency and accuracy of ICP matching.

Author Contributions: Conceptualization, S.D.; Data curation, S.D.; Formal analysis, S.D. and H.L.; Investigation, X.L.; Methodology, S.D.; Software, S.D.; Validation, S.D. and H.L.; Funding acquisition, G.L.; Supervision, X.L. and G.L.; Project administration, G.L.; Writing—original draft preparation, S.D.; Writing—review and editing, G.L., H.L. All authors have read and agreed to the published version of the manuscript.

References

1. Lin, Y.; Gao, F.; Qin, T.; Gao, W.; Liu, T.; Wu, W.; Yang, Z.; Shen, S. Autonomous aerial navigation using monocular visual inertial fusion. *IEEE Intell. Transp. Syst. Mag.* **2018**, *35*, 23–51. [CrossRef]
2. Pinies, P.; Lupton, T.; Sukkarieh, S; Tardos, J.D. Inertial aiding of inverse depth SLAM using a monocular camera. In Proceedings of the IEEE International Conference on Robotics and Automation, Roma, Italy, 10–14 April 2007; pp. 2797–2802.
3. Tong, C.H.; Barfoot, T.D. Gaussian process gaussnewton for 3d laser-based visual odometry. In Proceedings of the IEEE International Conference on Robotics and Automation, Karlsruhe, Germany, 6–10 May 2013; pp. 5024–5011.
4. Barfoot, T.D.; McManus, C.; Anderson, S; Dong, H.; Beerepoot, E.; Tong, C.H.; Furgale, P.; Gammell, J.D.; Enright, J. Into darkness: Visual navigation based on a lidar-intensity-image pipeline. *Robot. Res.* **2016**, *114*, 487–504.
5. Yang, J.; Li, H.; Campbell, D.; Jia, Y. Go-ICP: A Globally Optimal Solution to 3D ICP Point-Set Registration. *IEEE Trans. Pattern Anal. Mach. Intell.* **2016**, *38*, 2241–2254. [CrossRef] [PubMed]
6. Ren, Z.; Wang, L.; Bi, L. Robust GICP-Based 3D LiDAR SLAM for Underground Mining Environment. *Sensors* **2019**, *19*, 2915. [CrossRef] [PubMed]
7. Besl, P.J.; McKay, N.D. A Method for Registration of 3D Shapes. *IEEE Trans. Pattern Anal. Mach. Intell.* **1992**, *14*, 239–256. [CrossRef]
8. Segal, A.; Haehnel, D.; Thrun, S. Generalized-ICP. In Proceedings of the Robotics: Science and Systems, Zurich, Switzerland, 25–28 June 2009; Volume 2, p. 435.
9. Zhou, Q.Y.; Park, J.; Koltun, V. Fast global registration. In Proceedings of the European Conference on Computer Vision, Amsterdam, The Netherlands, 11–14 October 2016; pp. 766–782.
10. Zhang, J.; Singh, S. LOAM: Lidar Odometry and Mapping in Real-time. In Proceedings of the Robotics: Science and Systems, Cambridge, CA, USA, 12–16 July 2014; pp. 1–9.
11. Zaganidis, A.; Sun, L.; Duckett, T.; Cielniak, G. Integrating Deep Semantic Segmentation into 3D Point Cloud Registration. *IEEE Robot. Autom. Lett.* **2018**, *3*, 2942–2949. [CrossRef]
12. Stoyanov, T.; Magnusson, M.; Lilienthal, A.J. Point Set Registration through Minimization of the L2 Distance between 3D-NDT Models. In Proceedings of the 2012 IEEE International Conference on Robotics and Automation, Saint Paul, MN, USA, 14–18 May 2012; pp. 5196–5201.

13. Liu, H.; Ye, Q.; Wang, H.; Chen, L.; Yang, J. A Precise and Robust Segmentation-Based Lidar Localization System for Automated Urban Driving. *Remote Sens.* **2019**, *11*, 1348. [CrossRef]

14. Opromolla, R.; Fasano, G; Grassi, M.; Savvaris, A.; Moccia, A. PCA-Based Line Detection from Range Data for Mapping and Localization-Aiding of UAVs. *Int. J. Aerosp. Eng.* **2017**, *38*, 1–14. [CrossRef]

15. Durrant-Whyte, H.; Bailey, T. Simultaneous Localization and Mapping (SLAM): Part I the essential algorithms. *IEEE Robot. Autom. Mag.* **2006**, *13*, 99–110. [CrossRef]

16. Grisetti, G.; Kummerle, R.; Stachniss, C.; Burgard, W. A Tutorial on GraphBased SLAM. *IEEE Intell. Transp. Syst. Mag.* **2010**, *4*, 31–43. [CrossRef]

17. Rusu, R. ; Cousins, S. 3D is here: Point Cloud Library (PCL). In Proceedings of the 2011 IEEE International Conference on Robotics and Automation, Shanghai, China, 9–13 May 2011; pp. 1–4.

18. Na, K.; Byun, J.; Roh, M.; Seo, B. The ground segmentation of 3D LIDAR point cloud with the optimized region merging. In Proceedings of the 2017 IEEE/RSJ International Conference on Intelligent Robots and Systems, Las Vegas, NV, USA, 2–6 December 2013; pp. 445–450.

19. Luo, Z.; Mohrenschildt, M.; Habibi, S. A Probability Occupancy Grid Based Approach for Real-Time LiDAR Ground Segmentation. *IEEE Trans. Intell. Transp. Syst.* **2019**, 1–13. [CrossRef]

20. Shan, T. Englot, B. LeGO-LOAM: Lightweight and Ground-Optimized Lidar Odometry and Mapping on Variable Terrain. In Proceedings of the 2018 IEEE/RSJ International Conference on Intelligent Robots and Systems (IROS), Madrid, Spain, 1–5 October 2018; pp. 4758–4765.

21. Pomares, A.; Martínez, J.; Mandow, A.; Martinez, M.; Moran, M.; Morales, J. Ground Extraction from 3D Lidar Point Clouds with the Classification Learner App. In Proceedings of the 26th Mediterranean Conference on Control and Automation (MED), Zadar, Croatia, 19–22 June 2018; pp. 400–405.

22. Hackel, T.; Wegner, Jan D.; Schindler, K. Fast Semantic Segmentation of 3d Point Clouds with Strongly Varying Density. *ISPRS Ann. Photogramm. Remote Sens. Spat. Inf. Sci.* **2016**, *3*, 177–184. [CrossRef]

23. Velas, M.; Spanel, M.; Hradis, M.; Herout, A. CNN for Very Fast Ground Segmentation in Velodyne Lidar Data. In Proceedings of the 2018 IEEE International Conference on Autonomous Robot Systems and Competitions (ICARSC), Torres Vedras, Portugal, 25–27 April 2018; pp. 97–103.

24. Qi, C.R.; Su, H.; Mo, K.; Guibas, L.J. PointNet: Deep Learning on Point Sets for 3D Classification and Segmentation. In Proceedings of the IEEE Conference on Computer Vision and Pattern Recognition, Honolulu, HI, USA, 21–26 July 2017; pp. 652–660.

25. Huhle, B.; Magnusson, M.; Strasser, W.; Lilienthal, A.J. Registration of colored 3D point clouds with a Kernel-based extension to the normal distributions transform. In Proceedings of the 2008 IEEE International Conference on Robotics and Automation, Pasadena, CA, USA, 19–23 May 2008; pp. 4025–4030.

26. Nüchter, A.; Wulf, O.; Lingemann, K.; Hertzberg, J.; Wagner, B.; Surmann, H. 3D mapping with semantic knowledge. In *Robot Soccer World Cup*; Springer: Berlin, Germany, 2005; pp. 335–346.

27. Zaganidis, A.; Magnusson, M.; Duckett, T.; Cielniak, G. Semantic-assisted 3d normal distributions transform for scan registration in environments with limited structure. In Proceedings of the 2017 IEEE/RSJ International Conference on Intelligent Robots and Systems, Vancouver, BC, Canada, 24–28 September 2017.

28. PiniÉs, P.; TardÓs, J.D. Large-Scale SLAM Building Conditionally Independent Local Maps: Application to Monocular Vision. *IEEE Trans. Robot.* **2008**, *24*, 1094–1106. [CrossRef]

29. Grisetti, G.; Stachniss, C.; Burgard, W. Improved Techniques for Grid Mapping With Rao-Blackwellized Particle Filters. *IEEE Trans. Robot.* **2007**, *23*, 34–46. [CrossRef]

30. Olofsson, B.; Antonsson, J.; Kortier, H.G.; Bernhardsson, B.; Robertsson, A.; Johansson, R. Sensor fusion for robotic workspace state estimation. *IEEE/ASME Trans. Mechatronics* **2016**, *21*, 2236–2248. [CrossRef]

31. Wang, K.; Liu, Y.; Li, L. A Simple and Parallel Algorithm for Real-Time Robot Localization by Fusing Monocular Vision and Odometry/AHRS Sensors. *IEEE/ASME Trans. Mechatronics* **2014**, *19*, 1447–1457. [CrossRef]

32. Borrmann, D.; Elseberg, J.; Lingemann, K.; Nüchter, A.; Hertzberg, J. Globally consistent 3D mapping with scan matching. *Robot. Auton. Syst.* **2008**, *56*, 130–142. [CrossRef]

33. Elseberg, J.; Borrmann, D.; Nüchter, A. One billion points in the cloud—An octree for efficient processing of 3D laser scans. *Int. J. Photogramm. Remote. Sens.* **2013**, *76*, 76–88. [CrossRef]

34. Bogoslavskyi, I.; Stachniss, C. Fast range image-based segmentation of sparse 3D laser scans for online operation. In Proceedings of the 2016 IEEE/RSJ International Conference on Intelligent Robots and Systems (IROS), Daejeon, Korea, 9–14 October 2016; pp. 163–169.

35. Savitzky, A.; Golay, M.J.E. Smoothing and Differentiation of Data by Simplified Least Squares Procedures. *Anal. Chem.* **1964**, *36*, 1627–1639. [CrossRef]
36. Nüchter, A.; Lingemann, K. 6D SLAM Software. Available online: http://slam6d.sourceforge.net/ (accessed on 31 December 2019).
37. Geiger, A.; Lenz, P.; Urtasum, R. Are we ready for autonomous driving? the kitti vision benchmark suite. In Proceedings of the IEEE Conference on Computer Vision and Pattern Recognition (CVPR), Providence, RI, USA, 16–21 June 2012; pp. 3354–3361.
38. May, S.; Droeschel, D.; Holz, D.; Fuchs, S.; Malis, E.; Nüchter, A.; Hertzberg, J. Three dimensional mapping with time of flight cameras. *J. Field Robot.* **2009**, *26*, 934–965. [CrossRef]
39. Cvišić,Igor; Ćesić,Josip; Marković, Ivan; Petrović,Ivan. SOFT-SLAM: Computationally Efficient Stereo Visual SLAM for Autonomous UAVs. *J. Field Robot.* **2017**, *35*, 578–595.

3

Formation Control and Distributed Goal Assignment for Multi-Agent Non-Holonomic Systems

Wojciech Kowalczyk

Institute of Automation and Robotics, Poznań University of Technology (PUT), Piotrowo 3A, 60-965 Poznań, Poland; wojciech.kowalczyk@put.poznan.pl.

Abstract: This paper presents control algorithms for multiple non-holonomic mobile robots moving in formation. Trajectory tracking based on linear feedback control is combined with inter-agent collision avoidance. Artificial potential functions (APF) are used to generate a repulsive component of the control. Stability analysis is based on a Lyapunov-like function. Then the presented method is extended to include a goal exchange algorithm that makes the convergence of the formation much more rapid and, in addition, reduces the number of collision avoidance interactions. The extended method is theoretically justified using a Lyapunov-like function. The controller is discontinuous but the set of discontinuity points is of zero measure. The novelty of the proposed method lies in integration of the closed-loop control for non-holonomic mobile robots with the distributed goal assignment, which is usually regarded in the literature as part of trajectory planning problem. A Lyapunov-like function joins both trajectory tracking and goal assignment analyses. It is shown that distributed goal exchange supports stability of the closed-loop control system. Moreover, robots are equipped with a reactive collision avoidance mechanism, which often does not exist in the known algorithms. The effectiveness of the presented method is illustrated by numerical simulations carried out on the large formation of robots.

Keywords: formation of robots; non-holonomic robot; stability analysis; Lyapunov-like function; target assignment; goal exchange; path following; switching control

1. Introduction

The idea to use artificial potential fields to control manipulators and mobile robots was introduced by Khatib [1] in 1986. In this approach both attraction to the goal and repulsion from the obstacles are negated gradients of the artificial potential functions (APF). His paper not only presents the theory, but also a solution of the practical problem, implemented in the Puma 560 robot simulator. It is worth noting that much earlier, in 1977, Laitmann and Skowronski [2] investigated control of two agents avoiding collision with each other. This work was purely theoretical. The authors continued their work in the following years [3].

Since the 1990s, intensive research on the trajectory tracking control for non-holonomic mobile robots has been conducted [4–6]. The algorithm presented further in this paper is based on the method from [4]. This method considers a single, differentially-driven mobile robot moving in a free space. Its goal is to track a desired trajectory. The paper includes a stability analysis.

The last decade has seen a lot of publications on multiple mobile robot control. In [7], the goal of multiple mobile robots is to track desired trajectories avoiding inter-agent collisions. The same type of task is considered further in this paper. The tracking controller is different, as is the formula of APF, but the method of combining trajectory tracking and collision avoidance is the same. In [8], the same type of task is considered, but the dynamics of mobile platforms and uncertainties of its parameters are taken into account. Kowalczyk et al. [9] propose a vector-field-orientation algorithm for multiple mobile robots moving in the environment with circle shaped static obstacles. The dynamic properties of mobile platforms are also taken into account. The paper [10] presents a kinematic controller for the formation of robots that move in a queue. The goal is to keep desired displacements between robots and avoid collisions in the transient states. Hatanaka et al. [11] investigate a cooperative estimation problem for visual sensor networks based on multi-agent optimization techniques. The paper [12] addresses the formation control problem for fleets of autonomous underwater vehicles. Yoshioka et al. [13] deal with formation control strategies based on virtual structure for multiple non-holonomic systems.

Recent years have also seen a number of publications on barrier functions. In [14], coordination control for a group of mobile robots is combined with collision avoidance provided by a safety barrier. If the coordination control command leads to collision, the safety barrier dominates the controller and computes a safe control closest to coordination control law. In the method proposed in [15] control barrier functions are unified with performance objectives expressed as control Lyapunov functions. The authors of the paper [16] provide a theoretical framework to synthesize controllers using finite time convergence control barrier functions guided by linear temporal logic specifications for continuous time multi-agent systems.

The paper [17] addresses the problem of optimal goal assignment for the formation of holonomic robots moving on a plane. A linear bottleneck assignment problem solution is used to minimize the maximum completion time or maximum distance for any robot in the formation. The authors of [18] consider formation of non-holonomic mobile vehicles that has to change the geometrical shape of the formation. Goal assignment minimizes the total distance travelled by agents. The exemplary application indicated by the authors is reconfiguration of the formation when it approaches a narrow passage. In [19,20], goal assignment based on distances squared is proposed and tested for large formations, but the collision avoidance is resolved at the trajectory planning level. This makes this approach less robust in the case of unpredictable disturbances (which are natural in real applications). The second of these papers proposes a solution to the problem of collision avoidance with static and dynamic obstacles present in the environment. In [21], collision avoidance is obtained by applying safety constraints in optimal trajectory generation. The robots do not have non-holonomic constraints (they are quadrotors) and they have to change the shape of the formation using goal assignment that minimizes the total distance travelled by the robots. Turpin et al. [22] present concurrent assignment and planning of trajectories (CAPT) algorithms. The authors propose two variations of the algorithm: centralized and decentralized, and test them on a group of holonomic mobile robots moving in a three-dimensional space. The solution is based on the Hungarian assignment algorithm. The above-mentioned works do not deal with the problem of closed-loop control. For this reason stability issues were not considered there.

In comparison to the above two approaches, here the user or the higher level controller determines the locations of desired trajectories according to the needs (this can be considered as an expected feature in many applications). In addition, trajectory generation is decoupled from the closed-loop control (the system is modular). Such a solution is considered as a design pattern in robotics. The algorithm is responsible for tracking these trajectories and reacting to the risk of collisions between agents at the same time. Even if initial states of individual robots are far from the desired ones, the collision avoidance module works correctly. Furthermore, the algorithm characterizes conceptual and computational simplicity. The paper considers preserving data integrity during the goal exchange as it requires a simultaneous change of states in remote systems. This subject is omitted in the literature on goal assignment in multi-robot systems. To the best of the author's knowledge, no work has been published so far that proposes closed-loop control for multiple non-holonomic mobile robots combined with target assignment with analysis based on a Lyapunov-like function for both the tracking algorithm and target assignment. Panagou et al. [23] propose a similar method, but it assumes that agents are fully-actuated (modeled using integrators), and the analysis is based on multiple Lyapunov-like functions. This algorithm also uses a different criterion for goal exchange.

The algorithm proposed in this paper is applicable mainly to the homogeneous formations of non-holonomic mobile robots but also in the scenarios when two or more robots of the same type are involved in task execution. For a high number of applications of multiple mobile robots this situation occurs, e.g., exploration, mapping, safety, and surveillance.

In Section 2 a control algorithm for the formation of non-holonomic mobile robots is described. Section 3 analysis stability. The simulation results are presented in Section 4. The goal exchange algorithm is introduced in Section 5. Section 6 provides stability analysis of extended algorithm. Section 7 details distributed implementation. Some generalization of the proposed goal exchange rule is given in Section 8. Section 9 discusses the problem of maintaining data integrity in the process of the goal exchange. Section 10 offers simulation results for the goal exchange algorithm. Section 11 presents simulation results for limited wheel velocity controls. In the last Section concluding remarks are provided.

2. Control Algorithm

The kinematic model of the i-th differentially-driven mobile robot R_i ($i = 1 \ldots N$, N—number of robots) is given by the following equation:

$$
\dot{q}_i = \begin{bmatrix} \cos\theta_i & 0 \\ \sin\theta_i & 0 \\ 0 & 1 \end{bmatrix} u_i \tag{1}
$$

where vector $q_i = [x_i \ y_i \ \theta_i]^\top$ denotes the pose and x_i, y_i, θ_i are the position coordinates and orientation of the robot with respect to a global, fixed coordinate frame. Vector $u_i = \begin{bmatrix} v_i & \omega_i \end{bmatrix}^\top$ is the control vector with v_i denoting the linear velocity and ω_i denoting the angular velocity of the platform.

The task of the formation is to follow the virtual leader that moves with desired linear and angular velocities $[v_l \ \omega_l]^T$. The robots are expected to imitate the motion of the virtual leader. They should have the same velocities as the virtual leader. The position coordinates $[x_l \ y_l]^T$ of the virtual leader are used as a reference position for the individual robots but each of them has different displacement with respect to the leader:

$$
x_{id} = x_l + d_{ix} \qquad y_{id} = y_l + d_{iy}, \tag{2}
$$

where $[d_{ix} \ d_{iy}]^T$ is desired displacement of the i-th robot. As the robots position converge to the desired values their orientations θ_i converge to the orientation of the virtual leader θ_l.

The collision avoidance behaviour is based on the APF. This concept was originally proposed in [1]. All robots are surrounded by APFs that raise to infinity near objects border r_j (j—number of the robots/obstacles) and decreases to zero at some distance R_j, $R_j > r_j$.

One can introduce the following function [6]:

$$
B_{aij}(l_{ij}) = \begin{cases} 0 & \text{for} & l_{ij} < r_j \\ e^{\frac{l_{ij}-r_j}{l_{ij}-R_j}} & \text{for} & r_j \le l_{ij} < R_j \\ 0 & \text{for} & l_{ij} \ge R_j \end{cases} \tag{3}
$$

that gives output $B_{aij}(l_{ij}) \in \langle 0, 1\rangle$. The distance between the i-th robot and the j-th robot is defined as the Euclidean length $l_{ij} = \left\| [x_j \ y_j]^\top - [x_i \ y_i]^\top \right\|$.

Scaling the function given by Equation (3) within the range $\langle 0, \infty \rangle$ can be given as follows:

$$
V_{aij}(l_{ij}) = \frac{B_{aij}(l_{ij})}{1 - B_{aij}(l_{ij})}, \tag{4}
$$

that is used later to avoid collisions.

In further description, the terms 'collision area' or 'collision region' are used for locations fulfilling the condition $l_{ij} < r_j$. The range $r_j < l_{ij} < R_j$ is called 'collision avoidance area' or 'collision avoidance region' (Figure 1).

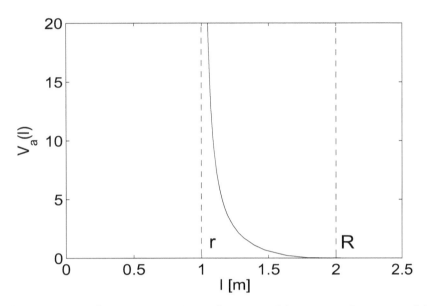

Figure 1. Artificial potential functions (APF) as a function of distance to the centre of the robot (indexes omitted for simplicity).

The goal of the control is to drive the formation along the desired trajectory avoiding collisions between agents. It is equivalent to bringing the following quantities to zero:

$$p_{ix} = x_{id} - x_i$$
$$p_{iy} = y_{id} - y_i$$
$$p_{i\theta} = \theta_l - \theta_i. \tag{5}$$

Assumption 1. $\forall\{i,j\}, i \neq j, ||[x_{id} \ y_{id}]^T - [x_{jd} \ y_{jd}]^T|| > R_j$.

Assumption 2. *If robot i gets into the collision avoidance region of any other robot j, $j \neq i$ its desired trajectory is temporarily frozen ($\dot{x}_{id} = 0$, $\dot{y}_{id} = 0$). If the robot leaves the avoidance area its desired coordinates are immediately updated. As long as the robot remains in the avoidance region, its desired coordinates are periodically updated at certain discrete instants of time. The time period t_u of this update process is relatively large in comparison to the main control loop sample time.*

Assumption 1 comes down to the statement that desired paths of individual robots are planned in such a way that in steady state all robots are out of the collision avoidance regions of other robots.

Assumption 2 means that the tracking process is temporarily suspended because collision avoidance has a higher priority. Once the robot is outside the collision detection region, it updates the reference to the new values. In addition, when the robot is in the collision avoidance region its reference is periodically updated. This low-frequency process supports leaving the unstable equilibrium points that occur e.g., when one robot is located exactly between the other robots and its goal.

The system error expressed with respect to the coordinate frame fixed to the robot is described below:

$$\begin{bmatrix} e_{ix} \\ e_{iy} \\ e_{i\theta} \end{bmatrix} = \begin{bmatrix} \cos(\theta_i) & \sin(\theta_i) & 0 \\ -\sin(\theta_i) & \cos(\theta_i) & 0 \\ 0 & 0 & 1 \end{bmatrix} \begin{bmatrix} p_{ix} \\ p_{iy} \\ p_{i\theta} \end{bmatrix}. \tag{6}$$

Using the above equations and non-holonomic constraint $\dot{y}_i \cos(\theta_i) - \dot{x}_i \sin(\theta_i) = 0$ the error dynamics between the leader and the follower are as follows:

$$\dot{e}_{ix} = e_{iy}\omega_i - v_i + v_l \cos e_{i\theta}$$
$$\dot{e}_{iy} = -e_{ix}\omega_i + v_l \sin e_{i\theta}$$
$$\dot{e}_{i\theta} = \omega_l - \omega_i. \tag{7}$$

One can introduce the position correction variables that consist of position error and collision avoidance terms:

$$P_{ix} = p_{ix} - \sum_{j=1,j\neq i}^{N} \frac{\partial V_{aij}}{\partial x_i}$$
$$P_{iy} = p_{iy} - \sum_{j=1,j\neq i}^{N} \frac{\partial V_{aij}}{\partial y_i}. \tag{8}$$

V_{aij} depends on x_i and y_i according to Equation (5). It is assumed that the robots avoid collisions with each other and there are no other obstacles in the taskspace. The correction variables can be transformed to the local coordinate frame fixed in the mass centre of the robot:

$$\begin{bmatrix} E_{ix} \\ E_{iy} \\ e_{i\theta} \end{bmatrix} = \begin{bmatrix} \cos(\theta_i) & \sin(\theta_i) & 0 \\ -\sin(\theta_i) & \cos(\theta_i) & 0 \\ 0 & 0 & 1 \end{bmatrix} \begin{bmatrix} P_{ix} \\ P_{iy} \\ p_{i\theta} \end{bmatrix}. \tag{9}$$

Differentiating the first two equations of (5) with respect to the p_{ix} and p_{iy} respectively one obtains:

$$\frac{\partial x_i}{\partial p_{ix}} = -1 \qquad \frac{\partial y_i}{\partial p_{iy}} = -1. \tag{10}$$

Using (10) one can write:

$$\frac{\partial V_{aij}}{\partial p_{ix}} = \frac{\partial V_{aij}}{\partial x_i} \frac{\partial x_i}{\partial p_{ix}} = -\frac{\partial V_{aij}}{\partial x_i}$$
$$\frac{\partial V_{aij}}{\partial p_{iy}} = \frac{\partial V_{aij}}{\partial y_i} \frac{\partial y_i}{\partial p_{iy}} = -\frac{\partial V_{aij}}{\partial y_i}. \tag{11}$$

Taking into account Equations (8) and (9) the gradient of the APF can be expressed with respect to the local coordinate frame fixed to the i-th robot:

$$\begin{bmatrix} \frac{\partial V_{aij}}{\partial e_{jx}} \\ \frac{\partial V_{aij}}{\partial e_{iy}} \end{bmatrix} = \begin{bmatrix} \cos \theta_i & \sin \theta_i \\ -\sin \theta_i & \cos \theta_i \end{bmatrix} \begin{bmatrix} \frac{\partial V_{aij}}{\partial p_{ix}} \\ \frac{\partial V_{aij}}{\partial p_{iy}} \end{bmatrix}. \tag{12}$$

Equation (12) can be verified easily by taking partial derivatives of $V_{aij}(d_{ix} - p_{ix}, d_{iy} - p_{iy}) = V_{aij}(d_{ix} - p_{ix}(e_{ix}, e_{iy}), d_{iy} - p_{iy}(e_{ix}, e_{iy}))$ with respect to e_{ix}, e_{iy} and taking into account the inverse transformation of the first two equations of Equation (6).

Using Equation (11), the above equation can be written as follows:

$$\begin{bmatrix} \frac{\partial V_{aij}}{\partial e_{jx}} \\ \frac{\partial V_{aij}}{\partial e_{iy}} \end{bmatrix} = \begin{bmatrix} -\cos \theta_i & -\sin \theta_i \\ \sin \theta_i & -\cos \theta_i \end{bmatrix} \begin{bmatrix} \frac{\partial V_{aij}}{\partial x_i} \\ \frac{\partial V_{aij}}{\partial y_i} \end{bmatrix}. \tag{13}$$

Equations (9) and (12) can be transformed to the following form:

$$E_{ix} = p_{ix}\cos(\theta_i) + p_{iy}\sin(\theta_i) + \sum_{j=1,j\neq i}^{N} \frac{\partial V_{aij}}{\partial e_{ix}}$$

$$E_{iy} = -p_{ix}\sin(\theta_i) + p_{iy}\cos(\theta_i) + \sum_{j=1,j\neq i}^{N} \frac{\partial V_{aij}}{\partial e_{iy}}$$

$$e_{i\theta} = p_{i\theta}, \tag{14}$$

where each derivative of the APF is transformed from the global coordinate frame to the local coordinate frame fixed to the robot. Finally, the correction variables expressed with respect to the local coordinate frame (Figure 2) are as follows:

$$E_{ix} = e_{ix} + \sum_{j=1,j\neq i}^{N} \frac{\partial V_{aij}}{\partial e_{ix}}$$

$$E_{iy} = e_{iy} + \sum_{j=1,j\neq i}^{N} \frac{\partial V_{aij}}{\partial e_{iy}}. \tag{15}$$

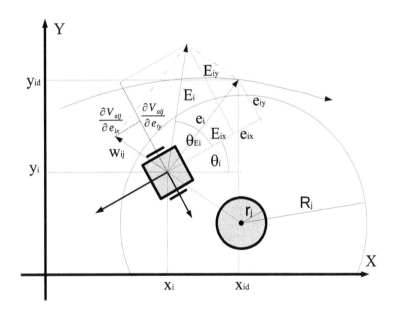

Figure 2. Robot in the environment with an obstacle.

Note the similarity of the structure of Equations (8) (updated by Equations (11)) and (15).

The trajectory tracking algorithm from [4] was chosen based on the author's experience. It is simple, easy to implement, and, above all, it is effective. Tracking control with persistent excitation [24] and the vector-field-orientation method [25] were also taken into account. The former gives much worse convergence time, the latter gives even better convergence but it is more difficult to implement on a real robot.

The control for N robots extended by the collision avoidance is as follows:

$$v_i = v_l\cos e_{i\theta} + k_1 E_{ix}$$
$$\omega_i = \omega_l + k_2\mathrm{sgn}(v_l)E_{iy} + k_3 e_{i\theta}, \tag{16}$$

where k_1, k_2 and k_3 are positive constant design parameters.

Assumption 3. *If the value of the linear control signal is less than considered threshold value v_t, i.e., $|v| < v_t$ (v_t-positive constant), it is replaced with a new value $\tilde{v} = S(v)v_t$, where*

$$S(v) = \begin{cases} -1 & \text{for} \quad v < 0 \\ 1 & \text{for} \quad v \geq 0 \end{cases}, \tag{17}$$

(indexes omitted for simplicity).

Substituting Equation (16) for (7) error dynamics is given by the following equations:

$$\begin{aligned}
\dot{e}_{ix} &= e_{iy}\omega_i - k_1 E_{ix} \\
\dot{e}_{iy} &= -e_{ix}\omega_i + v_l \sin e_{i\theta} \\
\dot{e}_{i\theta} &= -k_2 \mathrm{sgn}(v_l) E_{iy} - k_3 e_{i\theta}
\end{aligned} \tag{18}$$

Transforming (18) using (16) and taking into account Assumption 2 (when the robot gets into the collision avoidance region, in the collision avoidance state, velocities v_l and ω_l are replaced with 0 value) error dynamics can be expressed in the following form:

$$\begin{aligned}
\dot{e}_{ix} &= k_3 e_{iy} e_{i\theta} - k_1 E_{ix} \\
\dot{e}_{iy} &= -k_3 e_{i\theta} e_{ix} \\
\dot{e}_{i\theta} &= -k_3 e_{i\theta}
\end{aligned} \tag{19}$$

Orientation error $e_{i\theta}$ decreases exponentially to zero (refer to the last equation in (19)).

In Figure 3 a schematic diagram of the control system is presented. The following signal vectors are marked: $[x\ y]^T = [x_1 \ \dots \ x_N \ y_1 \ \dots \ y_N]^T$, $\theta = [\theta_1 \ \dots \ \theta_N]^T$, $[x_d\ y_d]^T = [x_{1d} \ \dots \ x_{Nd} \ y_{1d} \ \dots \ y_{Nd}]^T$, $[v\ \omega]^T = [v_1 \ \dots \ v_N \ \omega_1 \ \dots \ \omega_N]^T$, $[p_x\ p_y]^T = [p_{1x} \ \dots \ p_{Nx} \ p_{1y} \ \dots \ p_{Ny}]^T$, $[e_x\ e_y]^T = [e_{1x} \ \dots \ e_{Nx} \ e_{1y} \ \dots \ e_{Ny}]^T$, $[E_x\ E_y]^T = [E_{1x} \ \dots \ E_{Nx} \ E_{1y} \ \dots \ E_{Ny}]^T$.

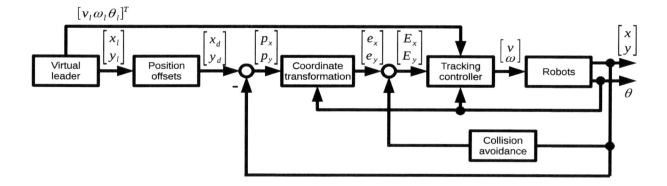

Figure 3. Control system.

3. Stability of the System

In this section stability analysis of the closed-loop system is presented. When the i-th robot is out of the collision regions of the other robots (APF takes the value zero) the analysis given in [4] is actual and will not be repeated here. Further the analysis for the situation in which the i-th robot is in the collision region of other robot is presented.

For further analysis a new variable is introduced: $\theta_{iE} = Atan2(-E_{iy}, -E_{ix})$ ($Atan2(\bullet, \bullet)$ is a version of the $Atan(\bullet)$ function covering all four quarters of the Euclidean plane)—auxiliary orientation variable.

Proposition 1. *The system (1) with controls (16) is stable if the desired trajectories fulfil the condition* $\theta_{iE} \notin \langle \frac{\pi}{2} \pm \theta_{E\Delta} \pm \pi d \rangle$ *(d = 0, ±1, ±2, ...), where* $\theta_{E\Delta}$ *is a small constant.*

As stated in [7], if $\theta_{iE} \in \langle \frac{\pi}{2} \pm \theta_{E\Delta} \pm \pi d \rangle$ (the combination of obstacle position and reference trajectory drive the robot into the neighbourhood of a singular configuration where the condition in Proposition 1 does not hold) one solution is to add perturbation to the desired signal. The system can also leave the neighbourhood of the singularity easily since the robot can reorient itself in place if the condition is not satisfied. This requires a special procedure to be implemented, which will not be discussed here.

Proof. Consider the following Lyapunov-like function:

$$V = \sum_{i=1}^{N} \left[\frac{1}{2}(e_{ix}^2 + e_{iy}^2 + e_{i\theta}^2) + \sum_{j=1, j\neq i}^{N} V_{aij} \right]. \tag{20}$$

When the robot is outside of the collision avoidance region, i.e., $l_{ij} \geq R_j$, the system is equivalent to the one presented in [4] (robot moving in a free space) and stability analysis presented in this paper still holds.

If the robot is in the collision avoidance region of the other robot time derivative of the Lyapunov-like function is calculated as follows:

$$\frac{dV}{dt} = \sum_{i=1}^{N} \left[e_{ix}\dot{e}_{ix} + e_{iy}\dot{e}_{iy} + e_{i\theta}\dot{e}_{i\theta} + \sum_{j=1, j\neq i}^{N} \left(\frac{\partial V_{aij}}{\partial e_{ix}}\dot{e}_{ix} + \frac{\partial V_{aij}}{\partial e_{iy}}\dot{e}_{iy} \right) \right]. \tag{21}$$

Taking into account Equation (15) the above formula can be transformed to the following form:

$$\frac{dV}{dt} = \sum_{i=1}^{N} \left[E_{ix}\dot{e}_{ix} + E_{iy}\dot{e}_{iy} + e_{i\theta}\dot{e}_{i\theta} \right]. \tag{22}$$

Next, using Equation (19) one obtains:

$$\dot{V} = \sum_{i=1}^{N} \left[k_3 E_{ix} e_{iy} e_{i\theta} - k_3 E_{iy} e_{ix} e_{i\theta} - k_3 e_{i\theta}^2 - k_1 E_{ix}^2 \right]. \tag{23}$$

Substituting $E_{ix} = D_i \cos \theta_{iE}$ and $E_{iy} = D_i \sin \theta_{iE}$, $D_i = \sqrt{E_{ix}^2 + E_{iy}^2}$ in the above equation one obtains:

$$\begin{aligned}
\dot{V} &= \sum_{i=1}^{N} \left[k_3 D_i \cos \theta_{iE} e_{iy} e_{i\theta} - k_3 D \sin \theta_{iE} e_{ix} e_{i\theta} - k_3 e_{i\theta}^2 - k_1 D_i^2 \cos^2 \theta_{iE} \right] \\
&= \sum_{i=1}^{N} \left[-\frac{1}{2}k_3 e_{i\theta}^2 + k_3 D_i \cos \theta_{iE} e_{iy} e_{i\theta} - \frac{1}{2}k_3 e_{i\theta}^2 + k_3 D_i \sin \theta_{iE} e_{ix} e_{i\theta} - k_1 D_i^2 \cos^2 \theta_{iE} \right] \\
&= \sum_{i=1}^{N} \left\{ -k_3 \left[(\frac{e_{i\theta}}{\sqrt{2}} - \frac{1}{\sqrt{2}}D_i \cos \theta_{iE} e_{iy})^2 - \frac{1}{2}D_i^2 \cos^2 \theta_{iE} e_{iy}^2 \right] \right. \\
&\quad \left. -k_3 \left[(\frac{e_{i\theta}}{\sqrt{2}} + \frac{1}{\sqrt{2}}D_i \sin \theta_{iE} e_{ix})^2 - \frac{1}{2}D_i^2 \sin^2 \theta_{iE} e_{ix}^2 \right] - k_1 D_i^2 \cos^2 \theta_{iE} \right\}.
\end{aligned}$$

To simplify further calculations, new scalar functions are introduced:

$$a_i = \frac{e_{i\theta}}{\sqrt{2}} - \frac{1}{\sqrt{2}} D_i \cos \theta_{iE} e_{iy}, \qquad b_i = \frac{e_{i\theta}}{\sqrt{2}} + \frac{1}{\sqrt{2}} D_i \sin \theta_{iE} e_{ix}. \tag{24}$$

Taking into account (24) $\dot{V}$ can be written as follows:

$$\dot{V} = \sum_{i=1}^{N} \left[-k_3 a_i^2 - k_3 b_i^2 - k_1 D_i^2 \cos^2 \theta_{iE} + \frac{1}{2} k_3 D_i^2 \cos^2 \theta_{iE} e_{iy}^2 + \frac{1}{2} k_3 D_i^2 \sin^2 \theta_{iE} e_{ix}^2 \right] \tag{25}$$

$$\dot{V} \leq \sum_{i=1}^{N} \left[-k_3 a_i^2 - k_3 b_i^2 - k_1 D_i^2 \cos^2 \theta_{iE} + \frac{1}{2} k_3 D_i^2 e_{iy}^2 + \frac{1}{2} k_3 D_i^2 e_{ix}^2 \right]. \tag{26}$$

The closed-loop system is stable ($\dot{V} \leq 0$) if the following condition is fulfilled:

$$\sum_{i=1}^{N} \left[k_1 \cos^2 \theta_{iE} - \frac{1}{2} k_3 (e_{ix}^2 + e_{iy}^2) \right] > 0. \tag{27}$$

As $\cos^2 \theta_{iE} > 0$ because of the assumption in Proposition 1 it cannot be arbitrarily small, the condition (27) can be met by setting a sufficiently high value of k_1 or by reducing k_3.

Note that the error dynamics (19) with frozen reference velocities can be decomposed into two subsystems (Figure 4). The origin of the system Σ_2 is exponentially stable if $k_3 > 0$. Each of the subsystems is input to state stable (ISS). Stability of the origin may be concluded invoking the small-gain theorem for ISS systems [26].

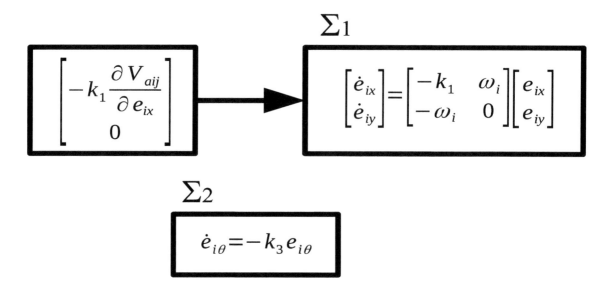

Figure 4. Diagram of the control system in the collision avoidance mode.

The boundedness of the output of the collision avoidance subsystem is necessary to prove stability. Taking the first equation in (19), one can state that if $\frac{\partial V_{aij}}{\partial e_{ix}}$ is sufficiently high (that happens if the robot is very close to the obstacle; there is no problem of boundedness in the other cases (refer to the properties of the APF, Figure 1)), i.e., $\frac{\partial V_{aij}}{\partial e_{ix}} \gg e_{ix}$, $\frac{\partial V_{aij}}{\partial e_{ix}} \gg e_{iy}$, and $\frac{\partial V_{aij}}{\partial e_{ix}} \gg e_{i\theta}$ the error dynamics can be approximated as follows:

$$\dot{e}_{ix} \cong -k_1 \frac{\partial V_{aij}}{\partial e_{ix}}. \tag{28}$$

From the Equation (28) it is clear that $\dot{e}_{ix}$ and $\frac{\partial V_{aij}}{\partial e_{ix}}$ have different signs and as a result $\frac{\partial V_{aij}}{\partial e_{ix}} \dot{e}_{ix} < 0$. To fulfil the condition that $\dot{V}_{aij} = \frac{\partial V_{aij}}{\partial e_{ix}} \dot{e}_{ix} + \frac{\partial V_{aij}}{\partial e_{iy}} \dot{e}_{iy}$ is less than zero the second term on the right hand side must be less than the first one taking their absolute values. This can be obtained by reducing k_3 parameter (refer to Equation (16)). The property $\dot{V}_{aij} \leq 0$ guarantees boundedness of both V_{aij} and $\frac{\partial V_{aij}}{\partial e_{ix}}$. Finally one can state that the collision avoidance block that is input to the system shown in Figure 4 also has bounded output and both error components e_{ix} and e_{iy} in Σ_1 are bounded.

The above is true if the robot is not located close to the boundary of more than one robot at a time. This situation is unlikely because it leads to high controls that increase the distance between the robots quickly and therefore this will not be considered further. ⊡

As shown in [7] collision avoidance is guaranteed if $V_{aij} \leq 0$ and $\lim_{\|[x_i \, y_i]^\top - [x_j \, y_j]^\top\| \to r^+} V_{aij} = +\infty$, $i \neq j$.

Each robot needs information about positions of other robots in its neighbourhood to avoid collision (their orientations are not needed). It can be obtained using on-board sensors with the range equal to or greater than R. In addition, robots need to know their position and orientation errors to calculate the tracking component of the control. This requirement imposes the use of a system allowing localization with respect to the global coordinate frame, because usually, the motion task is defined with respect to it. The author plans to conduct experiments on real robots in the near future. The OptiTrack motion capture system will be used to obtain coordinates of robots (positions and orientations) which is enough for control purposes.

4. Simulation Results

In this section a numerical simulation for a group of $N = 48$ mobile robots is presented. The initial coordinates (both positions and orientations) were random. The goal of the formation was to follow a circular reference trajectory at the same time avoiding collisions between agents. The formation had a shape of a circle. The assignments of robots to particular goal points were also random.

The following settings of the algorithm were used: $k_1 = 0.5$, $k_2 = 0.5$, $k_3 = 1.0$, $t_u = 1$ s, $r = 0.3$ m, $R = 1.2$ m.

Figure 5a shows paths of robots on the (x, y) plane. To make the presentation clearer in Figure 5a–g signals of 45 robots are grey while the 3 selected ones are highlighted in black. In Figure 5h the three selected inter-agent distances are highlighted in black. Figure 5b,c present graphs of x and y coordinates as a function of time. The robots converged to the desired values in 115 s. Figure 5d shows a time graph of the orientations. In Figure 5e,f linear and angular controls, respectively, are shown. Initially and in the transient state, their values were high, exceeding the maximum values of a typical mobile platform. In practical implementation, they should be scaled down to realizable values. Figure 5g presents a time graph of the freeze procedure (refer to Assumption 2) of all robots. Although the drawings are not easily readable (because they include the 'freeze' signal of all robots), one can interpret that the last collision avoidance interaction ends in 108 s. In Figure 5h relative distances between robots are shown. Important information that can be read from this drawing is that no pair of robots reaches the inter-agent distance lower than or equal to $r = 0.3$ m (dashed line). This means that no collision has occurred.

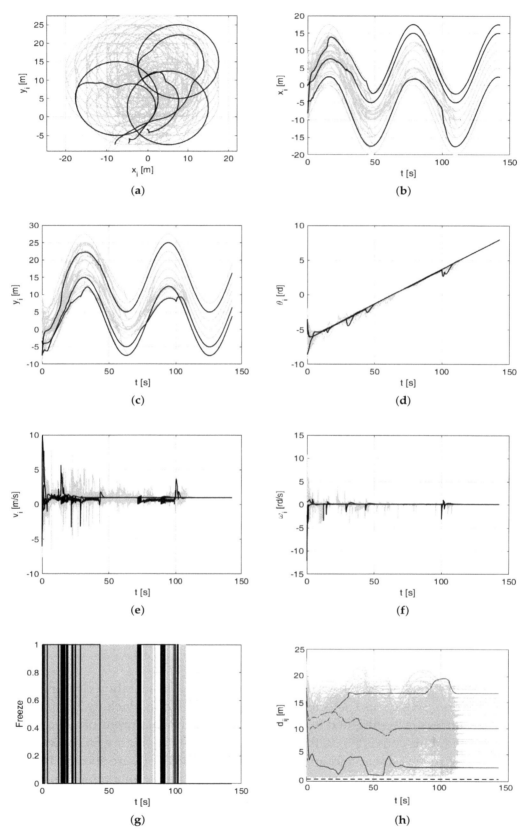

Figure 5. Numerical simulation 1: trajectory tracking for $N = 48$ robots. (**a**) locations of robots in xy-plane, (**b**) x coordinates as a function of time, (**c**) y coordinates as a function of time, (**d**) robot orientations as a function of time, (**e**) linear velocity controls, (**f**) angular velocity controls, (**g**) 'freeze' signals, (**h**) distances between robots.

5. Goal Exchange

This section presents a new control that introduces the ability to exchange goal between agents.

The block diagram of the new control is shown in Figure 6. Two new blocks are included: goal switching and permutation block.

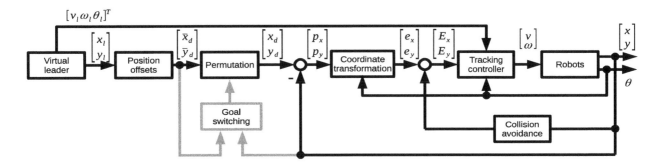

Figure 6. Control system with goal switching.

In the method proposed here Equation (2) is replaced as follows:

$$\bar{x}_{id} = x_l + d_{ix} \qquad \bar{y}_{id} = y_l + d_{iy}. \tag{29}$$

The new variables $\bar{x}_{id}$ and $\bar{y}_{id}$ are not representing the goal position of robot i, but the goal that can be assigned to any robot in the formation.

One can introduce the following aggregated goal coordinate vectors: $\bar{x}_d = [\bar{x}_{1d} \; \dots \; \bar{x}_{Nd}]^T$ and $\bar{y}_d = [\bar{y}_{1d} \; \dots \; \bar{y}_{Nd}]^T$ (numbers in lower index represent the numbers of the goals). The assignment of goals to particular robots is computed using $N \times N$ permutation matrix $P(t)$:

$$x_d = P(t)\bar{x}_d \qquad y_d = P(t)\bar{y}_d. \tag{30}$$

Resulting vectors contained goal coordinates assigned to particular robots $x_d = [x_{1d} \; \dots \; x_{Nd}]^T$ and $y_d = [y_{1d} \; \dots \; y_{Nd}]^T$ (number in lower index represents the number of the robot).

An additional control loop is introduced that acts asynchronously to the main control loop (Figure 6).

Let's assume that at some instant of time t_1 an arbitrary goal m is assigned to the robot k and another goal n is assigned to the robot l. This can be written as:

$$
\begin{aligned}
[x_{kd} \; y_{kd}]^T &= [\bar{x}_{md} \; \bar{y}_{md}]^T \\
[x_{ld} \; y_{ld}]^T &= [\bar{x}_{nd} \; \bar{y}_{nd}]^T.
\end{aligned} \tag{31}
$$

There are ones in permutation matrix $P(t_1)$ at element (m, k) and (n, l) and all other elements in rows m, n and columns k, l were zero.

At some discrete instant in time $t_s >= t_1$ for the pair of robots k and l and their goals m and n the following switching function was computed:

$$\sigma = \begin{cases} 1 & \text{if } ||p_{mk}||^2 + ||p_{nl}||^2 > ||p_{nk}||^2 + ||p_{ml}||^2 \\ 0 & \text{otherwise} \end{cases}, \tag{32}$$

where $p_{ij} = [\bar{x}_{id} - x_j \;\; \bar{y}_{id} - y_j]^T$.

If the switching function σ takes the value of 1 matrix P is changed as follows:

$$P(t) = S_{mn}P(t_s^-), \tag{33}$$

where $P(t_s^-)$ is the permutation matrix before switching.

The elementary matrix S_{mn} is a row-switching transformation. It swaps row m with row n and it takes the following form:

$$
S_{mn} =
\begin{bmatrix}
1 & 0 & \dots & 0 & \dots & 0 & \dots & 0 & 0 \\
0 & 1 & \dots & 0 & \dots & 0 & \dots & 0 & 0 \\
\vdots & \vdots & & \vdots & & \vdots & & \vdots & \vdots \\
0 & 0 & \dots & 0 & \dots & 1 & \dots & 0 & 0 \\
\vdots & \vdots & & \vdots & & \vdots & & \vdots & \vdots \\
0 & 0 & \dots & 1 & \dots & 0 & \dots & 0 & 0 \\
\vdots & \vdots & & \vdots & & \vdots & & \vdots & \vdots \\
0 & 0 & \dots & 0 & \dots & 0 & \dots & 1 & 0 \\
0 & 0 & \dots & 0 & \dots & 0 & \dots & 0 & 1
\end{bmatrix}
\begin{matrix}
\\ \\ \\ \leftarrow m\text{-th row} \\ \\ \leftarrow n\text{-th row} \\ \\ \\ \\
\end{matrix}.
\tag{34}
$$

$$
\begin{matrix}
\uparrow & & \uparrow \\
m\text{-th} & & n\text{-th} \\
\text{col.} & & \text{col.}
\end{matrix}
$$

Transformation (33) describes a process of the goal exchange between agents k and l at time t_s. After that goal m, was assigned to robot l and goal n was assigned to robot k that is equivalent to the following equalities:

$$
\begin{aligned}
[x_{kd} \; y_{kd}]^T &= [\bar{x}_{nd} \; \bar{y}_{nd}]^T \\
[x_{ld} \; y_{ld}]^T &= [\bar{x}_{md} \; \bar{y}_{md}]^T.
\end{aligned}
\tag{35}
$$

Note that the process of goal exchange is asynchronous with the main control loop. It operated at lower frequency because it required communication between remote agents, which is time consuming. The low frequency subsystem is highlighted in grey in Figure 6.

6. Stability of the System with Target Assignment

The goal exchange procedure significantly improves system convergence and reduces the number of collision avoidance interactions between agents. On the other hand, its execution time was not critical for the control of the system.

Stability analysis of the control system with goal switching was conducted using the same Lyapunov-like function (20) as in Section 3.

Proposition 2. *The procedure given by Equation (33) results in a decrease of the Lyapunov-like function Equation (20).*

Proof. A hypothetical position error p_{ij} can be expressed in a local coordinate frame by the following transformation

$$
e_{ij} =
\begin{bmatrix}
\cos(\theta_j) & \sin(\theta_j) \\
-\sin(\theta_j) & \cos(\theta_j)
\end{bmatrix}
p_{ij}.
\tag{36}
$$

that is invariant under scaling, and thus, the following equality holds true:

$$
||e_{ij}|| = ||p_{ij}||
\tag{37}
$$

(notice that index i is the number of the goal and index j is the number of the robot).

Using Equation (37) the switching function (32) can be rewritten as follows:

$$
\sigma =
\begin{cases}
1 & \text{if } ||e_{mk}||^2 + ||e_{nl}||^2 > ||e_{nk}||^2 + ||e_{ml}||^2 \\
0 & \text{otherwise}
\end{cases}.
\tag{38}
$$

To carry out further analysis the Lyapunov-like function (20) will be rewritten as follows:

$$
V = \overbrace{\sum_{i=1}^{N} \frac{1}{2} e_{i\theta}^2}^{V_\theta} + \overbrace{\sum_{i=1}^{N} \sum_{j=1,j\neq i}^{N} V_{aij}}^{V_a} + \frac{1}{2}(e_{1x}^2 + e_{1y}^2) + \dots
$$

$$
+ \underbrace{\frac{1}{2}(e_{kx}^2 + e_{ky}^2)}_{V_{pk}} + \dots + \underbrace{\frac{1}{2}(e_{lx}^2 + e_{ly}^2)}_{V_{pl}} + \dots + \frac{1}{2}(e_{Nx}^2 + e_{Ny}^2).
$$

(39)

Terms V_{pk} and V_{pl} are related to the position errors of robot k and l, respectively. Notice that other terms of the Lyapunov-like function V are invariant under goal assignment as V_a depends only on the distances between agents and V_θ remains constant because all agents share the same reference orientation θ_l. The position error terms related to robots that were not involved in goal exchange were also invariant under goal exchange.

Two cases will be considered further: case 1 at t_s^-, and case 2 at t_s.

In case 1 the sum of position terms of robots k and l can be transformed using (36) and (31) as follows:

$$
\begin{aligned}
V_{p1} = V_{pk} + V_{pl} &= \frac{1}{2}(e_{kx}^2 + e_{ky}^2) + \frac{1}{2}(e_{lx}^2 + e_{ly}^2) \\
&= \frac{1}{2}||[e_{kx}\ e_{ky}]^T||^2 + \frac{1}{2}||[e_{lx}\ e_{ly}]^T||^2 \\
&= \frac{1}{2}||[p_{kx}\ p_{ky}]^T||^2 + \frac{1}{2}||[p_{lx}\ p_{ly}]^T||^2 \\
&= \frac{1}{2}(p_{kx}^2 + p_{ky}^2) + \frac{1}{2}(p_{lx}^2 + p_{ly}^2) \\
&= \frac{1}{2}([x_{kd} - x_k\ y_{kd} - y_k])^2 + \frac{1}{2}([x_{ld} - x_l\ y_{ld} - y_l])^2 \\
&= \frac{1}{2}([\bar{x}_{md} - x_k\ \bar{y}_{md} - y_k])^2 + \frac{1}{2}([\bar{x}_{nd} - x_l\ \bar{y}_{nd} - y_l])^2 \\
&= \frac{1}{2}(||p_{mk}||^2 + ||p_{nl}||^2).
\end{aligned}
$$

In case 2 the sum of position terms of robot k and l, repeating the initial steps above and taking into account (35) is given by:

$$
\begin{aligned}
V_{p2} = V_{pk} + V_{pl} &= \\
&= \frac{1}{2}([x_{kd} - x_k\ y_{kd} - y_k])^2 + \frac{1}{2}([x_{ld} - x_l\ y_{ld} - y_l])^2 \\
&= \frac{1}{2}([\bar{x}_{nd} - x_k\ \bar{y}_{nd} - y_k])^2 + \frac{1}{2}([\bar{x}_{md} - x_l\ \bar{y}_{md} - y_l])^2 \\
&= \frac{1}{2}(||p_{nk}||^2 + ||p_{ml}||^2).
\end{aligned}
$$

Note that V_{p1} (omitting the constant multiplier $\frac{1}{2}$) is the left hand side of the inequality in the first condition of the switching function (32) while V_{p2} is the right hand side of this condition (also omitting the multiplier). This leads to the conclusion that as $V_{p1} > V_{p2}$, goal exchange results in a rapid decrease (discontinuous) of the Lyapunov-like function. $\square$

All other properties of the V still hold including $\dot{V} \leq 0$ if the condition given by Equation (27) is fulfilled.

Proposition 3. *The procedure given by Equation (33) results in a decrease of the sum of the position errors squared.*

Proof. As V_a and V_θ in Equation (39) are invariant under the goal assignment and the sum of all other terms represent the sum of the position errors squared (omitting the constant multiplier $\frac{1}{2}$), the proof of Proposition 3 comes directly from Proposition 2. $\square$

Notice that the sum of the position errors squared can be easily expressed in the global coordinate frame using equality $e_{ix}^2 + e_{iy}^2 = ||[e_{ix}\ e_{iy}]^T||^2 = ||[p_{ix}\ p_{iy}]^T||^2 = p_{ix}^2 + p_{iy}^2$ (refer to Equation (6)) to transform the position error of each robot in (39).

The tracking algorithm presented in Section 2 together with the goal exchange procedure resulted in time intervals continuous algorithm with discrete optimization that uses the Lyapunov-like function as a criterion to be minimized. The discontinuities occur in two situations: when the robot is in the collision avoidance region (the reference trajectory is temporarily frozen and then unfrozen) and when a pair of agents exchange the goals. The set of these discontinuity points was of zero measure. Note that including goal exchange in the control reinforces fulfilling condition (27) because it supports reduction of the component $\sum_{i=1}^{N} \frac{1}{2}k_3(e_{ix}^2 + e_{iy}^2)$.

The presented algorithm does not guarantee optimal solutions but each goal exchange improves the quality of the resulting motion. The total improvement depends significantly on the initial state of the system. In extreme cases, there is a situation in which the initial coordinates are close to optimal. The procedure may lead to no goal exchange, and thus no improvement, even though communication costs have been incurred. On the other hand, if the initial coordinates are not special, benefits of using the procedure are usually considerable.

7. Distributed Goal Exchange

The procedure described in Section 5 can be implemented in a distributed manner. Both key components of the goal switching algorithm; computation of the switching function (32) and permutation matrix transformation (33), involve only two agents. Reliable connection between them was needed as the process was conducted in a sequence of steps. After the agents establish the connection one of them (i-th) sends its position coordinates (x_i, y_i) and goal location to the other (j-th). The second robot computes switching function (32) which is the verdict on the goal exchange. If it is negative the robots disconnect and continue motion to their goals, otherwise robot j sends its position coordinates (x_j, y_j) and goal location to robot i. This part is critical in the process and must be designed carefully to ensure correct task execution.

To make distributed goal exchange possible, the robots must be equipped with the on-board radio modules allowing inter-agent communication. Even if not all pairs of agents were capable of communicating, the goal exchange algorithm improved the result. On the other hand, without communication between all pairs of agents the algorithm did not fail. In the vast majority of environments there was no problem with the communication range (current technology allows communication through many routers, base stations and peers). The exceptions may be space and oceanic applications. They will not be considered here. The author plans to conduct laboratory experiments where there are no restrictions on communication. From the practical point of view, robots need to know the network addresses of other robots and sequentially attempt to establish connection and, if successful, exchange the goals.

8. On Some Generalization

The condition (32) can be rewritten in more general form as follows:

$$\sigma = \begin{cases} 1 & \text{if } ||p_{mk}||^n + ||p_{nl}||^n > ||p_{nk}||^n + ||p_{ml}||^n \\ 0 & \text{otherwise} \end{cases} \tag{40}$$

Taking $n = 1$ leads to the shortest path criterion that seems to be natural in many cases because the shortest path induces lower motion cost. Unfortunately this observation may not be true in a cluttered case. This will be shown later in this section. Note that in the cases for $n \neq 2$ the Lyapunov analysis is much more complex. In [23] the similar method that results in the shortest total distance to the goals is presented.

Several specific scenarios for the simple case of two robots are analysed further. These scenarios should be treated as an approximation of a real case because typically paths of real robots are not straight lines in the case of the platforms that are not fully actuated (like the differentially driven mobile platform considered here).

In Figure 7 two robots are the same distance away from their goals. Initially the goals were assigned as marked with dashed arrows. The goal exchange procedure led to the assignment marked with continuous arrows. The resulting paths were less collisional or even non-collisional (as they are parallel). The new assignment was optimal using both the shortest path criterion ($n = 1$) and quadratic criterion ($n = 2$).

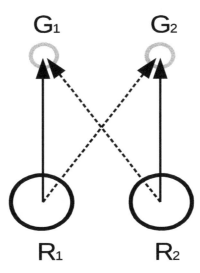

Figure 7. Two robot-goal assignment—case 1.

Figure 8 shows a case in which two robots and their goals lie on a straight line. Initially goal 2 is assigned to robot 1 and goal 1 is assigned to robot 2. This situation caused a saddle point because during the motion to the goal R_1 stays on the path of R_2. One can observe that the shortest path criterion ($n = 1$) produces exactly the same result for both possible goal assignments, while quadratic one ($n = 2$) produces the result marked by continuous arrows. By the assignment $R_1 - G_1$ and $R_2 - G_2$ the goals can be reached by the robots without bypass manoeuvre. This is one of the examples showing the significant advantage of the quadratic criterion over the shortest path criterion proposed in [23].

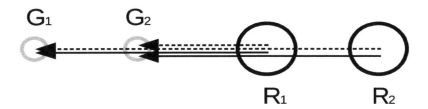

Figure 8. Two robot-goal assignment—case 2.

In Figure 9 R_1 is exactly at the goal G_2 assigned to it. G_1 was assigned to R_2. The quadratic criterion resulted in the opposite assignment (continuous arrows). Notice that for the shortest path criterion the collision avoidance interaction between R_1 and R_2 was possible. For this type of situation the quadratic criterion resulted in goal exchange for all locations in the hatched circle. The opposite situation is presented in Figure 10. Fields of squares shown in the figures represent values of cost functions for two possible goal assignments (compare two left-hand side squares with right-hand square). The quadratic criterion resulted (in comparison to the shortest path) in a higher cost function for assigning far goals to the robots. It favoured a larger number of short assignments instead of the lower number of farther ones. This promoted the reduction of collision interaction situations.

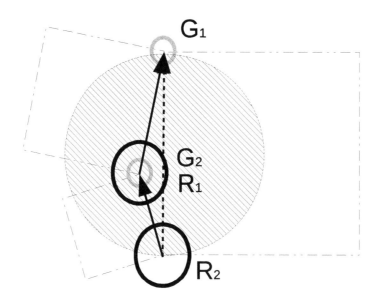

Figure 9. Two robot-goal assignment—case 3.

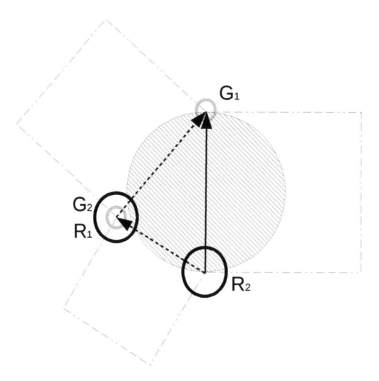

Figure 10. Two robot-goal assignment—case 4.

In Figure 11 certain positions of the robot R_1 and goals G_1 and G_2 have been assumed. If robot R_2 is located in the hatched area the quadratic criterion assigns it to G_2, otherwise it is assigned to G_1. Some initial configuration may have led to temporary collision interaction states (i.e., when R_2 initially is located to the left of the R_1) but the saddle point occurrence was not possible. Dashed circles on the sides represent examples of boundary locations (for goal exchange) of the robot R_2.

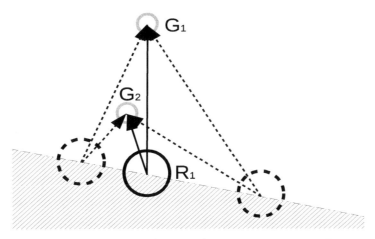

Figure 11. Two robot-goal assignment—case 5.

The considered cases do not cover all possible scenarios, but since there is no formal guarantee that the number of collision avoidance interactions between agents is reduced, they illustrate, together with simulation results, that in typical situations goal exchange procedure leads to the simplification of the control task.

9. Ensuring Integrity

As the goal exchange process involves two agents that are physically separated machines and they communicate through a wireless link the goal exchange process is at the risk of failure. Using a reliable communication protocol like Transmission Control Protocol (TCP) and dividing the process into a sequence of stages acknowledged by the remote host the fault-tolerance of the system can be increased. Assuming that the transmitted packets are encrypted (which is standard nowadays) and the implementation is relatively simple (the author believes that software bugs can be corrected) byzantine fault tolerance (BFT) [27] is not considered here.

The first stage of the goal exchange process is establishing the connection between agents. It is proposed to use the TCP connection because this protocol uses sequence numbers, acknowledgements, checksums and it is the most reliable, widely used communication protocol. Agents know network addresses of the other agents. They can be given in advance, provided by the higher-level system or obtained using a dedicated network broadcasting service. The attempt to connect to the agent that is already involved in the goal exchange process should be rejected. This can be easily and effectively implemented using TCP.

The second stage is transmission of the robot location coordinates and the goal from agent one to another. The receiver computes σ (Equation (32)) and sends back the obtained value. If $\sigma = 0$ agent closes the connection, otherwise they go to the third step.

The third step is a goal exchange that is the most critical part of the process. It must be guaranteed that no goal stays unassigned and no goal can be assigned to more than one robot in the case of agent/communication failure. To fulfil this condition the goal exchange must have all properties of database transaction: it must be atomic, consistent, isolated and durable. In practice this idealistic solution can be approximated by applying one of the widely used algorithms: two-phase commit protocol [28], three-phase commit protocol [29], or Paxos [30]. All of them introduce a coordinator block that is the central point of the algorithm. It can be run (for example as a separate process) on the one of the machines involved in the goal exchange procedure. Notice that in the case of failure (communication error, agent failure, etc.) the operation of goal exchange is aborted. This leads to slower convergence of robots to their desired values but is not critical for the task execution.

10. Simulation Results for Goal Exchange

This section presents numerical simulation of the algorithm extended with goal exchange procedure. The initial conditions are exactly the same as in Section 4 (results for the algorithm without goal exchange). The parameters of the controllers were also the same. The initial value of the permutation matrix was the identity matrix $P(0) = I$.

Figure 12a shows paths of robots on the (x, y) plane. As in the previous experiment signals of 3 robots (out of 48) are highlighted in black. Figure 12b,c present graphs of x and y coordinates as a function of time. The robots converge to the desired values in less than 20 s, a significantly better result than 115 s (without goal exchange). This experiment was completed at 28 s due to faster convergence. Figure 12d shows a time graph of the orientations. In Figure 12e,f linear and angular controls are plotted. They reach constant values in less than 20 s. Figure 12g presents a time graph of the freeze procedure (refer to Assumption 2). It can be seen that the last collision avoidance interaction ends at 13 s. In Figure 12h relative distances between robots are shown. No pair of robots reaches inter-agent distance lower than or equal to $r = 0.3$ m (dashed line). It means that no collision has occurred.

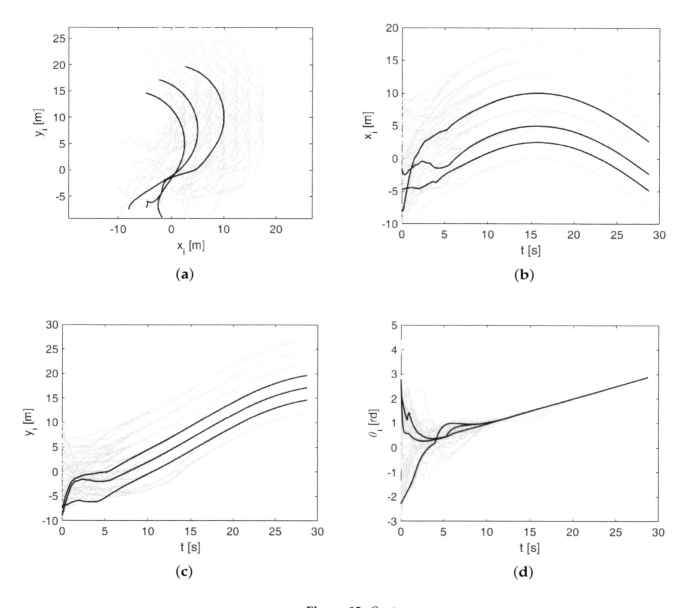

(a)

(b)

(c)

(d)

Figure 12. *Cont.*

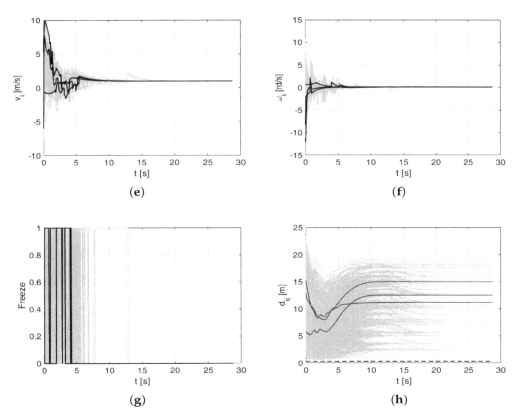

Figure 12. Numerical simulation 2: trajectory tracking for $N = 48$ robots with goal exchange. (a) locations of robots in xy-plane, (b) x coordinates as a function of time, (c) y coordinates as a function of time, (d) robot orientations as a function of time, (e) linear velocity controls, (f) angular velocity controls, (g) 'freeze' signals, (h) distances between robots.

Notice that in the presented simulation no communication delay of the goal exchange procedure was taken into account. Depending on the quality of the network single goal exchange may take even hundreds of milliseconds. On the other hand, as the procedure involves only a pair of agents, many of such pairs can execute goal exchange at the same time. Of course another limitation was the bandwidth of the communication network. These issues will be investigated by the author in the near future. In the presented numerical simulations the number of goal switchings was 238, which is a significant number.

Visualizations of the exemplary experiments are available on the website http://wojciech.kowalczyk.pracownik.put.poznan.pl/research/target-assignment/ts.html.

11. Simulation Results for Saturated Wheel Controls

As the APFs used to avoid collisions are unbounded, the algorithm should be tested for the limited wheel velocities (resulting from the motor velocity limits). This section presents a numerical simulation for the mobile robots with the wheel diameter of 0.0245 m, the distance between wheels amounting to 0.148 m and the maximum angular velocity of 48.6 rd/s.

A special scaling procedure was applied to the wheel controls. The desired wheel velocities were scaled down when at least one of the wheels exceeds the assumed limitation. The scaled control signal u_{iws} is calculated as follows:

$$u_{iws} = s_i u_{iw}, \tag{41}$$

where

$$s_i = \begin{cases} \frac{\omega_{max}}{\omega_{io}} & \text{if} \quad \omega_{io} > \omega_{max} \\ 1 & \text{otherwise} \end{cases}, \tag{42}$$

and

$$\omega_{io} = \max\{|\omega_{iR}|, |\omega_{iL}|\}, \tag{43}$$

where ω_{iR}, ω_{iL} denote right and left wheel angular velocity, ω_{max} is the predefined maximum allowed angular velocity for each wheel.

Figure 13a,b show time graphs of the right and left wheels of the platforms. As in the previous experiments, signals of three robots (out of 48) are highlighted in black. It can be clearly observed that both of them were limited to ±48.6 rd/s. Linear and angular velocities of the platform are shown in Figure 13c,d. Their velocities are lower in comparison to the non-limited case (refer to Figure 12e,f). Figure 13e presents relative distances between the robots. The area below dashed line represents the collision region. It can be seen that no pair of robots has reached it—no collision occurred during this experiment.

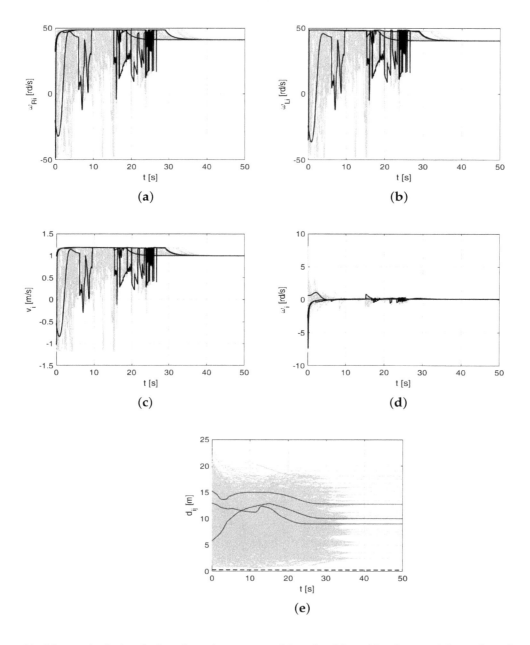

Figure 13. Numerical simulation 3: trajectory tracking for $N = 48$ robots with goal exchange and saturated wheel controls. (**a**) right wheel velocities, (**b**) left wheel velocities, (**c**) linear velocity controls, (**d**) angular velocity controls, (**e**) distances between robots.

12. Conclusions

This paper presents a new control algorithm for the formation of non-holonomic mobile robots. Inter-agent communication is used to check if exchange of goals between robots reduces the system's overall Lyapunov-like function and the sum of position errors squared. The procedure is verified by numerical simulations for large group of non-holonomic mobile robots moving in the formation. The simulations also include the case in which wheel velocity controls are limited. A significant improvement of system convergence is shown. The author plans to conduct extensive tests of the presented algorithm on real two-wheeled mobile robots in the near future.

References

1. Khatib, O. Real-time Obstacle Avoidance for Manipulators and Mobile Robots. *Int. J. Robot. Res.* **1986**, *5*, 90–98. [CrossRef]
2. Leitmann, G.; Skowronski, J. Avoidance Control. *J. Optim. Theory Appl.* **1977**, *23*, 581–591. [CrossRef]
3. Leitmann, G. Guaranteed Avoidance Strategies. *J. Optim. Theory Appl.* **1980**, *32*, 569–576. [CrossRef]
4. De Wit, C.C.; Khennouf, H.; Samson, C.; Sordalen, O.J. Nonlinear Control Design for Mobile Robots. *Recent Trends Mob. Rob.* **1994**, 121–156._0005. [CrossRef]
5. Morin, P.; Samson, C. Practical stabilization of driftless systems on Lie groups: The transverse function approach. *IEEE Trans. Autom. Control* **2003**, *48*, 1496–1508. [CrossRef]
6. Kowalczyk, W.; Michałek, M.; Kozłowski, K. Trajectory Tracking Control with Obstacle Avoidance Capability for Unicycle-like Mobile Robot. *Bull. Pol. Acad. Sci. Tech. Sci.* **2012**, *60*, 537–546. [CrossRef]
7. Mastellone, S.; Stipanovic, D.; Spong, M. Formation control and collision avoidance for multi-agent non-holonomic systems: Theory and experiments. *Int. J. Robot. Res.* **2008**, 107–126. [CrossRef]
8. Do, K.D. Formation Tracking Control of Unicycle-Type Mobile Robots With Limited Sensing Ranges. *IEEE Trans. Control Syst. Technol.* **2008**, *16*, 527–538. [CrossRef]
9. Kowalczyk, W.; Kozłowski, K.; Tar, J.K. Trajectory Tracking for Multiple Unicycles in the Environment with Obstacles. In Proceedings of the 19th International Workshop on Robotics in Alpe-Adria-Danube Region (RAAD 2010), Budapest, Hungary, 23–27 June 2010; pp. 451–456. [CrossRef]
10. Kowalczyk, W.; Kozłowski, K. Leader-Follower Control and Collision Avoidance for the Formation of Differentially-Driven Mobile Robots. In Proceedings of the MMAR 2018—23rd International Conference on Methods and Models in Automation and Robotics, Międzyzdroje, Poland, 27–30 August 2018.
11. Hatanaka, T.; Fujita, M.; Bullo, F. Vision-based cooperative estimation via multi-agent optimization. In Proceedings of the 49th IEEE Conference on Decision and Control (CDC), Atlanta, GA, USA, 15–17 December 2010; pp. 2492–2497. doi:10.1109/CDC.2010.5717384. [CrossRef]
12. Millan, P.; Orihuela, L.; Jurado, I.; Rubio, F. Formation Control of Autonomous Underwater Vehicles Subject to Communication Delays. *IEEE Trans. Control Syst. Technol.* **2014**, *22*, 770–777. [CrossRef]
13. Yoshioka, C.; Namerikawa, T. Formation Control of Nonholonomic Multi-Vehicle Systems based on Virtual Structure. *IFAC Proc. Vol.* **2008**, *41*, 5149–5154. [CrossRef]
14. Borrmann, U.; Wang, L.; Ames, A.D.; Egerstedt, M. Control Barrier Certificates for Safe Swarm Behavior. *IFAC-PapersOnLine* **2015**, *48*, 68–73. [CrossRef]
15. Ames, A.D.; Xu, X.; Grizzle, J.W.; Tabuada, P. Control Barrier Function Based Quadratic Programs for Safety Critical Systems. *IEEE Trans. Autom. Control* **2017**, *62*, 3861–3876. [CrossRef]
16. Srinivasan, M.; Coogan, S.; Egerstedt, M. Control of Multi-Agent Systems with Finite Time Control Barrier Certificates and Temporal Logic. In Proceedings of the 2018 IEEE Conference on Decision and Control (CDC), Miami Beach, FL, USA, 17–19 December 2018; pp. 1991–1996. doi:10.1109/CDC.2018.8619113. [CrossRef]
17. Akella, S. Assignment Algorithms for Variable Robot Formations. In Proceedings of the 12th International Workshop on the Algorithmic Foundations of Robotics, San Francisco, CA, USA, 18–20 December 2016.
18. Caldeira, A.; Paiva, L.; Fontes, D.; Fontes, F. Optimal Reorganization of a Formation of Nonholonomic Agents Using Shortest Paths. In Proceedings of the 2018 13th APCA International Conference on Automatic Control and Soft Computing (CONTROLO), Ponta Delgada, Azores, Portugal, 4–6 June 2018.
19. Alonso-Mora, J.; Baker, S.; Rus, D. Multi-robot formation control and object transport in dynamic environments via constrained optimization. *Int. J. Robot. Res.* **2017**, *36*, 1000–10217. [CrossRef]

20. Alonso-Mora, J.; Breitenmoser, A.; Rufli, M.; Siegwart, R.; Beardsley, P. Image and Animation Display with Multiple Mobile Robots. *Int. J. Robot. Res.* **2012**, *31*, 753–773. [CrossRef]

21. Desai, A.; Cappo, E.; Michael, N. Dynamically feasible and safe shape transitions for teams of aerial robots. In Proceedings of the 2016 IEEE/RSJ International Conference on Intelligent Robots and Systems (IROS), Daejeon Convention Center, Daejeon, Korea, 9–14 October 2016; pp. 5489–5494.

22. Turpin, M.; Michael, N.; Kumar, V. Capt: Concurrent Assignment and Planning of Trajectories for Multiple Robots. *Int. J. Robot. Res.* **2014**, *33*, 98–112. [CrossRef]

23. Panagou, D.; Turpin, M.; Kumar, V. Decentralized goal assignment and trajectory generation in multi-robot networks: A multiple Lyapunov functions approach. In Proceedings of the 2014 IEEE International Conference on Robotics and Automation (ICRA), Hong Kong, China, 31 May–5 June 2014; pp. 6757–6762. [CrossRef]

24. Loria, A.; Dasdemir, J.; Alvarez Jarquin, N. Leader—Follower Formation and Tracking Control of Mobile Robots Along Straight Paths. *IEEE Trans. Control Syst. Technol.* **2016**, *24*, 727–732. [CrossRef]

25. Michałek, M.; Kozłowski, K. Vector-Field-Orientation Feedback Control Method for a Differentially Driven Vehicle. *IEEE Trans. Control Syst. Technol.* **2010**, *18*, 45–65. [CrossRef]

26. Khalil, H.K. *Nonlinear Systems*, 3rd ed.; Prentice-Hall: New York, NY, USA, 2002.

27. Castro, M.; Liskov, B. Practical Byzantine Fault Tolerance. In Proceedings of the Third Symposium on Operating Systems Design and Implementation, New Orleans, LA, USA, 22–25 February 1999.

28. Bernstein, P.; Hadzilacos, V.; Goodman, N. *Concurrency Control and Recovery in Database Systems*; Addison Wesley Publishing Company: Boston, MA, USA, 1987; ISBN 0-201-10715-5.

29. Skeen, D.; Stonebraker, M. A Formal Model of Crash Recovery in a Distributed System. *IEEE Trans. Softw. Eng.* **1983**, *9*, 219–228. [CrossRef]

30. Lamport, L. The Part-Time Parliament. *ACM Trans. Comput. Syst.* **1998**, *16*, 133–169. [CrossRef]

4

Iterative Learning Method for In-Flight Auto-Tuning of UAV Controllers based on Basic Sensory Information

Wojciech Giernacki

Institute of Control, Robotics and Information Engineering, Electrical Department, Poznan University of Technology, Piotrowo 3a Street, 60-965 Poznan, Poland; wojciech.giernacki@put.poznan.pl.

Abstract: With an increasing number of multirotor unmanned aerial vehicles (UAVs), solutions supporting the improvement in their precision of operation and safety of autonomous flights are gaining importance. They are particularly crucial in transportation tasks, where control systems are required to provide a stable and controllable flight in various environmental conditions, especially after changing the total mass of the UAV (by adding extra load). In the paper, the problem of using only available basic sensory information for fast, locally best, iterative real-time auto-tuning of parameters of fixed-gain altitude controllers is considered. The machine learning method proposed for this purpose is based on a modified zero-order optimization algorithm (golden-search algorithm) and bootstrapping technique. It has been validated in numerous simulations and real-world experiments in terms of its effectiveness in such aspects as: the impact of environmental disturbances (wind gusts); flight with change in mass; and change of sensory information sources in the auto-tuning procedure. The main advantage of the proposed method is that for the trajectory primitives repeatedly followed by an UAV (for programmed controller gains), the method effectively minimizes the selected performance index (cost function). Such a performance index might, e.g., express indirect requirements about tracking quality and energy expenditure. In the paper, a comprehensive description of the method, as well as a wide discussion of the results obtained from experiments conducted in the AeroLab for a low-cost UAV (Bebop 2), are included. The results have confirmed high efficiency of the method at the expected, low computational complexity.

Keywords: UAV; auto-tuning; machine learning; iterative learning; extremum-seeking; altitude controller

1. Introduction

1.1. Auto-tuning of UAV Controllers—Context and Novelty

Common availability of low-cost, computationally efficient embedded systems and small size sensors directly influence the development of the construction of unmanned aerial vehicles and their applications, the number of which has been increasing in recent years [1–4]. In every UAV prototype, the need to ensure reliability and flight precision, both in manual and autonomous mode, are key aspects and depend directly on the selection of sensors [5], estimation methods [6], and the quality of position and orientation by controllers resulting from the applied control architecture [7,8]. In addition to the advanced control systems that often require precise models of UAV dynamics [9–12], due to their simplicity and versality, fixed-value controllers with a small number of parameters, are commonly and successfully used [13–16]. They determine the safety of operation, maximum flight duration and the UAV's in-flight behavior. That is why it is so important to learn and systematize the mechanisms of optimal self-tuning of their parameters for various environmental disturbances and for a radical change in the dynamics of the UAV itself due to a change in its total mass. Due to the attractive field of applications of such solutions in many areas (transportation and manipulation tasks performed by one or several UAV units [17–19], precision agriculture [20,21], missions requiring the sensory equipment to be re-armed, rescue operations [22], etc.), one is looking for fast solutions with low computational complexity that work in real-time mode.

While the state-of-the-art analysis shows several computationally complex approaches (requiring numerous repetitions and the use of the UAV model) to batch, optimal auto-tuning of controllers (via heuristic bio-inspired [23,24] and deterministic methods [25]), there has been no method reported to optimize the gains of fixed-value UAV controllers so far. No method has also been reported to do the latter in flight, iteratively and exclusively on the basis of available, periodic, basic sensory information (without using the UAV model)—to indirectly increase the flight duration by minimizing the energy expenditure through shaping a smooth flight characteristic. This issue has been selected as the core of the conducted research. The obtained result in the form of effective machine learning method for auto-tuning of gains of UAV controllers is a novelty presented in this article and thoroughly expands the concept of the method presented in [26] (using the weighted sum of the control error and control signal in predefining expectations for time courses and as a measure of tracking quality in the optimization algorithm). In addition, the most important added value also became:

- assessment and systematization (by means of simulation and experimental studies) of the influence of several environmental factors on the process of auto-tuning of UAV controllers during the flight by the proposed extremum-seeking method. The key issue here is the analysis of results in terms of assessing the quality of work of tuned controllers and the work of the optimization mechanism itself in the following test areas: presence of disturbances (wind gusts), UAV mass change, different sensory sources, flight dynamics/optimized performance index,

- outlining the rules for conducting the auto-tuning process of controllers, so that the automatic exploration of the gain space for individual controllers can be as safe as possible (one needs to keep in mind that the proposed method is not based on any stability criterion, which is its main limitation compared with numerous batch solutions based on models).

1.2. Motivation

In previous research [26], the author has drawn inspiration from the demanding problems of mobile robotics, which the world research centers have been coping with. Examples of such problems can be found in particular challenges of the Mohamed Bin Zayed International Robotics Challenge (http://www.mbzirc.com) [27], where the common denominator are tasks requiring the use of one or a group of UAV units to conduct autonomous flights with high precision in varied conditions (outdoor and indoor) and varied UAV mass. In preparation for the MBZIRC'2020 edition, it turned out that the only currently available auto-tuning algorithm on commercial auto-pilots (as Pixhawk, Naze32, Open Pilot, CC3D), named AutoTune, *"(...) uses changes in flight attitude input by the pilot to learn the key values for roll and pitch tuning. (...) While flying the pilot needs to input as many sharp attitude changes as possible so that the autotune code can learn how the aircraft responds"* [28]. Unfortunately, this solution is problematic due to the tuning safety (especially in prototyping UAV constructions) and control goal set: to provide the most smooth, feasible flight trajectories, which will reduce the control effort to reasonable level, and as a result will be maximally energy efficient. Therefore, in the method considered in this work, a gain tuning of UAV controllers based on dynamic behavior was replaced by more energy-efficient and automatic machine learning technique.

1.3. Related Work

Among numerous approaches to machine learning, and apart techniques using neural networks, which require many learning data sets, the mechanisms based on reward and punishment (as in the case of reinforcement learning approaches) are becoming increasingly common. In [29], Rodriguez-Ramos et al. have taught the control system to land autonomously on a moving vehicle, and in [30] Koch et al. trained a flight controller attitude control of a quadrotor through reinforcement learning. Despite the obviously large number of classic approaches to tuning of fixed-value controllers (Panda presents a whole array of such approaches, of which several dozen are practice-oriented [31]), the optimal techniques of iterative learning are invariably gaining on importance [32–34]. Iterative learning techniques have three desirable attributes, namely: automated tuning, low computational complexity (in optimization algorithms, a decision is made only on the basis of current, cyclic information from the selected performance index—cost function), and fast tuning speed [26,35] (in contrast to reinforcement learning approaches, which requires numerous experiments during the learning that makes it unpractical).

While the methods approximating the gradient of the cost function (first- and second-order optimization algorithms) presented in [25,36] can be quite problematic for UAV auto-tuning from noisy measurements (an aspect for careful comparisons in subsequent author's research), the zero-order optimization methods works efficiently because of the speed of calculations. However, it should be remembered and accepted that the obtained solution may be a local (there is no guarantee to obtain global solutions) or a value near it (depending on the declared level of expected accuracy of calculations ϵ).

Among the zero-order optimization methods presented by Chong & Zak in [25], such as Fibonacci-search, golden-search, equal division, and dichotomy algorithms, especially the first two of region elimination methods—developed by Kiefer [37] are effective in optimal control problems [38]. A broad description of the method based on Fibonacci numbers which was used for UAV altitude controller tuning can be found in the mentioned publication [26] of the author—especially mathematical basics and proofs for the region elimination mechanisms. Therefore, for undisturbed presentation of the proposed new method based on the modified golden-search algorithm used in the auto-tuning of the altitude controller during the UAV flight, only necessary mathematical description has been presented in the remaining part of the paper. Instead, the author paid more attention to the application aspects of the method (by placing the necessary pseudocodes) and a wide analysis of the results obtained from the conducted research experiments.

The paper is structured as follows: in Section 2 the UAV description as a control object and measurement system, as well as considered control system, is presented. Therein, the control purpose is highlighted, and the optimization problem is outlined. In the same Section, the proposed auto-tuning method is introduced, and its mathematical basics are explained. Furthermore, the experimental platform is shown. The comprehensive description of simulation and real-world experiments results with discussion are provided in Section 3. Finally, Section 4 presents conclusions and further work plans.

For a better understanding of the presented content, the most important symbols used in the paper are described in Table 1.

Table 1. Symbols used in this article.

Symbol of Variable	Explanation
α, β	weights (in cost function J)
θ, ϕ, ψ	*roll, pitch, yaw* angles
ρ	golden-search reduction factor
ϵ	expected accuracy in GLD method
$\mathcal{BF}$	body frame of reference
$\mathscr{D}^{(k)}$	considered range for the optimized parameter at k-th iteration
$\mathcal{EF}$	Earth frame of reference
$e(t)$	tracking error (in time domain)
$f(\cdot)$	cost function (in GLD method)
J	performance index (cost function in GLD procedure)

Table 1. *Cont.*

Symbol of Variable	Explanation
k_P, k_I, k_D	proportional/integral/derivative gains
N	minimal number of iterations required to ensure accuracy ϵ
N_b	number of the predefined bootstrap cycles
N_c	number of sampling periods necessary to calculate J at l-th iteration of the GLD
N_{max}	number of sampling periods related to the length of the tuning procedure
$\underline{p}_d$	vector of desired UAV position
$\underline{p}_m$	vector of measured UAV position
t_a	time of gathering information for calculation of J in GLD methods
t_h	flight time horizon
T_f	time constant of a low-pass filter of transfer function
T_p	sampling period for calculation of J
T_s	sampling period in low-pass filter
$u(t)$	control signal (in time domain)
$x^{(k^-)}$	lower bound for optimized parameter at k-th iteration
$x^{(k^+)}$	upper bound for optimized parameter at k-th iteration
$\hat{x}$	candidate point in the optimization procedure
$\hat{x}^*$	iterative estimate of the optimal solution
x_b, y_b, z_b	axes of the $\mathcal{BF}$
x_d, y_d, z_d	desired position coordinates
x_e, y_e, z_e	axes of the $\mathcal{EF}$
x_m, y_m, z_m	measured position of the UAV

2. Materials and Methods

2.1. Multirotor UAV as a Control Object and Its Measurement System

The multirotor UAV can be considered as a multidimensional control plant, being underactuated, strongly non-linear, and highly dynamic with (in general) non-stationary parameters. These features result from its physical structure—especially the use of several propulsion units mounted at the ends of the frame. In addition, measuring, processing, and communication systems are also attached to the middle of this frame—suited for a particular UAV construction. From the perspective of control, the appropriate selection of propulsion units (composed of: brushless direct current motors, electronic speed controllers and propellers) is a key aspect to ensure the expected flight dynamics expressed via thrust (T) and torque ($\underline{M}$) generated by the rotational movement of propellers [39]. By changing the rotational speed, it is possible to obtain the expected position and orientation of the UAV in 3D space, i.e., control of its 6 degrees of freedom (DOFs). The obtained control precision also depends on the quality of sensory information to a large extent. Presently, even in the simplest, low-cost UAVs (Figure 1), in order to determine current position and orientation estimates during the flight (e.g., based on more or less advanced modifications of Kalman filters [6]), the sensory data fusion is used (from 3-axes accelerometer, 3-axes gyroscope, 3-axes magnetometer, pressure sensor, optical-flow sensor, GPS, ultrasound sensor, etc.).

In the paper, two sources of measurements are used in the proposed auto-tuning procedure: on-board UAV avionics (for *roll* and *pitch* angles measurements) and external motion capture system (OptiTrack) (X, Y, Z position, and *yaw* angle). In the UAV autonomous control, to ensure unambiguous description of the UAV's position and orientation in 3D space, the North-East-Down (NED) configuration of the reference system is used, since the on-board measurements are expressed in local coordinate system ($\mathcal{BF}$—Body Frame), and the position control, as well as motion capture measurements are defined in the global one ($\mathcal{EF}$—Earth Frame). In the paper of Xia et al. [40], one may find a better known, basic information about the mechanisms of conversions, e.g., how the posture of the multirotor (its rotational and translational motion) can be described by the relative orientation between the $\mathcal{BF}$ and the $\mathcal{EF}$ with the use of the rotation matrix $R \in SO(3)$.

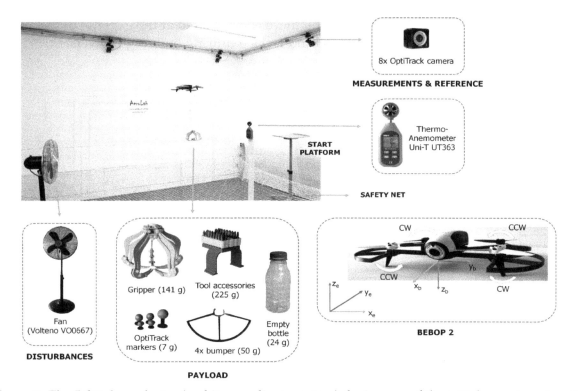

Figure 1. The *Bebop 2* quadrotor (and its coordinate system) during one of the initial experiments with the carrying of payload conducted in *AeroLab* of *Poznan University of Technology*.

2.2. Considered Control System and Control Purpose (Formulation of Optimization Problem)

The control system of multirotor UAV from Figure 2 considered here is based on cascaded control loops. There is control of angles *roll* (θ) and *pitch* (ϕ) around the x_b and y_b axes, according to the set (desired) position in the x_e and y_e axes in faster, internal control loops. Their control is performed in slower external loops. The control of θ and ϕ angles occurs indirectly in the realization of autonomous flight trajectory expressed using the vector of desired position trajectory $\underline{p}_d = (x_d, y_d, z_d)^T$ and desired angle of rotation *yaw* (ψ_d) around the z_e axis. The purpose of the autonomous control is then to ensure the smallest tracking errors $e(t)$ during the UAV flight, i.e., the difference in the values of the reference signals (desired) and output signals (actual/measured) [41]:

$$\underline{e}_p = \underline{p}_d - \underline{p}_m, \tag{1}$$

$$e_\psi = \psi_d - \psi_m, \tag{2}$$

where the m index refers to the measured values.

Bearing in mind that in UAVs the current tracking error information from (1) and (2) is used as the input of a given fixed-value controllers, in the commonly used proportional-derivative (PD) controller structure or proportional-integral-derivative (PID), it is proposed to use this information (as well as information from the output of a given type of controller with control signal $u(t)$) to formulate a measure of the tracking quality during UAV flight, i.e., the cost function/performance index $J(t)$ (see Figure 3), defined as follows:

$$J(t) = \int_0^{t_a} (\alpha\,|e(t)| + \beta\,|u(t)|)\,dt, \tag{3}$$

where t_a is the time of gathering information (to calculate new controller gains) in the optimization procedure. By introducing the penalty for excessive energy expenditure (expressed in the cost function through actual values of the control signal $u(t)$), it is possible to shape expectations towards transients and the controller's dynamics profile (providing smooth or dynamic flight trajectories). At small values of the β, the controller works aggressively, using more energy, often at the expense of the appearance of overshoot, which is undesirable in missions and tasks requiring high flight precision.

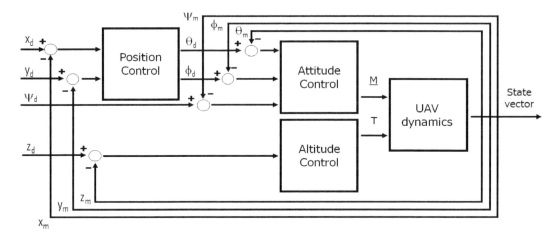

Figure 2. Diagram of considered control system.

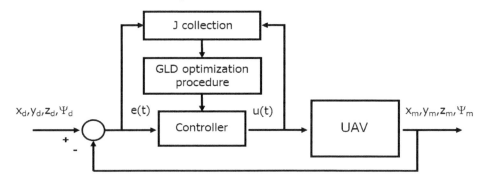

Figure 3. General block diagram of the control system with optimization.

Unconstrained control signal $u(t)$ is calculated from the controller's equation, which in the case of PID structure it is given by

$$u(t) = k_P e(t) + k_I \int_0^{t_h} e(t)\, dt + k_D \frac{d}{dt} e(t), \tag{4}$$

where t_h is a flight time horizon, k_P is the proportional gain, k_I represents the integral gain and k_D the derivative gain, respectively. Gains k_P and k_D are expected to be found using the proposed iterative learning method.

Remark 1. *In the article, when there is a reference to the PID controller, it should be remembered that only the k_P and k_D gains are tuned automatically, whereas the value k_I (used to eliminate the steady-state error) is selected in a manual manner. The proposed auto-tuning method can be used to optimize the gains of any type of controller with three (or even more) parameters; however, this will result in a longer tuning time. Therefore, from the application point of view, it is better to use the procedure presented further in the article.*

Recalling (4), this work deals with the search for the controller gains k_P and k_D, to minimize the cost function (3). That is, the current controller design procedure can be posed as an optimization problem where the solution to the following problem is sought:

$$\min_{k_1,k_2,\ldots,k_N} \quad J(t) = \int_0^{t_a} \left(\alpha \left| e(t) \right| + \beta \left| u(t) \right| \right) dt,$$

$$s.t. \quad \begin{array}{c} 0 \le k_1 \le k_1^{max} \\ 0 \le k_2 \le k_2^{max} \\ \ldots \\ 0 \le k_N \le k_N^{max} \end{array} \tag{5}$$

where k_1^{max}, k_2^{max}, ..., k_N^{max} are upper bounds of the predefined ranges of exploration in the optimization procedure of N controller parameters.

Remark 2. *In the numerical implementation of optimization problem from (5), to quantify the tracking quality by using the cost function (3), its discrete-time version is used (the integration operation is replaced with the sum of samples). Then the cost function is built from the weighted sum of the absolute values of the tracking error samples and the absolute values of the control signal samples (for a given sampling period T_p).*

2.3. Procedure for Tuning of Controllers

To increase the safety in the process of tuning UAV controller parameters during the flight, it is proposed to use the procedure from the flowchart (Figure 4), corresponding to the pyramid of subsequent expectations for the work of control system (Figure 5).

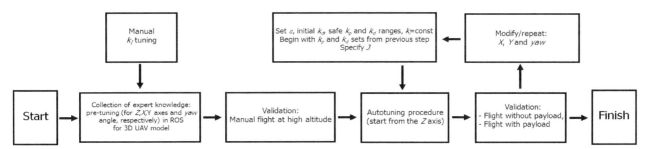

Figure 4. Flowchart for the proposed tuning strategy of the UAV controllers.

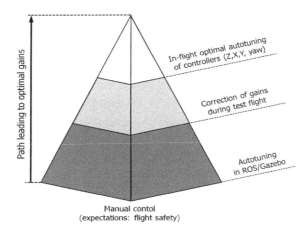

Figure 5. Following steps to obtain optimal gains of controllers.

Remark 3. *Manual tuning of UAV control system prototype is out of scope of this work (to focus on auto-tuning mechanisms). Some useful information regarding UAV controllers prototyping can be found at well-recognized by the UAV community webpages [42–44].*

2.4. Iterative Learning Method for In-Flight Tuning of UAV Controllers—General Idea

Bearing in mind that the search space for the J_{min} for all combinations of controller gains $\underline{k} = (k_1, k_2, \ldots, k_N)^T$ in predefined intervals (ranges) of gains, in the problem outlined in (5) is huge, one needs a fast, effective mechanism for search space exploration. It should be characterized by low computational complexity, and after checking the value of $J(t)$ in a maximum of dozen or several dozen of gain combinations, should be able to provide the value of J_{min} (locally best variant) or a significant improvement compared to the controller's original gains (expressed using e.g., the expected accuracy of the ϵ).

Recalling the publications cited in Section 1.3, iterative learning algorithms are characterized by fast convergence towards the minimum value—especially the region elimination methods (REMs). To be able to use them, one needs to refer to the general idea of iterative learning approaches (proposed by Arimoto et al. in [45]), i.e., minimization of the norm of error (here: cost function) in order to tune particular controller using the periodical repetitiveness of the trials (here: repetitions of the same, predefined trajectory primitives—see Figure 6). Then, to find locally best gains of a particular controller based on a given reference of x_d, y_d, z_d or ψ_d primitive, and corresponding measured value, the performance index $J(t)$ calculated during the flight with given safe ranges of controller parameters, enables the minimum-seeking procedure to find controller gains with respect to the preferred dynamics, and for a given tolerance of the solution.

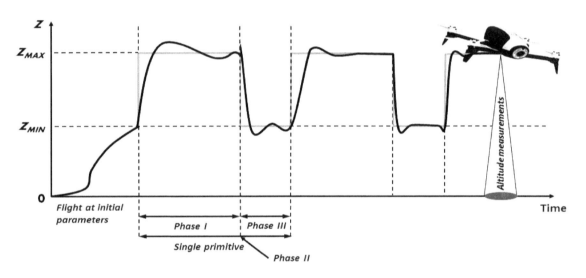

Figure 6. Following steps in iterative learning mechanism for altitude controller tuning.

Remark 4. *Since the method is based solely on the cyclic collection of measurement data to determine $J(t)$, the need to use the UAV model is reduced. However, its knowledge is advantageous when in the simulation conditions it is possible to roughly estimate/determine the maximum values of the elements of the $\underline{k}$ vector, for which the UAV does not lose its stability.*

For every single primitive being used in the optimization procedure, three phases can be distinguished (see Figure 6):

1. Acquisition of measurement data (current, sampled values: x_m, y_m, z_m or ψ_m) for set controller gains with the assumed T_p and the assumed form of the $J(t)$ function,
2. Determination of new controller gains based on the estimated value of the cost function from the phase no. 1,
3. Adjusting the controller according to iteratively corrected gains and waiting for the time necessary to stop the transient processes caused by it.

Determining the sequence of controller gains is possible by systematically narrowing the search space. For this purpose, the use of region elimination method based on the zero-order deterministic optimization algorithm (GLD), is proposed.

2.5. Region Elimination Method Based on GLD Algorithm

Let us consider the problem of iterative searching for a particular controller's gain as a problem of reducing the set range of this gain, in which the criterion of stopping the algorithm is the proximity of following solutions, i.e., the value of the cost function in subsequent solutions (for subsequent controller gains), and the convergence of the algorithm ensures the use of the mechanism based on golden-section search from [25] used in REMs.

Principles and assumptions in the GLD method are similar to those in the modified Fibonacci-search method (FIB) proposed in [26]. The most important are: the unimodality assumption of optimized cost function $f(\cdot)$, lack of knowledge about the global minimum (which gave rise to formulation of stopping criteria in the iterative tuning algorithm, e.g., given tolerance to find the minimizer), successively narrowing the range of values inside which the extremum is known to exist according to the definition (2.1) of the fundamental rule for REMs.

Definition 1. *Let us consider an optimization problem of a one-argument unimodal cost function $f : \mathcal{R} \to \mathcal{R}$ within in the predefined range $[x^{(0^-)}, x^{(0^+)}]$ in initial 0th iteration, where $x^{(0^-)} < x^{(0^+)}$ of a unimodal function f. The argument x of this function can be interpreted as a gain of controller (here: k_P or k_D), and the value of f can be understood as the J value (within some horizon) corresponded to it.*

Now, for a pair of two arguments $x^{(1^-)}$ i $x^{(1^+)}$, which lie in the range $[x^{(0^-)}, x^{(0^+)}]$, and which satisfy $x^{(0^-)} < x^{(1^-)} < x^{(1^+)} < x^{(0^+)}$, it is true that:

- *If $x^{(1^-)} > x^{(1^+)}$, then the minimum $\hat{x}^*$ does not lie in $(x^{(0^-)}, x^{(1^-)})$,*
- *If $x^{(1^-)} < x^{(1^+)}$, then the minimum $\hat{x}^*$ does not lie in $(x^{(1^+)}, x^{(0^+)})$,*
- *If $x^{(1^-)} = x^{(1^+)}$, then the minimum $\hat{x}^*$ does not lie in $(x^{(0^-)}, x^{(1^-)})$ and $(x^{(1^+)}, x^{(0^+)})$.*

A region elimination fundamental rule is used to find the $\hat{x}^*$ with the minimum value of f within predefined range, based on repeatedly selection of two arguments from the current range according to symmetrically reduction the range of possible arguments:

$$x^{(1^-)} - x^{(0^-)} = x^{(1^+)} - x^{(0^+)} = \rho(x^{(0^+)} - x^{(0^-)}), \tag{6}$$

where $\rho = \frac{3-\sqrt{5}}{2} = 0.381966$ is a golden-search reduction factor.

Remark 5. *An advantage of using the golden-search reduction factor (according to Algorithm 1) is the fast exploration of the interval, because following values of x (controller gains) are selected to use one of the values of the cost function calculated in the previous iteration. For this purpose, the interval is divided regarding the golden ratio. As a result, of the use of the golden-search reduction factor for a given interval, two new sub-intervals are obtained. For the new intervals, the ratio of the longer length to the shorter length is equal to the ratio of the length of the divided interval to the length of the longer interval.*

Due to this mechanism, and by using the golden-search reduction factor, the time of range exploration is shortened (through the reduction the number of points for which f function needs to be evaluate) or alternatively the f function values can be averaged for the same x in following iterations, i.e., $x^{((k+1)^+)}$ and $x^{(k^-)}$, which is useful in order to reduce the impact of measurement disturbances during the UAV outdoor flight.

Based on predefined initial range $x \in \mathcal{D}^{(0)} = \left[x^{(0^-)}, x^{(0^+)} \right]$, the golden-search algorithm can be implemented according to the pseudo-code presented below (Algorithm 1).

Algorithm 1 Golden-search algorithm.

Step 1. Evaluate the minimal number N of iterations required to provide the sufficient (predefined) value of the ϵ:

$$|x^* - \hat{x}^*| \leq \epsilon(x^{(0^+)} - x^{(0^-)}), \qquad (7)$$

where $|x^* - \hat{x}^*|$ is the absolute value of the difference between the true (unknown minimum x^*) and iterative solution $\hat{x}^*$ (which is assumed to be in the center of $\mathscr{D}^{(N)}$).

Step 2. For iteration $k = 1, \ldots, N$,

1) select a pair of intermediate points $\hat{x}^{(k^-)}$ and $\hat{x}^{(k^+)}$ ($\hat{x}^{(k^-)} < \hat{x}^{(k^+)}$, $\left\{ \hat{x}^{(k^-)}, \hat{x}^{(k^+)} \right\} \in \mathscr{D}^{(k-1)}$),

2) reduce the range to $\mathscr{D}^{(k)}$ based on REM fundamental rule:

   a) $x^{(k+1)} \in \mathscr{D}^{(k)} = \left[x^{(k-1^-)}, \hat{x}^{(k^+)} \right]$ for $f(\hat{x}^{(k^-)}) < f(\hat{x}^{(k^+)})$,

   b) $x^{(k+1)} \in \mathscr{D}^{(k)} = \left[\hat{x}^{(k^-)}, x^{(k-1^+)} \right]$ for $f(\hat{x}^{(k^-)}) \geq f(\hat{x}^{(k^+)})$,

   c) start next iteration $k := k + 1$.

Step 3. Stop the algorithm; put $\hat{x}^* = \frac{1}{2}(x^{(N^+)} + x^{(N^-)})$.

For the given value of ϵ, the minimum number N of iteration in the GLD algorithm can be calculated according to:

$$(1 - \rho)^N \leq \epsilon, \qquad (8)$$

and for $k = 1, \ldots, N$ one may find the pair of intermediate points using

$$\hat{x}^{(k^-)} = x^{(k-1^-)} + \rho(x^{(k-1^+)} - x^{(k-1^-)}), \qquad (9)$$
$$\hat{x}^{(k^+)} = x^{(k-1^-)} + (1 - \rho)(x^{(k-1^+)} - x^{(k-1^-)}). \qquad (10)$$

2.6. Optimal Gain Tuning of a Two-Parameter Controller Based on Bootstrapping Mechanism

In a two-dimensional space of parameters, the vector of parameters $\underline{x} = \begin{bmatrix} x_1 & x_2 \end{bmatrix}^T$ for the cost function $f(\underline{x})$ (calculated from in-flight measurements) can be interpreted as controller gains (here: k_P and k_D). For fast exploration of this space and to give a global character the GLD extremum-seeking procedure, Algorithm 2 is proposed. It is based on the bootstrapping mechanism (see Table 2), for the predefined bootstrap cycles N_b. In considered two-parameter controller tuning, in every single bootstrap, two launch of GLD algorithm (for each of controller gains) are executed to obtain expected value of the ϵ. Firstly, the gain no.1 is tuned (while the gain no. 2 is fixed), and then, the gain no. 2 (for fixed value of the no. 1).

Algorithm 2 Two-parameter controller tuning.

Step 0. Put the bootstrap cycles counter to $l = 0$; for initial $\mathscr{D}_i^{(l)}$ ($i = 1, 2$) define ϵ, N_b, and initial value of the second parameter $x_2^{(l)}$ (take $\hat{x}_2^{(l)^*} = x_2^{(l)}$), set $l := l + 1$.

Step 1. Find the optimal $\hat{x}_1^{(l)^*}$ using the GLD algorithm, with the second parameter fixed at $\hat{x}_2^{(l-1)^*}$.

Step 2. Calculate the optimal $\hat{x}_2^{(l)^*}$ analogously to the method from the Step 1, keeping the first parameter fixed at $\hat{x}_1^{(l)^*}$.

Step 3. If $l < N_b$, increase the bootstrap cycles counter $l := l + 1$, and proceed to Step 1, otherwise stop the algorithm—the optimal solution $\underline{\hat{x}}^* = \begin{bmatrix} \hat{x}_1^{(l)^*}, & \hat{x}_2^{(l)^*} \end{bmatrix}^T$ has been obtained after N_b bootstrap cycles, as desired.

Table 2. Steps in the bootstrapping mechanism.

Bootstrap No.	Gain No.1	Gain No.2
1	Tuning (according to the GLD REM)	Kept constant
1	Kept constant	Tuning
2	Tuning	Kept constant
2	Kept constant	Tuning
...	...	...
N_b	Tuning	Kept constant
N_b	Kept constant	Tuning

To ensure high effectiveness of the proposed method of auto-tuning, one should remember about several important aspects (in configuration and implementation):

- The proposed method requires predefining the initial, admissible ranges for $\underline{x}$, i.e., $\mathscr{D}_i^{(0)} = \left[x_i^{(0^-)}, \ x_i^{(0^+)} \right]$ for $i = 1, 2$. It is a crucial choice from the perspective of ensuring the safety of autonomous flight. If there is a such a possibility, it is strongly recommended to use the expert knowledge about the controller gains (from initial flights on the base of analysis of a rise time and the maximum overshoot, prototyping in virtual environment, default settings of on-board controller, detailed analysis of the UAV feedback control system, etc.),

- For the expected tolerance ϵ, the number N is calculated. $2N$ calculations of f are needed in the tuning of a pair of controller parameters of a single bootstrap,

- The algorithm's execution time depends on: N_b, N, and the time of a single reference primitive, which must be correlated with the expected UAV dynamics and its natural inertia,

- Recalling the most important principles of the zero-optimization method from [26], one needs to have in mind that the proposed method "(...) is iterative-based and collects information about the performance index (on incremental cost function value) at sampling time instants, equally spaced every T_p seconds" during the tuning experiments. Thus, for sampling period T_p, a single evaluation of f value according to Step 2 of Algorithm 1 with a change of a single parameter of controller is performed using Procedure 1 (for symbols from the Table 1).

- The performance index is calculated as

$$\Delta J^{(n)} = J^{(n+1)} + \Delta J^{(n)}, \tag{11}$$

where $\Delta J^{(n)}$ can be obtained from the discrete-time version of Equation (3), which for n-th sample (tracking error and control signal) at time $t = nT_p$ is given by

$$\Delta J^{(n)} = \alpha \left| e_n \right| + \beta \left| u_n \right|. \tag{12}$$

Algorithm 3 Evaluation of performance index (with single change of controller parameter) [26]

Recalling defined N_c, N_{max}, and n for $f(\cdot)$. Then:

- for $n = 1, \ldots, N_c - 1$ with the controller parameters are updated in the previous iteration, the performance index is evaluated using (11) by adding (12); set $J^{(0)} = 0$;
- for $n = N_c$ a single iteration of GLD algorithm is initialized, cost function is stored, and if possible—reduce the range for controller parameters or perform the bootstrap; it results in a transient behavior of the dynamical signal;
- for $n = N_c + 1, \ldots, N_{max}$ tuning is not performed; the controller parameters have been updated; no performance index is collected; transient behavior should decay.

2.7. Signals Acquisition and Their Filtration in the Proposed Method

Bearing in mind that in general to determine the performance index, sensory information is used from sources with different precision of estimation of the position and orientation of an UAV, therefore in the auto-tuning procedure it is proposed to use:

- the signals from the UAV odometry—processed using commonly used Kalman filtration. Thanks to that, it is possible to fuse data from several standard UAV on-board sensors,
- low-pass filtration (presented and tested primarily in [26]), expressed by a transfer function of first-order inertia type

$$G(s) = \frac{k}{1 + T_f s},$$ (13)

where k is its gain, and T_f is a chosen time constant (here: $k = 1$, $T_f = 0.1$ sec.

For the implementation of the GLD method, the discretized, recursive version of the low-pass filter (13) for the chosen sampling period T_s, is used:

$$y(n) = a(n-1) + (1-a)u(n-1),$$ (14)

where

$$a = \exp\left(-T_s/T_f\right),$$ (15)

and $y(n)$ and $u(n)$ are filtered and pure errors at sample n, respectively.

- (optional) measurement information from an external high-precision measurement system—for example, the motion capture system (for indoor flights) or GNSS (outdoor), treated as the ground truth in estimating the difference to UAV avionics measurements.

2.8. Experimental Platform

In the real-world experiments, the low-cost, micro quadrotor Bebop 2 from Parrot company, was used (see Figure 1 and [46]). Since it is equipped in P7 dual-core CPU Cortex 9 processor, 1 GB RAM memory, and 8 GB of flash memory, it is possible to perform on-board state estimation of the UAV using Extended Kalman Filter (EKF) for the data gathered from its on-board sensors listed in Table 3. The Bebop 2 uses the Busybox Linux operating system. Compact sizes of the UAV ($33 \times 38 \times 3.6$ cm with hull) and efficient propulsion units (4×1280 KV BLDC Motor, 7500-12000 rpm), in combination with 2700 mAh battery provide maximum flight time up to 25 minutes and maximum load capacity up to 550 g (which gives a maximum takeoff mass equal to 1050 g, since the UAV weighs 500 g).

Table 3. General characteristic of Bebop 2 sensors.

Parameter	Value
accelerometer & gyroscope	3-axes MPU 6050
pressure sensor (barometer)	MS5607 (analyses the flight altitude beyond 4.9 m)
ultrasound sensor	analyses the flight altitude up to 8 m
magnetometer	3-axes AKM 8963
geolocalization	Furuno GN-87F GNSS module (GPS+GLONASS+Galileo)
Wi-Fi Aerials	2.4 and 5 GHz dual dipole
vertical stabilization camera	photo every 16 ms
camera	14 Mpx 3-axis Full HD 1080p with Sunny 180 fish-eye lens: 1/2.3″

All experimental studies discussed in the article were carried out in `AeroLab` [47], the research space created at the *Institute of Control, Robotics and Information Engineering of Poznan University of Technology* for testing solutions in the field of UAVs flight autonomy, where ground truth is the OptiTrack motion capture system equipped with 8 Prime 13W cameras (with markers placed on the UAV), and a processing unit (PC) equipped with `Motive`—OptiTrack's unified motion capture software platform. The measurement program (Robot Operating System (ROS) node) is executed with the frequency of 100 Hz, control actions with 30 Hz, whereas the tuning methods with 5 Hz. The system is connected to the ground station (Figure 7) to which information about the current position and orientation of the UAV (from motion capture system and UAV) are transmitted. The ground station is the Lenovo Legion Y520 notebook, equipped with Intel Core i7-7700HQ (2.8 GHz frequency), 32 GB DDR4 RAM memory, SSD hard drive and GeForce GTX 1050 2048 MB under Linux Kinetic 16.04 LTS operating system. Such a powerful computer was proposed for the autonomous control of the Bebop 2 UAV, to conduct all necessary calculations at the ground station, including: path planning, data (measurements) processing, autonomous control, auto-tuning of controllers, safety control, etc.

The ground station was also used for tests of the proposed GLD auto-tuning method in simulation environment. These tests were carried out under the control of the ROS, using the open-source flight simulator `Sphinx` [48] and `bebop_autonomy` library [49] extended by models of cascade control system enabling simulation of autonomous flights in x_e and y_e axes(flight for the given coordinates). In the external position control loops, the PID-type controllers have been used.

During the flights, to ensure the safety, Bebop 2 was equipped with 4 bumpers (12.5 g each, made in 3D printing technology) protecting propellers, and in `AeroLab` an additional horizontal safety net was installed to protect it against hard crashes to the ground level. In addition, for security reasons, the priority over the autonomous flight of the drone was allocated to the operator equipped with SkyController 2, enabling manual flight control. Furthermore, a safety button was introduced to cut off the UAV power supply in a situation of imminent danger. It supported initial experiments, where additional safety rope was used.

In experiments on variable mass flights, the UAV was also equipped with a plastic bottle and a gripper (made in 3D printing technology), or alternatively with tool accessories mounted directly on the Bebop (see Figure 1). Additionally, in studies on the influence of environmental disturbances on the auto-tuning process, the UT363 thermo-anometer from Uni-T company was used to measure the air flow speed generated from the Volteno VO0667 fan.

For the simulation and experimental results presented in the next section of the article, a movie clips (available at the webpage http:www.uav.put.poznan.pl), were prepared.

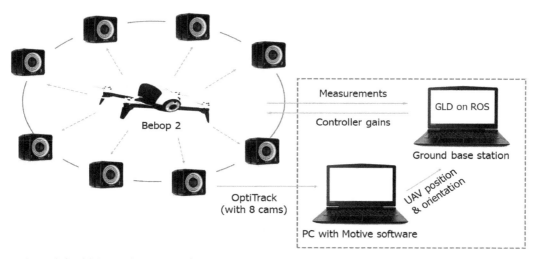

Figure 7. Simplified block diagram of measurement and control signal architecture used during the experiments with in-flight tuning of controllers.

3. Results and Discussion

3.1. Simulation Experiments

Let us consider the problem of searching locally best gains of the altitude PID-type controller of Bebop 2 unmanned aerial vehicle. Default gains are not made available by the Parrot company, hence the problem of finding the best gains (summarized in Table 4) has been treated at the prototyping stage. After development of the 3D model of this UAV (with bumpers) in the Blender software, it was implemented in the ROS/Gazebo environment, giving the physical dimensions, mass and moments of inertia from the real flying robot to its virtual counterpart embedded in the virtual `Aerolab` scenery. This enabled reliable preliminary experiments to be conducted in the simulator.

The research purposes were set as:

- recognizing the nature of optimized function $J = f(k_P, k_D)$ for its various structures (α = var, β = var),

- validation if given gain ranges of k_P and k_D (for a constant, very small value of $k_I = 0.0003$) are safe (i.e., if the closed-loop control system is stable),

- comparative analysis of the effectiveness of GLD and FIB methods.

Table 4. Gains of Bebop's controllers used in experiments.

	X-axis	Y-axis	Z-axis	θ	ϕ	ψ
k_P	0.69	0.69	1.32	default	default	0.07
k_I	0.00015	0.00015	0.0003	default	default	0.00001
k_D	50	50	10.2	default	default	0.9

In the first phase of the research, more than 33 hours of simulation tests were conducted. The results are presented in Figure 8. The same dynamics of the desired reference signal was set as in [26] for the FIB method. Every 12 seconds the UAV changed periodically the flight altitude (1.2→1.9→1.2 m). The value of J was being recorded for 10 sec repeatedly. For each combination of k_P and k_D gains, the J value was averaged from 5 trials. The results of 400 combinations of (k_P, k_D) were recorded for three various J functions. In none of the 2000 trials, the UAV model showed dangerous behavior, and as expected: higher values of k_P correspond to a better quality of reference signal tracking (lower values of J).

In the second phase of the research, the effectiveness of the GLD and FIB methods was compared for three initial values of the k_D and three J function structures. The very promising results are presented in Figure 8 and Table 5. Both methods effectively explore the gains space (k_P, k_D) in search for smaller values of J, avoiding the local minima (they do not "get stuck" in there)—see Figure 8 (right column). Depending on the set k_{Dinit} gain value, both the methods yield in similar k_P values, but various k_D, slowing down the expected tracking dynamics respectively (for larger values of β). It is particularly noteworthy to compare the signals for subsequent set values of β (Figure 9). Bearing in mind the diversity of UAV applications, it is possible to shape the "energy policy", i.e., through an introduction of larger values of β, one obtain a smooth, slower trajectory of the altitude signal, with a smaller control signal amplitudes (for which the β is punishing), which is conducive to extend the flight time.

In relation to the FIB method, an additional time of 96 sec (corresponding to 8 iterations of the auto-tuning algorithm), allows the GLD method in subsequent iterations only to slightly improve the value of the J performance index (respectively by 1.62%, 0.78%, and 2.10%). The introduction of the second bootstrap is justified in the FIB method (improvement by respectively: 11.92%, 4.95%, 1.40%), while in the case of the GLD method, just only one bootstrap provides similar results. The listings from the altitude controller auto-tuning process are available for both methods in the supplementary materials at the `AeroLab` webpage.

Table 5. Results of simulation experiments.

	FIB	GLD	FIB	GLD	FIB	GLD
α	1.0	1.0	0.9	0.9	0.8	0.8
β	0.0	0.0	0.1	0.1	0.2	0.2
k_{Dinit}	2.0	2.0	10.0	10.0	18.0	18.0
k_P range	[0.5,5.0]	[0.5,5.0]	[0.5,5.0]	[0.5,5.0]	[0.5,5.0]	[0.5,5.0]
k_D range	[1.0,20.0]	[1.0,20.0]	[1.0,20.0]	[1.0,20.0]	[1.0,20.0]	[1.0,20.0]
No. of bootstrap cycles	2	2	2	2	2	2
No. of main iterations	48	56	48	56	48	56
Tuning time [sec]	576	672	576	672	576	672
Low-pass filtration	yes	yes	yes	yes	yes	yes
ϵ	0.05	0.05	0.05	0.05	0.05	0.05
Best k_P and k_D values	3.56/8.70	3.70/8.18	4.63/5.99	4.49/8.58	4.23/10.51	4.39/15.44
J_1 (after the 1st bootstrap)	6.8235	5.4575	5.7596	5.4865	5.9641	5.8197
J_{48} (after 48 iter.)	6.0969	5.5024	5.4882	5.4492	5.8815	5.9059
J_{end} (after the tuning proc.)	6.0969	5.4149	5.4882	5.4068	5.8815	6.0298
J_{avg} (average for tuning proc.)	6.7264	5.4754	5.7730	5.5643	6.0345	5.9376

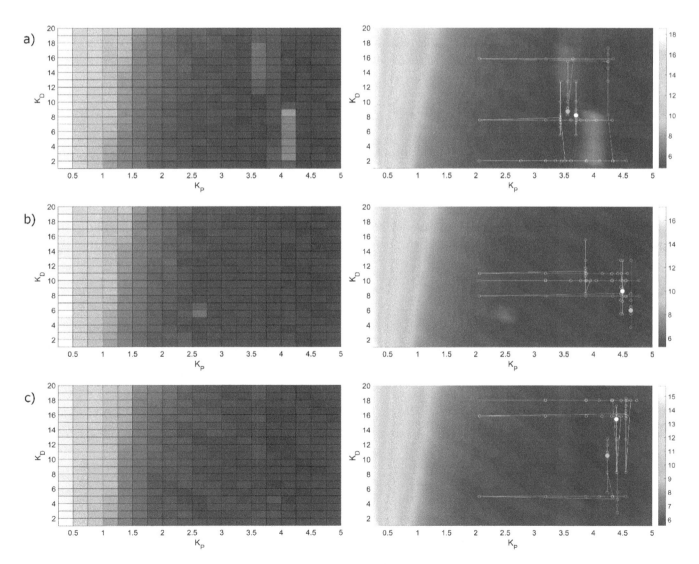

Figure 8. Obtained values of the J performance index for k_P and k_D combinations (left column) and $J = f(k_P, k_D)$ approximations (right column) for: (**a**) $\alpha = 1.0$, $\beta = 0.0$, (**b**) $\alpha = 0.9$, $\beta = 0.1$, (**c**) $\alpha = 0.8$, $\beta = 0.2$. FIB (green) and GLD (white) tuning results for: (**a**) $k_{Dinit} = 2$, (**b**) $k_{Dinit} = 10$, (**c**) $k_{Dinit} = 18$ (marked in red).

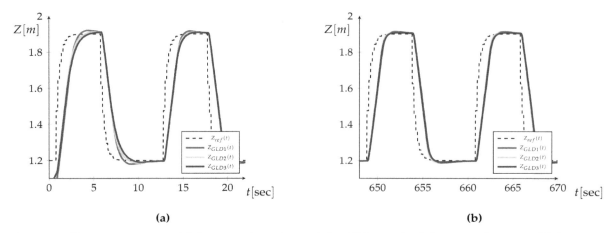

(a)   **(b)**

Figure 9. Time courses for: (**a**) first two iterations of the GLD method (mistuned gains), (**b**) last two iterations (well-tuned gains) for Z_{GLD1} ($\alpha = 1.0$, $\beta = 0.0$), Z_{GLD2} ($\alpha = 0.9$, $\beta = 0.1$), Z_{GLD3} ($\alpha = 0.8$, $\beta = 0.2$).

3.2. Experiments in Flight Conditions

The GLD method was verified in real-world experiments on the same UAV and for the same parameter configuration as in simulation tests. The method was tested with great attention paid to the efficiency of obtaining altitude controller gains and the tracking quality. From variety of conducted experiments, the author decided to present and discussed, a few, which are the most representative. Supplementary materials (video and listings) are available at: http://www.uav.put.poznan.pl.

3.2.1. Uncertainty of Altitude Measurements. Change of Sensory Information Sources

The aim of the experiment was to verify how imprecise and non-stationary the altitude measurements of the UAV flight are in the building, based on its basic on-board avionics only. The motion capture system was used as a ground truth. The results are shown in Figure 10. The task for the UAV was to fly to a fixed altitude of 1 m and hover in the air.

As the average error from registered trials is only 0.80%, the range of actual/instantaneous values ranges from 0.85 m to 1.14 m and increases with the passage of time. Such a dispersion of measurements is a problem and major difficulty in the proposed machine learning procedure used in real-world conditions for altitude controller tuning. Therefore, the motion capture system was used for further estimation of the UAV flight altitude. This eliminates the measurement error as a source of additional errors during the J calculation.

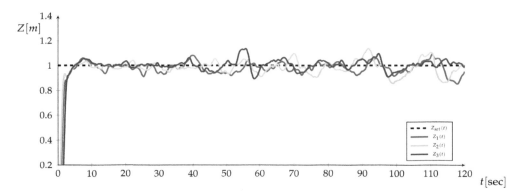

Figure 10. Tracking of the reference altitude Z_{set} by the Bebop 2 UAV in three trials (Z_1–Z_3).

3.2.2. Comparison of the Tuning Effectiveness: FIB vs. GLD Method Used in Real-World Conditions

In the Figure 11, the altitude controller gains during auto-tuning procedure using GLD and FIB methods, are presented. Based on simulation results it was decided to terminate both methods after 48 iterations. Final and average values of J (see Figure 12), are lower for the GLD method: 43.97% and

3.39%, for which the tracking quality is better (Figure 13), e.g., lower overshoots were recorded during the tuning time.

Furthermore, it is worth mentioning that both methods here shown convergence in the vicinity of the two local minima of the $J = f(k_P, k_D)$ function, which were estimated based on the preliminary simulation experiments.

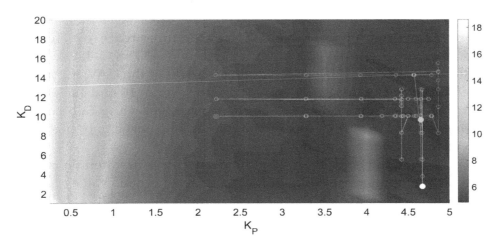

Figure 11. The altitude controller gains and $J = f(k_P, k_D)$ values during auto-tuning process using GLD (white color) and FIB (green) methods; $k_{Dinit} = 10$ (marked in red).

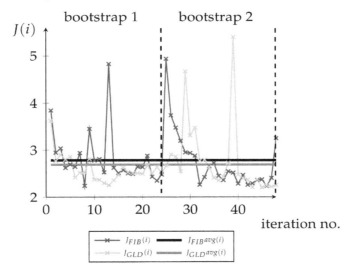

Figure 12. Time courses for the GLD and FIB tuning process in real-world conditions.

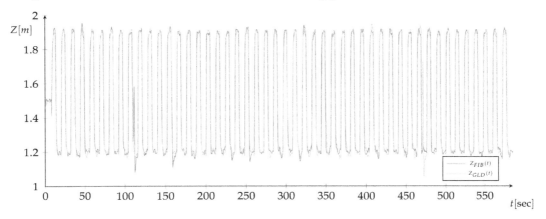

Figure 13. Values of $J(i)$ in consecutive steps (i) of GLD and FIB tuning.

3.2.3. Analysis of the Impact of Environmental Disturbances (Wind Gusts) on the Auto-Tuning Procedure

Usually in scientific world literature presented results of research on UAV flights under wind gust conditions concern the case when the air stream is directed towards the UAV frontally. In real flight conditions, this direction is usually random and variable in time. Thus, it was decided to verify the effectiveness of the GLD auto-tuning method with a low-pass filtration, during the UAV flight, in the stream of the air generated from the rotating fan (1.2 m high), at a distance of 1.8 m, behind the UAV on the left, as in Figure 14.

In the auto-tuning procedure, the disturbances were introduced twice (see Figure 15). In the first phase, the maximum air flow speed was 2.7 m/s, in the second—3.7 m/s. It is a severe disturbance referring to the ratio of physical dimensions to the small weight of the UAV. A complete 56-iterative tuning cycle was conducted. The results are summarized in Figure 16 and Table A1 (see Appendix A), and compared with the results of the auto-tuning from the previous Subsection. Very similar, promising final values of the J performance index were obtained—even only slightly smaller for the case of impact of a wind gust during the GLD procedure.

Determinism of the method is illustrated by the results of 10 first iterations in both trials and iterations no. 29-38, where for different values of J, the calculated k_P gain values are identical. Similar behavior can be observed in the presence of wind gusts (iterations no. 15–20, and 43–48). In the future research, it is worth considering an approach in which two or several UAV units (agents) could be used to parallel measurements and averaging computations during the auto-tuning procedure, resulting in better tuning precision.

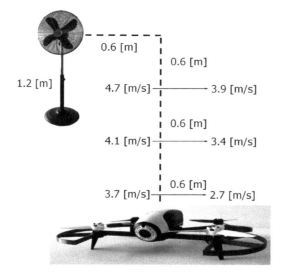

Figure 14. Test bed for research on wind gusts impact on the GLD method.

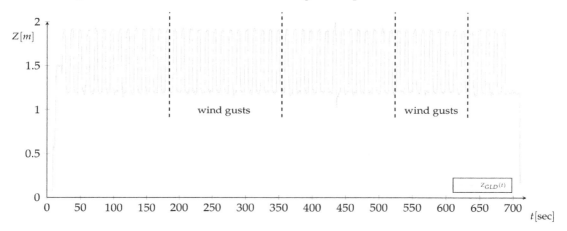

Figure 15. Time course for the GLD tuning process in the presence of wind gusts.

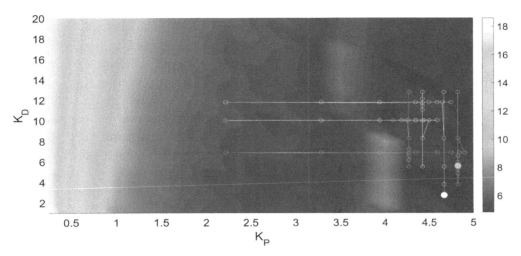

Figure 16. The altitude controller gains and $J = f(k_P, k_D)$ values during auto-tuning process using GLD method for nominal case (white color) and at the presence of wind gusts (green); $k_{Dinit} = 10$ (marked in red).

3.2.4. Flights and Auto-Tuning in UAV Mass Change Conditions

The last interesting aspect of the conducted research was to provide knowledge about the quality of the obtained gains in the context of transport tasks and the use of the GLD method to tune the gains of the altitude controller after changing the total takeoff weight of the UAV. A series of experimental studies was conducted for this purpose.

In the simulation experiments, the efficiency of tuning of the UAV altitude controller using the GLD method was verified in conditions of lifting of the additional payload (jar on the gripper and tool accessories attached to the UAV). The gains of the other controllers (for X and Y axes, and for yaw angle control), were adopted from Table 4. In subsequent simulations, the values α and β of the J function were changed. The results are presented in Table 6, and the search process for the controller's gains is illustrated in the attached video material. Based on the obtained results, it can be noticed that in the case of both payloads tested, the values of k_P were smaller than in the nominal case (flight without payload), and k_D values were larger. Increased starting mass of the UAV forces the use of more thrust to lift the UAV and at the same time—to provide its effective balance, so as not to cause any overshoots (exceeding the given/reference altitude). In the qualitative evaluation of the results of the auto-tuning procedure, the obtained controller using a similar gain value of the proportional part, compensates with a larger gain of k_D the nervous behavior of the UAV (which for particular J function tries to match the dynamics to higher UAV inertia).

Table 6. Results of simulation experiments—flying with: gripper & jar (GRIP), and tool accessories (TOOL); for 2 bootstrap cycles, 56 iterations of the GLD method with low-pass filtration and $\epsilon = 0.05$.

	GRIP	TOOL	GRIP	TOOL	GRIP	TOOL
α	1.0	1.0	0.9	0.9	0.8	0.8
β	0.0	0.0	0.1	0.1	0.2	0.2
k_{Dinit}	10.0	10.0	10.0	10.0	10.0	10.0
k_P range	[0.5,5.0]	[0.5,5.0]	[0.5,5.0]	[0.5,5.0]	[0.5,5.0]	[0.5,5.0]
k_D range	[1.0,20.0]	[1.0,20.0]	[1.0,20.0]	[1.0,20.0]	[1.0,20.0]	[1.0,20.0]
Best k_P and k_D values	2.30/15.11	2.39/18.94	2.20/17.88	1.08/18.94	0.82/15.77	0.67/13.40
J_1 (after the 1st bootstrap)	7.5143	7.9741	7.6119	7.6654	7.5773	7.5454
J_{end} (after the tuning proc.)	7.5150	7.8334	7.5473	7.6591	7.7385	7.4198
J_{avg} (average for tuning proc.)	7.4306	7.9191	7.7877	7.5485	7.5906	7.5454

In the first real-world experiment (Figure 17), the task of the UAV was to start the autonomous flight from a platform with a plastic bottle attached; then, to fly to the point where the GLD auto-tuning procedure begins; finally, to perform 56 iterations of the algorithm in the presence of wind gusts. The drone, using its on-board avionics (including the optical-flow and ultrasonic sensors) moved vertically after stabilizing the position of the gripper, since it recognized its position as altitude equal to 0, and in effect moved upwards, which created a danger. The same behavior was observed in the second experiment, where the UAV task was to compensate its position in the X, Y, and Z axis (refer to supplementary video material). A decision was made to change the type and manner of payload attachment as shown in Figure 18, which played its role, both in the GLD auto-tuning experiments with additional mass, as well as in transportation tasks at designated nominal gains (see Figure 19). In every conducted trial (Table 7), for subsequent J functions, similar behavior was observed as in the case of simulation tests. For example, let us consider the results obtained for $\alpha = 1.0$ (Figure 20). It can be noticed that the time courses with large overshoots (when the controller forces too hard the UAV, wanting to overcome its increased inertia), result in an increase in the value of J and are effectively rejected in the procedure of seeking the smallest value of this performance index. In addition, by analyzing the subsequent values of this index (Figure 21), it can be seen that the selection of the gain value k_D directly implies the UAV vertical flight dynamics profile. This is particularly seen in the first bootstrap (marked in Figure 20).

Auto-tuning in UAV mass change conditions will be the subject of a separate article, while it is worth stressing that the second problem encountered—mentioned at the beginning of the article—i.e., lack of stability criterion based on which it would be possible to estimate safe gains ranges of k_P and k_D for their exploration in the GLD method. Despite its high efficiency and safe operation in tuning of controllers of UAVs with nominal mass or with low extra mass, in case of large payloads (see Figure 22 for the case of 282 g) one can find examples of unstable flights. Then it is strongly recommended to use preliminary simulation tests based on the model. The introduction of the stability criterion into the proposed GLD method is in the area of further research interest of the author [50].

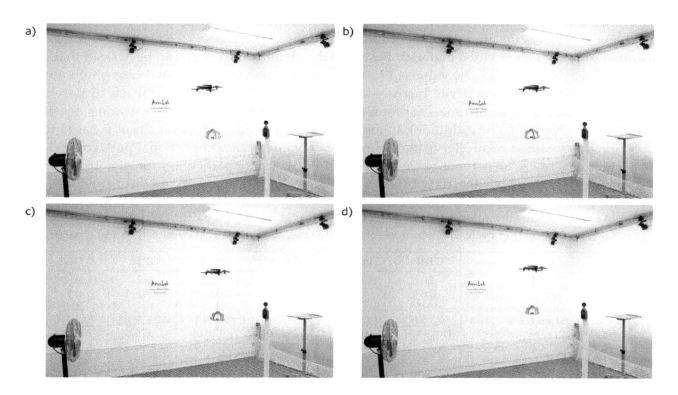

Figure 17. Snapshots from one of the initial research experiments with the auto-tuning of the altitude controller during the flight in the presence of wind gusts and with the mass attached to the UAV on a flexible joint.

Figure 18. The Bebop 2 with additional payload used for in-flight auto-tuning experiments.

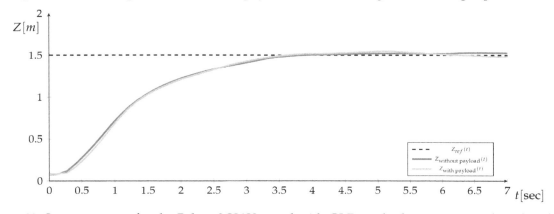

Figure 19. Step responses for the Bebop 2 UAV tuned with GLD method—variants: with and without addition mass (225 g tool accessories).

Table 7. Results of real-world experiments—flying with the payload (tool accessories); for 2 bootstrap cycles, 56 iterations of the GLD method with low-pass filtration and $\epsilon = 0.05$.

	Exp.1	**Exp.2**	**Exp.3**
α	1.0	0.9	0.8
β	0.0	0.1	0.2
k_{Dinit}	10.0	10.0	10.0
k_P range	[0.5,5.0]	[0.5,5.0]	[0.5,5.0]
k_D range	[1.0,20.0]	[1.0,20.0]	[1.0,20.0]
Best k_P and k_D values	3.92/7.85	4.02/9.57	3.20/11.68
J_1 (after the 1st bootstrap)	2.5070	2.4992	2.6096
J_{end} (after the tuning proc.)	1.7583	2.5779	2.5255
J_{avg} (average for tuning proc.)	2.8091	2.4566	2.6553

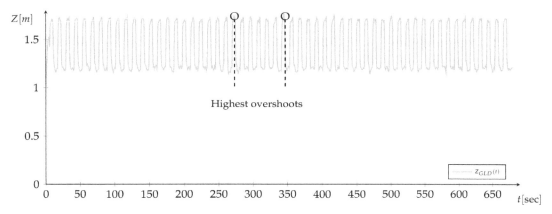

Figure 20. Time course from the tuning experiment via GLD method—variant: tuning of the altitude controller during the UAV flight with an additional (heavy) mass (225 g); the experiment interrupted due to the loss of stability.

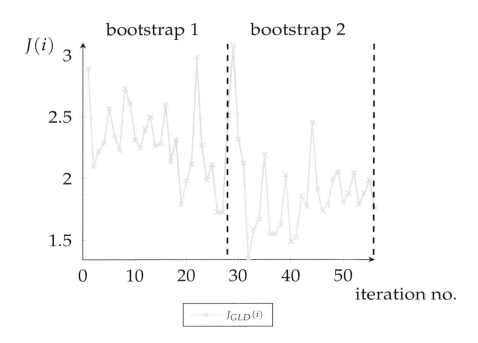

Figure 21. Values of $J(i)$ in consecutive steps (i) of the GLD method (Exp. no. 1)—flying with the payload.

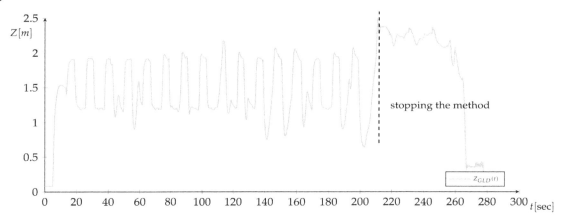

Figure 22. Time course from the real-world experiment (Exp. no. 1): tuning of the altitude controller during the UAV flight with an additional (heavy) mass (225 g).

4. Conclusions and Further Work

In the paper, a new and efficient real-time auto-tuning method for fixed-parameters controllers based on the modified golden-search (zero-order) optimization algorithm and bootstrapping technique, has been presented. The method ensures fast, iterative behavior, and as a result—returns in the worst case the locally best gains of controller, in the best case—globally optimal. The GLD method is fully automated, and uses a low-pass filtration while working in a stochastic environment. It is a model-free approach, but as it has been articulated in the paper, it is good to combine its advantages with initial model-based prototyping, since the method does not use any stability criterion. The author is interested and looking for the mathematical solutions, i.e., in the area of stochastic analysis and probability, which can be easily adapted into the proposed GLD procedure—without increase of its computational complexity. It will be useful in a context of solving mentioned transportation tasks and problems (especially when flying near to the lifting capacity of the UAV).

Acknowledgments: The author would like to thank Bartłomiej Kulecki for his help with software configuration.

Abbreviations

The following abbreviations are used in this manuscript:

BF	Body Frame
CCW	Counter-Clockwise
CW	Clockwise
DOF	Degrees of Freedom
EF	Earth Frame
FIB	Fibonacci-search Method
GLD	Golden-search Method
GNSS	Global navigation satellite system
GPS	Global Positioning System
MBZIRC	Mohamed Bin Zayed International Robotics Challenge
NED	North-East-Down
PD	Proportional-Derivative Controller
PID	Proportional-Integral-Derivative Controller
REM	Region Elimination Method
ROS	Robot Operating System
UAV	Unmanned Aerial Vehicle

Appendix A

Table A1. Comparison of the results of auto-tuning of the UAV's altitude controller using the GLD method—variants: nominal and at the presence of wind gusts.

	Nominal	Disturbed	Nominal	Disturbed	Nominal	Disturbed
No. of Iter.	k_P	k_P	k_D	k_D	J	J
1	2.2190	2.2190	10.0000	10.0000	3.6232	3.9781
2	3.2810	3.2810	10.0000	10.0000	2.7718	2.9733
3	3.2813	3.2813	10.0000	10.0000	2.9025	3.0650
4	3.9377	3.9377	10.0000	10.0000	2.7974	2.9231
5	3.9379	3.9379	10.0000	10.0000	2.8608	2.9099
6	4.3435	4.3435	10.0000	10.0000	2.4185	2.6717
7	4.3436	4.3436	10.0000	10.0000	2.5228	2.7708
8	4.5943	4.5943	10.0000	10.0000	2.5281	3.2745
9	4.1886	4.1886	10.0000	10.0000	2.8080	2.6899
10	4.3435	4.3435	10.0000	10.0000	2.3742	2.8161
11	4.3436	4.0928	10.0000	10.0000	2.3683	3.0077
12	4.4393	4.1886	10.0000	10.0000	2.3004	2.6560
13	4.4393	4.1886	10.0000	10.0000	2.2584	2.7008
14	4.4985	4.2478	10.0000	10.0000	2.3600	2.4583
15	4.4210	4.2661	8.2580	8.2580	2.4742	2.4305
16	4.4210	4.2661	12.7420	12.7420	2.6279	2.5428
17	4.4210	4.2661	5.4854	5.4854	2.5291	2.8694
18	4.4210	4.2661	8.2566	8.2566	2.5097	2.3611
19	4.4210	4.2661	8.2574	8.2574	2.5057	2.3465
20	4.4210	4.2661	9.9700	9.9700	2.4944	2.7500
21	4.4210	4.2661	9.9705	7.1985	2.6712	2.4564
22	4.4210	4.2661	11.0289	8.2569	2.5066	2.5012
23	4.4210	4.2661	11.0292	6.5441	2.6319	2.5666
24	4.4210	4.2661	11.6833	7.1982	2.5153	2.7464
25	4.4210	4.2661	11.6835	6.1397	2.6646	2.6194
26	4.4210	4.2661	12.0877	6.5439	2.9049	2.4165
27	4.4210	4.2661	11.4336	6.5441	2.8642	2.4360
28	4.4210	4.2661	11.6834	6.7939	2.5492	2.3116
29	2.2190	2.2190	11.7607	6.8711	4.6672	4.4838
30	3.2810	3.2810	11.7607	6.8711	3.3072	2.7008

Table A1. *Cont.*

No. of Iter.	Nominal k_P	Disturbed k_P	Nominal k_D	Disturbed k_D	Nominal J	Disturbed J
31	3.2813	3.2813	11.7607	6.8711	3.4729	2.9624
32	3.9377	3.9377	11.7607	6.8711	2.7758	2.4287
33	3.9379	3.9379	11.7607	6.8711	2.7904	2.5296
34	4.3435	4.3435	11.7607	6.8711	2.6357	2.3066
35	4.3436	4.3436	11.7607	6.8711	2.4105	4.4358
36	4.5943	4.5943	11.7607	6.8711	2.3535	2.6047
37	4.5943	4.5943	11.7607	6.8711	2.4362	2.5906
38	4.7493	4.7493	11.7607	6.8711	2.7252	2.4460
39	4.4986	4.7493	11.7607	6.8711	5.4026	2.5969
40	4.5943	4.8450	11.7607	6.8711	2.4769	2.3667
41	4.5943	4.8451	11.7607	6.8711	2.4032	2.3662
42	4.6535	4.9042	11.7607	6.8711	2.3701	2.9292
43	4.6718	4.8268	8.2580	8.2580	2.2139	2.3028
44	4.6718	4.8268	12.7420	12.7420	2.3797	2.5986
45	4.6718	4.8268	5.4854	5.4854	2.2050	2.1790
46	4.6718	4.8268	8.2566	8.2566	2.2468	2.3559
47	4.6718	4.8268	3.7720	3.7720	2.2244	2.4715
48	4.6718	4.8268	5.4846	5.4846	2.2563	2.1976
49	...	4.8268	...	5.4851	...	2.1532
50	...	4.8268	...	6.5435	...	2.5722
51	...	4.8268	...	4.8307	...	2.3696
52	...	4.8268	...	5.4848	...	2.3328
53	...	4.8268	...	5.4850	...	2.2445
54	...	4.8268	...	5.8892	...	2.7051
55	...	4.8268	...	5.2350	...	2.2805
56	...	4.8268	...	5.4848	...	2.2393

References

1. Valavanis, K.; Vachtsevanos, G.J. (Eds.) *Handbook of Unmanned Aerial Vehicles*; Springer: Dordrecht, The Netherlands, 2015.
2. Jordan, S.; Moore, J.; Hovet, S.; Box, J.; Perry, J.; Kirsche, K.; Lewis, D.; Tsz Ho Tse, Z. State-of-the-art technologies for UAV inspections. *IET Radar Sonar Navig.* **2018**, *12*, 151–164. [CrossRef]
3. Hinas, A.; Roberts, J.M.; Gonzalez, F. Vision-Based Target Finding and Inspection of a Ground Target Using a Multirotor UAV System. *Sensors* **2017**, *17*, 2929. [CrossRef] [PubMed]
4. Sandino, J.; Gonzalez, F.; Mengersen, K.; Gaston, K.J. UAVs and Machine Learning Revolutionising Invasive Grass and Vegetation Surveys in Remote Arid Lands. *Sensors* **2018**, *18*, 605. [CrossRef] [PubMed]
5. Dziuban, P.J.; Wojnar, A.; Zolich, A.; Cisek, K.; Szumiński, W. Solid State Sensors—Practical Implementation in Unmanned Aerial Vehicles (UAVs). *Procedia Eng.* **2012**, *47*, 1386–1389. [CrossRef]
6. Gośliński, J.; Giernacki, W.; Królikowski, A. A nonlinear Filter for Efficient Attitude Estimation of Unmanned Aerial Vehicle (UAV). *J. Intell. Robot. Syst.* **2018**. [CrossRef]
7. Urbański, K. Control of the Quadcopter Position Using Visual Feedback. In Proceedings of the 18th International Conference on Mechatronics (Mechatronika), Brno, Czech Republic, 5–7 December 2018; pp. 1–5.
8. Ebeid, E.; Skriver, M.; Terkildsen, K.H.; Jensen, K.; Schultz, U.P. A survey of Open-Source UAV flight controllers and flight simulators. *Microprocess. Microsyst.* **2018**, *61*, 11–20. [CrossRef]
9. Lozano, R. (Ed.) *Unmanned Aerial Vehicles: Embedded Control*; John Wiley & Sons: New York, NY, USA, 2010.
10. Santoso, F.; Garratt, M.A.; Anavatti, S.G. State-of-the-Art Intelligent Flight Control Systems in Unmanned Aerial Vehicles. *IEEE Trans. Autom. Sci. Eng.* **2018**, *15*, 613–627. [CrossRef]
11. Mahony, R.; Kumar, V.; Corke, P. Multirotor aerial vehicles: Modeling, estimation, and control of quadrotor. *IEEE Robot. Autom. Mag.* **2012**, *19*, 20–32. [CrossRef]

12. Ren, B.; Ge, S.; Chen, C.; Fua, C.; Lee, T. *Modeling, Control and Coordination of Helicopter Systems*; Springer: New York, NY, USA, 2012. [CrossRef]

13. Pounds, P.; Bersak, D.R.; Dollar, A.M. Stability of small-scale UAV helicopters and quadrotors with added payload mass under PID control. *Auton. Robots* **2012**, *33*, 129–142. [CrossRef]

14. Li, J.; Li, Y. Dynamic Analysis and PID Control for a Quadrotor. In Proceedings of the 2011 IEEE International Conference on Mechatronics and Automation (ICMA), Beijing, China, 7–10 August 2011; pp. 573–578. [CrossRef]

15. Espinoza, T.; Dzul, A.; Llama, M. Linear and nonlinear controllers applied to fixed-wing UAV. *Int. J. Adv. Robot. Syst.* **2013**, *10*, 1–10. [CrossRef]

16. Lee, K.U.; Kim, H.S.; Park, J.-B.; Choi, Y.-H. Hovering Control of a Quadrotor. In Proceedings of the 2012 12th International Conference on Control, Automation and Systems (ICCAS), JeJu Island, South Korea, 17–21 October 2012; pp. 162–167.

17. Pounds, P.E.; Dollar, A.M. Aerial Grasping from a Helicopter UAV Platform, Experimental Robotics. *Springer Tracts Adv. Robot.* **2014**, *79*, 269–283. [CrossRef]

18. Kohout, P. A System for Autonomous Grasping and Carrying of Objects by a Pair of Helicopters, Master's Thesis, Czech Technical University in Prague, Prague, Czech Republic, 2017.

19. Spica, R.; Franchi, A.; Oriolo, G.; Bülthoff, H.H.; Giordano, P.R. Aerial grasping of a moving target with a quadrotor UAV. In the Proceedings of the 2012 IEEE/RSJ International Conference on Intelligent Robots and Systems, Vilamoura, Portugal, 7–12 October 2012; pp. 4985–4992. [CrossRef]

20. Yang, F.; Xue, X.; Cai, C.; Sun, Z.; Zhou, Q. Numerical Simulation and Analysis on Spray Drift Movement of Multirotor Plant Protection Unmanned Aerial Vehicle. *Energies* **2018**, *11*, 2399. [CrossRef]

21. Rao Mogili, U.M.; Deepak, B.B.V.L. Review on Application of Drone Systems in Precision Agriculture. *Procedia Comput. Sci.* **2018**, *133*, 502–509. [CrossRef]

22. Imdoukh, A.; Shaker, A.; Al-Toukhy, A.; Kablaoui, D., El-Abd, M. Semi-autonomous indoor firefighting UAV. In Proceedings of the 2017 18th International Conference on Advanced Robotics (ICAR), Hong Kong, China, 10–12 July 2017; pp. 310-315. [CrossRef]

23. Duan, H.; Li, P. *Bio-inspired Computation in Unmanned Aerial Vehicles*; Springer: Berlin, Germany, 2014. [CrossRef]

24. Giernacki, W.; Espinoza Fraire, T.; Kozierski, P. Cuttlesh Optimization Algorithm in Autotuning of Altitude Controller of Unmanned Aerial Vehicle (UAV). In Proceedings of the Third Iberian Robotics Conference (ROBOT 2017), Seville, Spain, 22–24 November 2017; pp. 841–852. [CrossRef]

25. Chong, E.K.P.; Zak, S.H. *An Introduction to Optimization*, 2nd ed.; John Wiley & Sons: Hoboken, NJ, USA, 2001.

26. Giernacki, W.; Horla, D.; Báča, T.; Saska, M. Real-time model-free optimal autotuning method for unmanned aerial vehicle controllers based on Fibonacci-search algorithm. *Sensors* **2018**, *19*, 312. [CrossRef]

27. Spurný, V.; Báča, T.; Saska, M.; Pěnička, R.; Krajník, T.; Loianno, G.; Thomas, J.; Thakur, D.; Kumar, V. Cooperative Autonomous Search, Grasping and Delivering in Treasure Hunt Scenario by a Team of UAVs. *J. Field Robot.* **2018**, 1–24. [CrossRef]

28. Automatic Tuning with AUTOTUNE. Ardupilot.org. Available online: http://ardupilot.org/plane/docs/automatic-tuning-with-autotune.html (accessed on 12 November 2018).

29. Rodriguez-Ramos, A.; Sampedro, C.; Bavle, H.; de la Puente, P.; Campoy, P. A Deep Reinforcement Learning Strategy for UAV Autonomous Landing on a Moving Platform. *J. Intell. Robot. Syst.* **2018**, 1–16. [CrossRef]

30. Koch, W.; Mancuso, R.; West, R.; Bestavros, A. Reinforcement Learning for UAV Attitude Control. Available online: https://arxiv.org/abs/1804.04154 (accessed on 12 November 2018).

31. Panda, R.C. *Introduction to PID Controllers—Theory, Tuning and Application to Frontier Areas*; In-Tech: Rijeka, Croatia, 2012. [CrossRef]

32. Rios, L.; Sahinidis, N. Derivative-free optimization: A review of algorithms and comparison of software implementations. *J. Glob. Optim.* **2013**, *56*, 1247–1293. [CrossRef]

33. Spall, J.C. *Introduction to Stochastic Search and Optimization: Estimation, Simulation, and Control*; Wiley: New York, NY, USA, 2003. [CrossRef]

34. Hjalmarsson, H.; Gevers, M.; Gunnarsson, S.; Lequin, O. Iterative feedback tuning: Theory and applications. *IEEE Control Syst. Mag.* **1998**, *18*, 26–41. [CrossRef]

35. Reza-Alikhani, H. PID type iterative learning control with optimal variable coefficients. In the Proceedings of the 2010 5th IEEE International Conference Intelligent Systems, London, UK, 7–9 July 2010; pp. 1–6. [CrossRef]

36. Ghadimi, S.; Lan, G. Stochastic first-and zeroth-order methods for nonconvex stochastic programming. *SIAM J. Optim.* **2013**, *23*, 2341–2368. [CrossRef]

37. Kiefer, J. Sequential minimax search for a maximum. *Proc. Am. Math. Soc.* **1953**, *4*, 502–506. [CrossRef]

38. Brasch, T.; Byström, J.; Lystad, L.P. *Optimal Control and the Fibonacci Sequence*; Statistics Norway, Research Department: Oslo, Norway, 2012; pp. 1–33. Available online: https://www.ssb.no/a/publikasjoner/pdf/DP/dp674.pdf (accessed on 28 December 2018).

39. Theys, B.; Dimitriadis, G.; Hendrick, P.; De Schutter, J. Influence of propeller configuration on propulsion system efficiency of multi-rotor Unmanned Aerial Vehicles. In Proceedings of the 2016 International Conference on Unmanned Aircraft Systems (ICUAS), Arlington, TX, USA, 7–10 June 2016; pp. 195–201. [CrossRef]

40. Xia, D.; Cheng, L.; Yao, Y. A Robust Inner and Outer Loop Control Method for Trajectory Tracking of a Quadrotor. *Sensors* **2017**, *17*, 2147. [CrossRef] [PubMed]

41. Wang, Y.; Gao, F.; Doyle, F. Survey on iterative learning control, repetitive control, and run-to-run control. *J. Process Control* **2009**, *10*, 1589–1600. [CrossRef]

42. Multicopter PID Tuning Guide. Available online: https://docs.px4.io/en/config_mc/pid_tuning_guide_multicopter.html (accessed on 19 November 2018).

43. How to Tune PID I-Term on a Quadcopter. Available online: https://quadmeup.com/how-to-tune-pid-i-term-on-a-quadcopter/ (accessed on 18 November 2018).

44. Quadcopter PID Explained. Available online: https://oscarliang.com/quadcopter-pid-explained-tuning/ (accessed on 18 November 2018).

45. Arimoto, S.; Kawamura, S.; Miyazaki, F. Bettering operation of robots by learning. *J. Robot. Syst.* **1984**, *1*, 123–140. [CrossRef]

46. Parrot BEBOP 2. The Lightweight, Compact HD Video Drone. Available online: https://www.parrot.com/us/drones/parrot-bebop-2 (accessed on 26 November 2018).

47. AeroLab Poznan University of Technology Drone Laboratory Webpage. Available online: http://uav.put.poznan.pl/AeroLab (accessed on 2 December 2018).

48. What is Sphinx. Available online: https://developer.parrot.com/docs/sphinx/whatissphinx.html (accessed on 18 November 2018).

49. bebop_autonomy—ROS Driver for Parrot Bebop Drone (quadrocopter) 1.0 & 2.0. Available online: https://bebop-autonomy.readthedocs.io/en/latest/ (accessed on 18 November 2018).

50. Giernacki, W.; Horla, D.; Sadalla, T.; Espinoza Fraire, T. Optimal Tuning of Non-integer Order Controllers for Rotational Speed Control of UAV's Propulsion Unit Based on an Iterative Batch Method. *J. Control Eng. Appl. Inform.* **2018**, *24*, 22–31.

5

A Graph Representation Composed of Geometrical Components for Household Furniture Detection by Autonomous Mobile Robots

Oscar Alonso-Ramirez [1,*], Antonio Marin-Hernandez [1,*], Homero V. Rios-Figueroa [1], Michel Devy [2], Saul E. Pomares-Hernandez [3] and Ericka J. Rechy-Ramirez [1]

[1]   Artificial Intelligence Research Center, Universidad Veracruzana, Sebastian Camacho No. 5, Xalapa 91000, Mexico; hrios@uv.mx (H.V.R.-F.); erechy@uv.mx (E.J.R.-R.)

[2]   CNRS-LAAS, Université Toulouse, 7 avenue du Colonel Roche, F-31077 Toulouse CEDEX, France; devy@laas.fr

[3]   Department of Electronics, National Institute of Astrophysics, Optics and Electronics, Luis Enrique Erro No. 1, Puebla 72840, Mexico; spomares@inaoep.mx

[*]   Correspondence: oscalra_820@hotmail.com (O.A.-R.); anmarin@uv.mx (A.M.-H.);

Abstract: This study proposes a framework to detect and recognize household furniture using autonomous mobile robots. The proposed methodology is based on the analysis and integration of geometric features extracted over 3D point clouds. A relational graph is constructed using those features to model and recognize each piece of furniture. A set of sub-graphs corresponding to different partial views allows matching the robot's perception with partial furniture models. A reduced set of geometric features is employed: horizontal and vertical planes and the legs of the furniture. These features are characterized through their properties, such as: height, planarity and area. A fast and linear method for the detection of some geometric features is proposed, which is based on histograms of 3D points acquired from an RGB-D camera onboard the robot. Similarity measures for geometric features and graphs are proposed, as well. Our proposal has been validated in home-like environments with two different mobile robotic platforms; and partially on some 3D samples of a database.

Keywords: service robot; graph representation; similarity measure

1. Introduction

Nowadays, the use of service robots is more frequent in different environments for performing tasks such as: vacuuming floors, cleaning pools or mowing the lawn. In order to provide more complex and useful services, robots need to identify different objects in the environment; but also, they must understand the uses, relationships and characteristics of objects in the environment.

The extraction of an object's characteristics and its spatial relationships can help a robot to understand what makes an object useful. For example, the largest surfaces of a table and a bed differ in planarity and height; then, modeling object's characteristics can help the robot to identify their differences. A robot with a better understanding of the world is a more efficient service robot.

A robot can reconstruct the environment geometry through the extraction of geometrical structures on indoor scenes. Wall, floor or ceiling extractions are already widely used for environment characterization; however, there are few studies on extracting the geometric characteristics of furniture. Generally, the extraction is performed on very large scenes, composed of multiple scans and points of view, while we extract the geometric characteristics of furniture from only a single point of view (Figure 1).

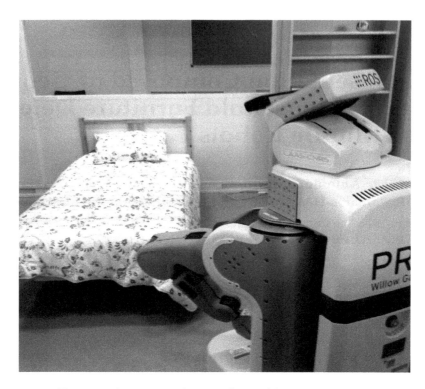

Figure 1. A service robot in a home-like environment.

Human environments are composed of many type of objects. Our study focuses on household furniture that can be moved by a typical human and that is designed to be moved during normal usage.

Our study omits kitchen furniture, fixed bookcases, closets or other static or fixed objects; because they can be categorized in the map as fixed components, therefore, the robot will always know their position. The frequency each piece of furniture is repositioned will vary widely: a chair is likely to move more often than a bed. Furthermore, the magnitude of repositioning a piece of furniture will vary widely: the intentional repositioning of a chair will likely move farther than the incidental repositioning of a table.

When an object is repositioned, if the robot does not extract the new position of the object and update its knowledge of the environment, then the robot's localization will be less accurate.

The main idea is to model these pieces of furniture (offline) in order to detect them on execution time and add them to the map simultaneously, not as obstacles, but as objects with semantic information that the robot could use later.

In this work, the object's horizontal or quasi-horizontal planes are key features (Figure 2). These planes are common in home environments: for sitting at a table, for lounging, or to support another object. The main difference between various planes is whether the horizontal plane is a regular or irregular flat surface. For example, the horizontal plane of a dining table or the horizontal plane of the top of a chest of drawers is different from the horizontal plane of a couch or the horizontal plane of a bed.

Modeling these planes can help a robot to detect those objects and to understand the world. It can help a robot, for example, to know where it can put a glass of water.

This work proposes that each piece of furniture is modeled with graphs. The nodes represent geometrical components, and the arcs represent the relationships between the nodes. Each graph of a piece of furniture has a main node representing a horizontal plane, generally the horizontal plane most commonly used by a human. Furthermore, this principle vertex is normally the horizontal plane a typical human can view from a regular perspective (Figure 2). In our framework, we take advantage of the fact that the horizontal plane most easily viewed by a typical human is usually the horizontal plane most used by a human. The robot has cameras positioned to provide the robot with a point of view similar to the point of view of a typical human.

Figure 2. Common scene of a home environment. Each object's horizontal plane is indicated with green.

To recognize the furniture, the robot uses an RGB-D camera to acquire three-dimensional data from the environment. After acquiring 3D data, the robot extracts the geometrical components and creates a graph for each object in the scene. Because the robot maintains the relationships between all coordinates of its parts, the robot can transform the point cloud to a particular reference frame, which simplifies the extraction of geometric components; specifically, to the robot's footprint reference frame.

From transformed point clouds of a given point of view, the robot extracts graphs corresponding to partial views for each piece of furniture in the scene. The graphs generated are later compared with the graphs of the object's models contained in a database.

The main contributions of this study are:

- a graph representation adapted to detect pieces of furniture using an autonomous mobile robot;
- a representation of partial views of furniture models given by sub-graphs;
- a fast and linear method for geometric feature extraction (planes and poles);
- metrics to compare the partial views and characteristics of geometric components; and
- a process to update the environment map when furniture is repositioned.

2. Related Work

RGB-D sensors have been widely used in robots; therefore, they are excellent for extracting information about diverse tasks in diverse environments. These sensors have been employed to solve many tasks; e.g., to construct 3D environments, for object detection and recognition and in human-robot interaction. For many tasks, but particularly when mobile robots must detect objects or humans, the task must be solved in real time or near real time; therefore, the rate at which the robot processes information is a key factor. Hence, an efficient 3D representation of objects that can quickly and accurately detect them is important.

Depending on the context or the environment, there are different techniques to detect and represent 3D objects. The most common techniques are based on point features.

The extraction of some 3D features is already available in libraries like Point Cloud Library (PCL) [1], including: spin images or fast point feature histograms. Those characteristics provide good results as the quantity of points increases. An increase in points, however, increases the computational time. A large object, such as furniture, requires computational times that are problematic for real-time tasks. A comparative evaluation of PCL 3D features on point clouds was given in [2].

The use of RGB-D cameras for detecting common objects (e.g., hats, cups, cans, etc.) has been accomplished by many research teams around the world. For example, a study [3] presented an approach based on depth kernel features that capture characteristics such as size, shape and edges. Another study [4] detected objects by combining sliding window detectors and 3D shape.

Others works ([5,6]) have followed a similar approach using features to detect other types of free-form objects. For instance, in [5], 3D-models were created and objects detected simultaneously by using a local surface feature. Additionally, local features in a multidimensional histogram have been combined to classify objects in range images [6]. These studies used specific features extracted from the objects and then compared the extracted features with a previously-created database, containing the models of the objects.

Furthermore, other studies have performed 3D object-detection based on pairs of points from oriented surfaces. For example, Wahl et al. [7] proposed a four-dimensional feature invariant to translation and rotation, which captures the intrinsic geometrical relationships; whereas Drost et al. [8] have proposed a global-model description based on oriented pairs of points. These models are independent from local surface-information, which improves search speed. Both methods are used to recognize 3D free-form objects in CAD models.

In [9], the method presented in [8] was applied to detect furniture for an "object-oriented" SLAM technique. By detecting multiple repetitive pieces of furniture, the classic SLAM technique was extended. However, they used a limited range of types of furniture, and a poor detection of furniture was reported when the furniture was distant or partially occluded.

On the other hand, Wu et al. [10] proposed a different object representation to recognize and reconstruct CAD models from pieces of furniture. Specifically, they proposed to represent the 3D shape of objects as a probability distribution of binary variables on a 3D voxel grid using a convolutional deep belief network.

As stated in [11], it is reasonable to represent an indoor environment as a collection of planes because a typical indoor environment is mostly planar surfaces. In [12], using a 3D point cloud and 2D laser scans, planar surfaces were segmented, but those planes are used only as landmarks for map creation. In [13], geometrical structures are used to describe the environment. In this work, using rectangular planes and boxes, a kitchen environment is reconstructed in order to provide to the robot a map with more information about, i.e., how to use or open a particular piece of furniture.

A set of planar structures to represent pieces of furniture was presented in [14], which stated that their planar representations "have a certain size, orientation, height above ground and spatial relation to each other". This method was used in [15] to create semantic maps of furniture. This method is similar to our framework; however, their method used a set of rules, while our method uses a probabilistic framework. Our method is more flexible, more able to deal with uncertainty, more able to process partial information and can be easily incorporated into many SLAM methods.

In relation to plane extraction, a faster alternative than the common methods for plane segmentation was presented in [16]. They used integral images, taking advantage of the structured point cloud from RGB-D cameras.

Another option is the use of semantic information from the environment to improve the furniture detection. For example in [17], geometrical properties of the 3D world and the contextual relations between the objects were used to detect objects and understand the environment. By using a Conditional Random Field (CRF) model, they integrated object appearance, geometry and relationships with the environment. This tackled some of the problems with feature-based approaches, including pose variation, object occlusion or illumination changes.

In [18], the use of the visual appearance, shape features and contextual relations such as object co-occurrence was proposed to semantically label a full 3D point cloud scene. To use this information, they proposed a graphical model isomorphic to a Markov random field.

The main idea of our approach is to improve the understanding of the environment by identifying the pieces of furniture, which is still a very challenging task, as stated in [19]. These pieces of furniture will be represented by a graph structure as a combination of geometrical entities.

Using graphs to represent environment relations was done in [20,21]. Particularly, in [20], a semantic model of the scene based on objects was created; where each node in the graph represented an object and the edges represented their relationship, which were also used to improve the object's detection. The work in [21] used graphs to describe the configuration of basic shapes for the detection of features over a large point cloud. In this case, the nodes represent geometric primitives, such as planes, cylinders, spheres, etc.

Our approach uses a representation similar to [21]. Each object is decomposed into geometric primitives and represented by a graph. However, our approach differs because it processes only one point cloud at a time, and it is a probabilistic framework.

3. Furniture Model Representation and Similarity Measurements

Our proposal uses graphs to represent furniture models. Specifically, our approach focuses on geometrical components and relationships in the graph instead of a complete representation of the shape of the furniture.

Each graph contains the furniture's geometrical components as nodes or a vertex. The edges or arcs represent the adjacency of the geometrical components. These geometrical components are described using a set of features that characterize them.

3.1. Furniture Graph Representation

A graph is defined as an ordered pair $G = (V, E)$ containing a set V of vertices or nodes and a set E of edges or arcs. A piece of furniture F^i is represented by a graph as follows:

$$F^i = (V^i, E^i), \qquad \text{with } i = [1, ..., N_f]$$
(1)

where F^i is an element from the set of furniture models $\mathcal{F}$; the sets V^i and E^i contain the vertices and edges associated with the ith class; and N_f is the number of models in the set $\mathcal{F}$, i.e., $|\mathcal{F}| = N_f$.

The set of vertices V^i and the set of edges E^i are described using lists as follows:

$$V^i = \{v_1^i, v_2^i, ..., v_{n_v^i}^i\}$$
$$E^i = \{e_1^i, e_2^i, ..., e_{n_e^i}^i\}$$
(2)

where n_v^i and n_e^i are the number of vertices and edges, respectively, for the ith piece of furniture.

The functions $V(F^i)$ and $E(F^i)$ are used to recover the corresponding lists of vertex and edges of the graph F^i.

An edge e_j^i is the jth link in the set, joining two nodes for the ith piece of furniture. As connections between nodes are a few, we use a simple list to store them. Thus, each edge e_j^i is described as:

$$e_j^i = (a, b)$$
(3)

where a and b correspond to the linked vertices $v_a^i, v_b^i \in V^i$, such that $a \neq b$; and since the graph is an undirected graph, $e_j^i = (a, b) = (b, a)$.

3.2. Geometric Components

The vertices on a furniture graph model are geometric components, which roughly correspond to the different parts of the furniture. For instance, a chair has six geometric components: one horizontal plane for sitting, one vertical plane for the backrest and four tubes for the legs. Each component has different characteristics to describe it.

Generally, a geometric component Gc^k is a non-homogeneous set of characteristics:

$$Gc^k = \{ft_1^k, ft_2^k, ..., ft_{n_{ft}^k}^k\}$$
(4)

where k designates an element from the set $\mathcal{G}c$, which contains N_k different types of geometric components, and n_{ft}^k is the number of characteristics of the kth geometric component. Characteristics or features of a geometrical component can be of various types or sources. A horizontal plane, for example, includes characteristics of height, area and relative measures.

Each vertex v_j^i is then a geometric component of type k. The function $G_c(v_j^i)$ returns then the type and the set of features of geometric component k.

Figure 3 shows an example of a furniture model F^i. It is composed of four vertices ($n_v^i = 4$) and three edges ($n_e^i = 3$). There are three different types of vertices because each geometric component ($N_k = 3$) is represented graphically with a different shape of the node.

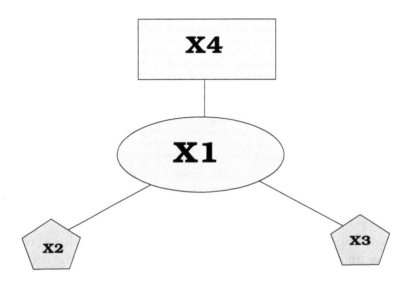

Figure 3. Example of a furniture graph with four nodes and three edges. The shape of the node correspond to a type of geometric component.

Despite the simplified representation of the furniture, the robot cannot see all furniture components from a single position.

3.3. Partial Views

At any time, the robot's perspective limits the geometric components the robot can observe. For each piece of furniture, the robot has a complete graph including every geometric component of the piece of furniture, then some subgraphs should be generated corresponding to different views that a robot can have. Each sub-graph contains the geometric components the robot would observe from a hypothetical perspective.

Considering the following definition, a graph H is called a subgraph of G, such that $V(H) \subseteq V(G)$ and $E(H) \subseteq E(G)$, then a partial view Fp^i for a piece of furniture is a subgraph from F^i, which is described as:

$$Fp^i = (\tilde{V}^i, \tilde{E}^i) \tag{5}$$

such that $\tilde{V}^i \subseteq V^i$ and $\tilde{E}^i \subseteq E^i$. The number of potential partial views is equal to the number of possible subsets in the set F^i; however, not all partial views are useful. See Section 4.3.

In order to match robot perception with the generated models, similarity measurements for graphs and geometric components should be defined.

3.4. Similarity Measurements

3.4.1. Similarity of Geometric Components

Generally, a similarity measure s_{Gc} of two type k geometric components Gc^k and $Gc^{k\prime}$ is defined as:

$$s_{Gc}^k(Gc^k, Gc^{k\prime}) = 1 - \sum_i^{n_{ft}^k} w_{Gi}^k d(ft_i^k, ft_i^{k\prime}) \tag{6}$$

where k represents the type of geometric component, w_{Gi} are weights and $d(ft_i^k, ft_i^{k'})$ is a function of the difference of the ith feature of the geometric components Gc^k and $Gc^{k'}$, defined as follows:

$$\delta\varphi = \frac{|ft_i^k - ft_i^{k'}| - \epsilon_{ft_i}}{ft_i^k} \tag{7}$$

then:

$$d(ft_i^k, ft_i^{k'}) = \begin{cases} 0, & \delta\varphi < 0 \\ \delta\varphi, & 0 \le \delta\varphi \le 1 \\ 1, & \delta\varphi > 1 \end{cases} \tag{8}$$

where ϵ_{ft_i} is a measure of the uncertainty of the ith characteristic. ϵ_{ft_i} is considered a small value that specifies the tolerance between two characteristics. Equation (7) normalizes the difference; whereas Equation (8) equals zero if the difference of two characteristics is within the acceptable uncertainty ϵ_{ft_i} and equals one when they are totally different.

3.4.2. Similarity of Graphs

Likewise, the similarity s_F of two furniture graphs (or partial graphs) F^i and $F^{i'}$ is defined as:

$$s_F(F^i, F^{i'}) = \sum_j^{n_v^i} w_{Fj} s_{Gc}^k(Gc_j^k, Gc_j^{k'}) \tag{9}$$

where w_{Fj} are weights, corresponding to the contribution of the similarity s_{Gc} between the corresponding geometric components j to the graph model F^i.

It is important to note that:

$$\sum_i^{n_{ft}^k} w_{Gi}^k = 1$$
$$\sum_j^{n_v^i} w_{Fj} = 1 \tag{10}$$

In the next section, values for Equations (6) and (9), in a specific context and environment, will be provided.

4. Creation of Models and Extraction of Geometrical Components

In order to generate the proposed graphs, 3D models per furniture are required; so that geometrical components can be extracted.

Nowadays, it is possible to find online 3D models for a wide variety of objects and in many diverse formats. However, those 3D models contain many surfaces and components, which are never visible from a human (or a robot) perspective (e.g., the bottom of a chair or table). It could be possible to generate the proposed graph representation from those 3D models; nevertheless, it would be necessary to make some assumptions or eliminate components not commonly visible.

At this stage of our proposal, the particular model of each piece of furniture is necessary is necessary, not a generic model of the type of furniture. Furthermore, it was decided to construct the model for each piece of furniture from real views taken by the robot; consequently, visible geometrical components of the furniture can be extracted, and then, an accurate graph representation can be created.

Furniture models were generated from point clouds obtained with an RGB-D camera mounted on the head of the robot, in order to have a similar perception to a human being. The point clouds were merged together with the help of an Iterative Closest Point (ICP) algorithm. In order to make an accurate registration, the ICP algorithm finds and uses the rigid transformation between two point clouds. Finally, a downsample was performed to get an even distribution of the points on the 3D model. This is achieved by dividing the 3D space into voxels and combining the points that lie within into one output point. This allow reducing the number of points in the point cloud while maintaining the characteristics as a whole. Both the ICP and the downsampling algorithm were used from the PCL library [1] In Figure 4, some examples are shown of the 3D point cloud models.

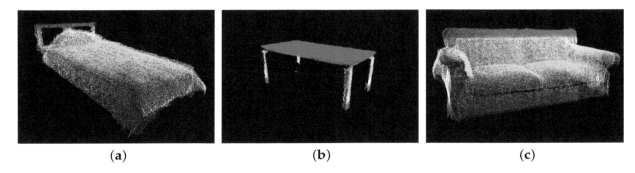

(a)   (b)   (c)

Figure 4. Example of the created models, for (**a**) a bed, for (**b**) a table and for (**c**) a couch.

4.1. Extraction of Geometrical Components

Once 3D models of furniture are available, the type of geometric components should be chosen and then extracted. At this stage of the work, it has been decided to use a reduced set of geometrical components, which is composed of horizontal and vertical planes and legs (poles). As can been seen, most of the furniture is composed of flat surfaces (horizontal or vertical) and legs or poles. Conversely, some surfaces are not strictly flat (e.g., the horizontal surface of a bed or the backrest of a coach); however, many of them can be roughly approximated to a flat surface with some relaxed parameters. For example, the main planes for a bed and a table can be approximated by a plane equation, but with different dispersion; a small value for the table and a bigger value for the bed. Currently, curve vertical surfaces have not been considered in our study. Nevertheless, they can be incorporated later.

4.1.1. Horizontal Planes' Extraction

Horizontal plane detection and extraction is achieved using a method based on histograms. For instance, a table and a bed have differences in their horizontal planes. In order to capture the characteristics of a wide variety of horizontal planes, three specific tasks are performed:

1. Considering that a robot and its sensors are correctly linked (i.e., between all reference frames, fixed or mobile), it is possible to obtain a transformation for a point cloud coming from a sensor in the robot's head into a reference frame to the base or footprint of the robot (see Figure 5). The TF package in ROS (Robotic Operation System) performs this transformation at approximately 100 Hz.
2. Once transformation between corresponding references frames is performed, a histogram of heights of the points in the cloud with reference to the floor is constructed.

    Over this histogram, horizontal planes (generally composed of a wide set of points) generate a peak or impulse. By extracting all the points that lie over those regions (peaks), the horizontal planes can be recovered. The form and characteristics of the peak or impulse in the histogram refer to the characteristics of the plane. For example, the widths of the peaks in Figure 6 are different; these correspond to: (1) a flat and regular surface for a chest of drawers (Figure 6b) and (2) a rough surface of a bed (Figure 6d).

    Nevertheless, there can be several planes merged together in a peak in a scene, i.e., two or more planes with the same height.
3. To separate planes merged together in a peak in a scene, first, all the points of the peak are extracted, and then, they are projected to the floor plane. A clustering algorithm is then performed in order to separate points corresponding to each plane, as shown in Figure 7.

Figure 5. Visualization of a PR2 robot with its coordinate frames and the point cloud from the scene transformed to the world reference frame.

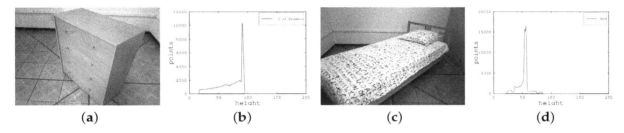

(a)   (b)   (c)   (d)

Figure 6. Example of a height histogram of two different pieces of furniture. In (**a**), the RGB image of a chest of drawers, and in (**b**), its height histogram. In (**c,d**), the RGB image of a bed and its height histogram, respectively.

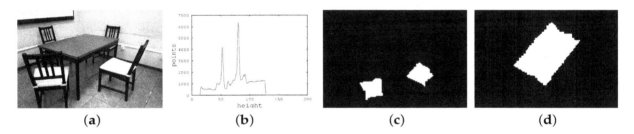

(a)   (b)   (c)   (d)

Figure 7. Example of a height histogram and the objects segmentation. In (**a**), the RGB from the scene, in (**b**), the height histogram, and (**c,d**) show the projections for each peak found on the histogram.

4.1.2. Detection of Vertical Planes and Poles

Vertical planes' and poles' extraction follows a similar approach. Specifically, they are obtained as follows:

1. In these cases, the distribution of the points is analyzed on a 2D histogram generated by projecting all the points into the floor plane.
2. Considering this 2D histogram as a grayscale image, all the points from a vertical plane will form a line on the image, and the poles will form a spot; then extracting the lines and spots, and their corresponding projected points, will provide the vertical planes and poles (Figure 8).

Finally, image processing algorithms can be applied in order to segment those lines on the 2D histogram and then recover points corresponding to those vertical planes.

In the cases where two vertical planes are projected to the same line, they can be separated by a similar approach mentioned in the previous section for segmenting horizontal planes; however in this case, the points are projected to a vertical plane. Then, it is possible to cluster them and perform some calculations on them like the area of the projected surface. For the current state of the work, we are more interested in the form of the furniture rather than its parts, so adjacent planes belonging to the same piece of furniture, i.e., the drawers of a chest of drawers, are not separated. If the separation of these planes were necessary, an approach similar to the one proposed in [22] could be used; where they segmented the drawers from a kitchen.

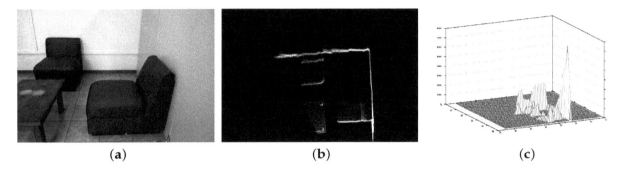

(a)   **(b)**   **(c)**

Figure 8. Example of a floor projection and its 2D histogram. In (**a**), the RGB image from the scene, in (**b**), the floor projection, and in (**c**), the 2D histogram.

With this approach, it is also possible to detect small regions or dots that correspond to the legs of the chairs or tables. Despite the small sizes of the regions or dots, they are helpful to characterize the furniture.

Our proposal can work with full PCD scenes without any requirement of a downsampling, which must be performed in algorithms like RANSAC in order to maintain a low computational cost. Moreover, histograms are computed linearly.

4.2. Characteristics of the Geometrical Components

Every geometrical component must be characterized in order to complete the graph for every piece of furniture. For simplicity, at this point, all the geometrical components have the same characteristics. However, more geometrical components can be added or replaced in the future. The following features or characteristics of the geometrical components are considered:

- Height: the average height of the points belonging to the geometric component.
- Height deviation: standard height deviation of the points in the peak or region.
- Area: area covered by the points.

Figure 9 shows the values of each characteristic for the main horizontal plane of some furniture models. The sizes of the boxes were obtained by a min-max method. More details will be given in Section 5.2. The parallelepiped represents the variation (uncertainty) for each variable.

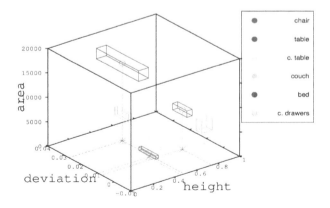

Figure 9. Characteristics of the main geometrical component (horizontal plane) for diverse pieces of furniture.

4.3. Examples of Graph Representation

Once geometric components have been described, it is possible to construct a graph for each piece of furniture.

Let F^* be a piece of furniture (e.g., a table); therefore, as stated in Equation (1), the graph is described as:

$$F^* = (V^*, E^*)$$

where V^* has five vertices and E^* four edges (i.e., $n_v^* = 5$ and $n_e^* = 4$). The five vertices correspond to: one vertex for the horizontal surface of the table and one for each of the four legs. The main vertex is the horizontal plane, and an edge will be added whenever two geometrical components are adjacent, so in this particular case, the edges correspond to the union between each leg and the horizontal plane.

Figure 10a presents the graph corresponding to a dinning table. Similarly, Figure 10b shows the graph of a chair. It can be seen that the chair's graph has one more vertex than the table's graph. This extra vertex is a vertical component corresponding to the backrest of the chair.

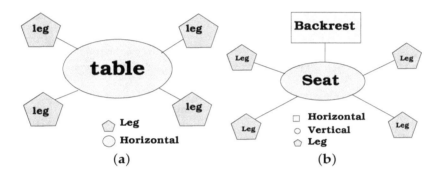

(a)   (b)

Figure 10. Example of complete graph models for: (**a**) a dinning table and (**b**) a chair; different shapes represent different types of geometric components.

As mentioned earlier, a graph serves as a complete model for each piece of furniture because it contains all its geometrical components. However, as described in Section 3.3, the robot cannot view all the components of a given piece of furniture because of the perspective. Therefore, subgraphs are created in order to compare robot perception with models.

To generate a small set of sub-graphs corresponding to partial views, several points of view have been grouped into four quadrants. These are: two subgraphs for the front left and right views and two more for the back view left and right (Figure 11). However, due to symmetry and occlusions, the set of subgraphs can be reduced. For example, in the case of a dining table, there is only one graph without sub-graphs because its four legs can be seen from many viewpoints.

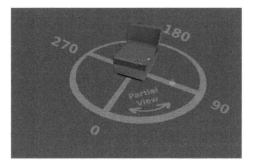

Figure 11. Visualizing the different points of view used to generate partial views.

Consequently, graphs require also to specify which planes are on opposite sides (if there are any), because this information is important to specify which components are visible from every view.

The visibility of a given plane is encoded at the vertex. For example, for a chest of drawers, it is not possible to see the front and the back at the same time.

Figure 12 shows an example of a graph model and some subgraphs for a couch graph model. It can be seen from Figure 12a that the small rectangles to the side of the nodes indicate the opposite nodes. The sub-graphs (Figure 12b,c) represent two sub-graphs of the left and right frontal views, respectively. The reduction of the number of vertex of the graph is clear. Specifically, the backrest and the front nodes are shown, whereas the rear node is not presented because it is not visible for the robot from a frontal view of the furniture. Thus, a subgraph avoids comparing components that are not visible from a given viewpoint.

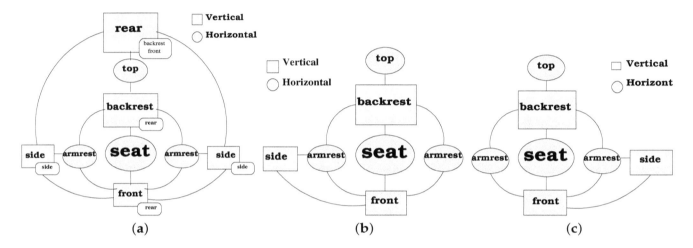

Figure 12. In (**a**), the graph corresponding to the complete couch graph, and in (**b**,**c**), two sub-graphs for the front left and right views.

5. Determination of Values for Models and Geometric Components

In order to validate our proposal, a home environment has been used. This environment is composed of six different pieces of furniture ($N_f = 6$), which are:

$$\mathcal{F} = \{dinning\ table, chair, couch, center\ table,$$
$$bed, chest\ of\ drawers\}$$

To represent the components of those pieces of furniture, the following three types of geometrical components have been selected:

$$\mathcal{G}c = \{horizontal\ plane, vertical\ plane, legs\}$$

and the features of the geometrical components are described in Section 4.2.

5.1. Weights for the Geometrical Components' Comparison

In order to compute the proposed similarity s_{Gc} between two geometrical components, it is necessary to determine the corresponding weights w_{Gi} in Equation (6). Those values have been determined empirically, as follows:

From a set of scenes taken by the robot, a set of them where each piece of furniture was totally visible. Then, geometrical components were extracted following the methodology proposed in

Section 4.1. The weights were selected according to the importance of each feature to a correct classification of the geometrical component.

Table 1 shows the corresponding weights for the three geometrical components.

Table 1. Weights for similarity estimation.

	w_{G1}^k (Height)	w_{G2}^k (h. Deviation)	w_{G3}^k (Area)
w_{Gi}^H (horizontal)	0.65	0.15	0.25
w_{Gi}^V (vertical)	0.5	0.2	0.3
w_{Gi}^L (legs)	0.5	0.2	0.3

5.2. Uncertainty

Uncertainty values in Equation (7) were estimated using an empirical process. From some views selected for each piece of furniture fully observable, the difference with its correspondent model was calculated; in order to have an estimation of the variation of corresponding values, with the complete geometrical component. Then, the highest difference for each characteristic was selected as the uncertainty.

As can be seen from Figure 9, the use of characteristics (height, height deviation and area) is sufficient to classify the main horizontal planes. Moreover, over this space, characteristics and uncertainty from each horizontal plane make regions fully classifiable.

There are other features of the geometrical components that are useful to define the type of geometrical component or their relations, so they have been added to the vertex structure. These features are:

- Center: the 3D point center of the points that makes the geometrical component.

- PCA eigenvectors and eigenvalues: eigenvector and eigenvalues resulting from a PCA analysis for the region points.

The center is helpful to establish spatial relations, and the PCA values help to discriminate between vertical planes and poles. By finding and applying an orthogonal transformation, PCA converts a set of possible correlated variables to a set of linearly uncorrelated variables called principal components. Since, in PCA, the first principal component has the largest possible variance, then, in the case of poles, the first component should be aligned with the vertical axis, and the variance of other components should be significantly smaller. This not so in the case of planes, where two first components can have similar variance values. Components are obtained by the eigenvector and eigenvalues from PCA.

5.3. Weights for the Graphs' Comparison

Conversely, as weights were determined using Equation (7), the weights for the similarity between graphs (Equation (9)) were calculated based on the total area of models for each piece of furniture.

Given the projected area of each geometrical component of the graph model, the total area is calculated. Then, the weights for each vertex (geometrical component) have been defined as the percentage of its area in comparison to the total area. Moreover, when dealing with a subgraph from a model, the total area is determined by the sum of areas from the nodes from that particular view (subgraph). Thus, there is a particular weight vector for each graph and subgraph in the environment.

Table 2 shows the values of computed areas for the chest of drawers corresponding to the graph and subgraph of the model (Figure 13).

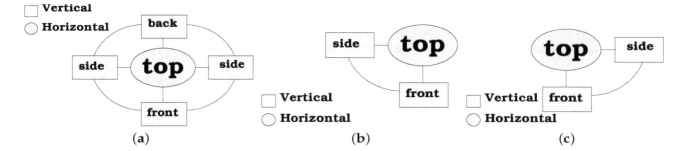

Figure 13. Graph models for the chest of drawers: in (**a**), the graph for the full model, and in (**b,c**), graphs for partial front views, left and right, respectively.

Table 2. Example of the weights for the comparison of the chest of drawers graph, based on the area of each geometrical component.

	Side	Front	Side	Back	Top
area	1575.5	9155.5	1575.5	9155.5	1914
Model %	6.74	39.16	6.74	39.16	8.19
Partial View %	12.45	72.41			15.13
Partial View %		72.41	12.45		15.13

Table 3 shows the weights in the case of the dinning table, which does not have sub-graphs (Figure 10a).

Table 3. Example of the weights for the comparison of the dining table graph, based on the area of each geometrical component.

	Table	Leg	Leg	Leg	Leg
area	8159	947	947	947	947
Model %	68.31	7.92	7.92	7.92	7.92
Partial View %	68.31	7.92	7.92	7.92	7.92

6. Evaluations

Consider an observation of a scene, where geometrical components have been extracted, by applying the methods described in Section 4.1.

Let O be the set of all geometrical components observed on a scene, then:

$$O = \{O^1, ..., O^{N_k}\} \tag{11}$$

where O^k is the subset of geometrical components of the type k.

In this way, observed horizontal geometrical components found on the scene are in the same subset, lets say O^*. Consequently, it is possible to extract each one of them in the subset and then compare them to the main nodes for each furniture graph.

Once the similarity between the horizontal components on the scene and the models has been calculated, all the categories with a similarity higher than a certain threshold are chosen as probable models for each horizontal component. A graph is then constructed for each horizontal component, where adjacent geometrical components are merged with it. Then, this graph is compared with the subgraphs of the probable models previously selected.

The first column in Figure 14 shows some scenes from the environment, where different pieces of furniture are present. Only four images with the six types of furniture are shown. The scene in Figure 14a is composed of a dinning table and a chair. After the geometrical components are extracted (Figure 14b), two horizontal planes corresponding to the dinning table and the chair are selected.

A comparison of those horizontal components to each one of the main nodes of the furniture graphs is performed. This results in two probable models (table and chest of drawers) for the plane labeled as "H0", which actually corresponds to the table, and three probable models (chair, center table and couch) for the plane labeled "H01" (which corresponds to a chair). The similarities computed can be observed as "Node Sim." in Table 4.

Next, graphs are constructed for each horizontal plane (the main node) and adding its adjacent components. In this case, both graphs have only one adjacent node.

Figure 15a shows the generated Graph ("G0") from the scene and the partial-views graphs from the selected probable models. It can be observed that "G0" has an adjacent leg node, so it can only be a sub-graph for the table graph since the chest of drawers graph has only adjacent vertical nodes.

For "G1" (Figure 16a), its adjacent node is a vertical node with a higher height than the main node, so it is matched to the backrest node of the chair and of the couch (Figure 16b,d). Moreover, there is no match with the center table graph (Figure 16c).

The similarity for adjacent nodes is noted in Table 4. Graph similarity is calculated with Equation (9) and shown in the last column of the table. "G0" is selected as a table and "G1" as a chair (Figure 14c).

Figure 14 shows the results of applying the described procedure to different scenes with different types of furniture. The first column (Figure 14a,d,g,j) shows the point clouds from the scenes. The column at the center (Figure 14b,e,h,k) shows the geometrical components found on the corresponding scene. Finally, the last column (Figure 14c,f,i,l) shows the generated graphs classified correctly.

Table 4. Example for graph classification.

Main Node	Main Node Sim.	Adjacent Nodes	Adjacent Node Sim.	Graph Similarity
H00	Table 0.8743	V02	Table leg 0.7015	0.0.6670
H00	Chest of Drawers 0.8014	V02	No Match	X
H01	Chair 0.9891	V01	Backrest 0.8878	0.5740
H01	Center Table 0.8895	V01	No match	X
H01	Couch 0.7277	V01	Backrest 0.7150	0.3204

As our approach is probabilistic, it can deal with the noise from the sensor; as well as partially occluded views, at this time, with occlusions no greater than 50% of the main horizontal plane, as this plane is the key factor in the graph.

While types of furniture can have the same graph structure, values in their components are particular for a given instance. Therefore, it is not possible to recognize with the same graph different instances of a type of furniture. In other words, a particular graph for a bed cannot be used for beds with different sizes; however, the graph structure could be the same.

Additionally, to test the approach on more examples, we have tested on selected images from the dataset SUNRGB-D [23]. Figure 17 shows the results of the geometrical components' extraction and the graph generation for the selected images. The color images corresponding to house environments similar to the test environments scenes are presented at the top of the figure; the corresponding geometrical component extraction is shown at the center of the figure; and the corresponding graphs are presented at the bottom of the figure.

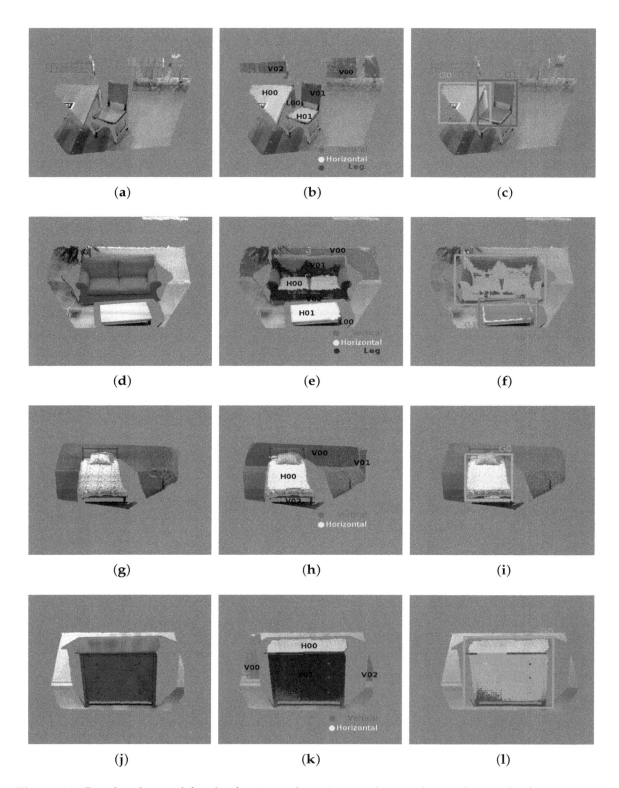

Figure 14. Results obtained for the furniture detection; each row shows the results for one scene. In (**a,d,g,j**), the original point clouds from the scenes. In (**b,e,h,k**) are shown the geometrical components found in the scene; the vertical planes are in green color, the horizontals in yellow and the legs in red. The bounding boxes in (**c,f,i,l**) show the graphs generated that were correctly identified as a piece of furniture.

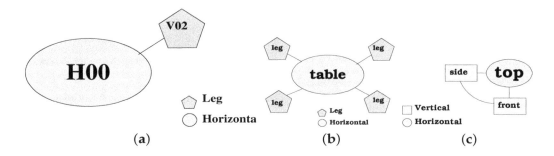

Figure 15. Graph comparison: in (**a**), one of the graphs generated from the scene in Figure 14b; in (**b**,**c**), partial graphs selected for matching with the graph in (**a**), corresponding to the table and the chest of drawers.

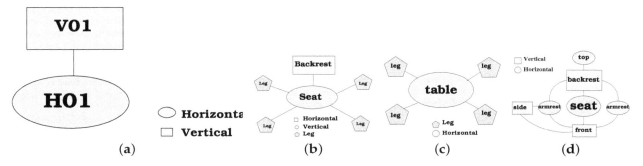

Figure 16. Graph comparison: in (**a**), one of the graphs generated from the scene in Figure 14b, and in (**b**–**d**), partial graphs selected for matching corresponding to the chair, the center table and the couch.

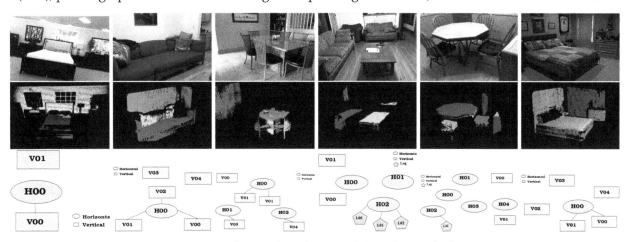

Figure 17. Tests over images from the dataset SUNRGB-D [23]: at the top, the RGB images, in the middle, the corresponding extracted geometrical components, and at the bottom, the corresponding graphs.

It is important to note that to detect the geometrical components over these scenes, the point cloud from each selected scene was treated using the intrinsic and extrinsic parameters provided in the dataset. Consequently, a similar footprint reference frame is obtained such as the frame provided by the robots in this work. This manual adjustment was necessary to simulate the translation needed to align the coordinate frame of the image with the environment, as described in Section 4.1.

Since the proposed approach requires a complete model for each piece of furniture, which were not provided by the dataset, it was not possible to match the graphs generated to a given piece of furniture. However, the approach has worked as expected (Figure 17).

7. Conclusions and Future Work

This study proposed a graph representation useful for detecting and recognizing furniture using autonomous mobile robots. These graphs are composed of geometrical components, where each geometrical component corresponds roughly to a different part of a given piece of furniture. At this stage of our proposal, only three different types of geometrical components were considered; horizontal planes, vertical planes and the poles. A fast and linear method for extraction of geometrical components has been presented to deal with 3D data from the scenes. Additionally, similarity metrics have been proposed in order to compare the geometrical components and the graphs between the models in the learned set and the scenes. Our graph representation allows performing a directed search when comparing the graphs. To validate our approach, evaluations were performed on house-like environments.

Two different environments were tested with two different robots provided with different brands of RGB-D cameras. The environments presented the same type of furniture, but not completely the same instances; and robots had their RGB-D camera in the head to provide a similar point of view to a human. These preliminary evaluations have proven the efficiency of the presented approach.

As future work, other geometrical components (spheres and cylinders) and other characteristics will be added and assessed to improve our proposal. Furthermore, other types of furniture will be evaluated using our proposal.

It will be desirable to bring the robots to a real house environment to test their performance.

Author Contributions: Conceptualization, O.A.-R. and A.M.-H. Investigation, O.A.-R., Supervision, A.M.-H., H.V.R.-F. and M.D. Formal analysis, H.V.-R.F. and M.D. Validation, S.E.P.-H. and E.J.R.-R.

Acknowledgments: The lead author thanks CONACYT for the scholarship granted during his Ph.D.

References

1. PCL—Point Cloud Library (PCL). Available online: http://www.pointclouds.org (accessed on 10 October 2017).
2. Alexandre, L.A. 3D descriptors for object and category recognition: A comparative evaluation. In Proceedings of the Workshop on Color-Depth Camera Fusion in Robotics at the IEEE/RSJ International Conference on Intelligent Robots and Systems (IROS), Vilamoura, Portugal, 7–12 October 2012; Volume 1, p. 7.
3. Bo, L.; Ren, X.; Fox, D. Depth kernel descriptors for object recognition. In Proceedings of the IEEE/RSJ International Conference on Intelligent Robots and Systems, San Francisco, CA, USA, 25–30 September 2011; pp. 821–826. [CrossRef]
4. Lai, K.; Bo, L.; Ren, X.; Fox, D. Detection-based object labeling in 3D scenes. In Proceedings of the IEEE International Conference on Robotics and Automation, Saint Paul, MN, USA, 14–18 May 2012; pp. 1330–1337. [CrossRef]
5. Guo, Y.; Bennamoun, M.; Sohel, F.; Lu, M.; Wan, J. An Integrated Framework for 3-D Modeling, Object Detection, and Pose Estimation From Point-Clouds. *IEEE Trans. Instrum. Meas.* **2015**, *64*, 683–693. [CrossRef]
6. Hetzel, G.; Leibe, B.; Levi, P.; Schiele, B. 3D object recognition from range images using local feature histograms. In Proceedings of the IEEE Computer Society Conference on Computer Vision and Pattern Recognition (CVPR 2001), Kauai, HI, USA, 8–14 December 2001; Volume 2, pp. 394–399. [CrossRef]
7. Wahl, E.; Hillenbrand, U.; Hirzinger, G. Surflet-pair-relation histograms: A statistical 3D-shape representation for rapid classification. In Proceedings of the Fourth International Conference on 3-D Digital Imaging and Modeling, Banff, AB, Canada, 6–10 October 2003; pp. 474–481. [CrossRef]
8. Drost, B.; Ulrich, M.; Navab, N.; Ilic, S. Model globally, match locally: Efficient and robust 3D object recognition. In Proceedings of the IEEE Computer Society Conference on Computer Vision and Pattern Recognition (CVPR'10), San Francisco, CA, USA, 13–18 June 2010; pp. 998–1005. [CrossRef]

9. Salas-Moreno, R.F.; Newcombe, R.A.; Strasdat, H.; Kelly, P.H.J.; Davison, A.J. SLAM++: Simultaneous Localisation and Mapping at the Level of Objects. In Proceedings of the IEEE Conference on Computer Vision and Pattern Recognition (CVPR), Long Beach, CA, USA, 15–21 June 2013; pp. 1352–1359. [CrossRef]

10. Wu, Z.; Song, S.; Khosla, A.; Yu, F.; Zhang, L.; Tang, X.; Xiao, J. 3D ShapeNets: A deep representation for volumetric shapes. In Proceedings of the IEEE Conference on Computer Vision and Pattern Recognition (CVPR), Long Beach, CA, USA, 15–21 June 2015; pp. 1912–1920. [CrossRef]

11. Swadzba, A.; Wachsmuth, S. A detailed analysis of a new 3D spatial feature vector for indoor scene classification. *Robot. Auton. Syst.* **2014**, *62*, 646–662. [CrossRef]

12. Trevor, A.J.B.; Rogers, J.G.; Christensen, H.I. Planar surface SLAM with 3D and 2D sensors. In Proceedings of the IEEE International Conference on Robotics and Automation (ICRA'12), Saint Paul, MN, USA, 14–18 May 2012; pp. 3041–3048. [CrossRef]

13. Rusu, R.B.; Marton, Z.C.; Blodow, N.; Dolha, M.; Beetz, M. Towards 3D Point cloud based object maps for household environments. *Robot. Auton. Syst.* **2008**, *56*, 927–941. [CrossRef]

14. Günther, M.; Wiemann, T.; Albrecht, S.; Hertzberg, J. Building semantic object maps from sparse and noisy 3D data. In Proceedings of the IEEE/RSJ International Conference on Intelligent Robots and Systems, Tokyo, Japan, 3–7 November 2013; pp. 2228–2233. [CrossRef]

15. Günther, M.; Wiemann, T.; Albrecht, S.; Hertzberg, J. Model-based furniture recognition for building semantic object maps. *Artif. Intell.* **2017**, *247*, 336–351. [CrossRef]

16. Holz, D.; Holzer, S.; Rusu, R.B.; Behnke, S. Real-Time Plane Segmentation Using RGB-D Cameras. In *RoboCup 2011: Robot Soccer World Cup XV*; Röfer, T., Mayer, N.M., Savage, J., Saranlı, U., Eds.; Springer: Berlin/Heidelberg, Germany, 2012; pp. 306–317.

17. Lin, D.; Fidler, S.; Urtasun, R. Holistic Scene Understanding for 3D Object Detection with RGBD Cameras. In Proceedings of the International Conference on Computer Vision (ICCV), Seoul, Korea, 27 October–3 November 2013; pp. 1417–1424. [CrossRef]

18. Koppula, H.S.; Anand, A.; Joachims, T.; Saxena, A. Semantic Labeling of 3D Point Clouds for Indoor Scenes. In *Advances in Neural Information Processing Systems 24*; Shawe-Taylor, J., Zemel, R.S., Bartlett, P.L., Pereira, F., Weinberger, K.Q., Eds.; Curran Associates, Inc.: Dutchess County, NY, USA, 2011; pp. 244–252.

19. Wittrowski, J.; Ziegler, L.; Swadzba, A. 3D Implicit Shape Models Using Ray Based Hough Voting for Furniture Recognition. In Proceedings of the International Conference on 3D Vision—3DV 2013, Seattle, WA, USA, 29 June–1 July 2013; pp. 366–373. [CrossRef]

20. Chen, K.; Lai, Y.K.; Wu, Y.X.; Martin, R.; Hu, S.M. Automatic Semantic Modeling of Indoor Scenes from Low-quality RGB-D Data Using Contextual Information. *ACM Trans. Graph.* **2014**, *33*, 208:1–208:12. [CrossRef]

21. Schnabel, R.; Wessel, R.; Wahl, R.; Klein, R. Shape recognition in 3d point-clouds. In Proceedings of the 16th International Conference in Central Europe on Computer Graphics, Visualization and Computer Vision in co-operation with EUROGRAPHICS, Pilsen, Czech Republic, 4–7 February 2008; pp. 65–72.

22. Rusu, R.B.; Marton, Z.C.; Blodow, N.; Holzbach, A.; Beetz, M. Model-based and learned semantic object labeling in 3D point cloud maps of kitchen environments. In Proceedings of the IEEE/RSJ International Conference on Intelligent Robots and Systems, St. Louis, MO, USA, 10–15 October 2009; pp. 3601–3608. [CrossRef]

23. Song, S.; Lichtenberg, S.P.; Xiao, J. SUN RGB-D: A RGB-D scene understanding benchmark suite. In Proceedings of the IEEE Conference on Computer Vision and Pattern Recognition (CVPR), Long Beach, CA, USA, 15–21 June 2015; Volume 5, p. 6.

6

Numerical Evaluation of Sample Gathering Solutions for Mobile Robots

Adrian Burlacu [1,†], Marius Kloetzer [1,*,†] and Cristian Mahulea [2,†]

[1] Department of Automatic Control and Applied Informatics, "Gheorghe Asachi" Technical University of Iasi, 700050 Iasi, Romania; aburlacu@ac.tuiasi.ro

[2] Department of Computer Science and Systems Engineering, University of Zaragoza, 50018 Zaragoza, Spain; cmahulea@unizar.es

* Correspondence: kmarius@ac.tuiasi.ro
† The authors contributed equally to this work.

Abstract: This paper applies mathematical modeling and solution numerical evaluation to the problem of collecting a set of samples scattered throughout a graph environment and transporting them to a storage facility. A team of identical robots is available, where each robot has a limited amount of energy and it can carry one sample at a time. The graph weights are related to energy and time consumed for moving between adjacent nodes, and thus, the task is transformed to a specific optimal assignment problem. The design of the mathematical model starts from a mixed-integer linear programming problem whose solution yields an optimal movement plan that minimizes the total time for gathering all samples. For reducing the computational complexity of the optimal solution, we develop two sub-optimal relaxations and then we quantitatively compare all the approaches based on extensive numerical simulations. The numerical evaluation yields a decision diagram that can help a user to choose the appropriate method for a given problem instance.

Keywords: sample gathering problem; mobile robots; mathematical modeling; numerical evaluation; centralized architecture; optimization

1. Introduction

Much robotics research develops automatic planning procedures for autonomous agents such that a given mission is accomplished under an optimality criterion. The missions are usually related to standard problems such as navigation, coverage, localization, and mapping [1,2]. Some works provide strategies directly implementable on particular robots with complicated dynamics and multiple sensors [3]. Other research aims to increase the task expressiveness [4], e.g., starting from Boolean-inspired specifications [5,6] up to temporal logic ones [7–10], even if the obtained plans may be applied only to simple robots.

It is often common to construct discrete models for the environment and robot movement capabilities, by using results from multiple areas such as systems theory, computational geometry [11,12], and discrete event systems [13,14].

The current research is focused on solving a sample gathering problem. The considered task belongs to the general class of optimal assignment problems [15], since the sample can correspond to jobs and the robots to machines. The minimization of the overall time for gathering all samples (yielded by the "slowest" agent) thus translates to so-called min-max problems [16] or bottleneck assignment problems [15] (Chapter 6.2). However, these standard frameworks do not consider different numbers of jobs and machines, nor machines (agents) with limited energy amounts.

A broad taxonomy of allocations in multi-robot teams is presented in [17,18], according to which our problem belongs to the class of assignments of single-robot tasks (one task requiring one robot)

in multi-task robot systems (a robot can move, pick up, and deposit samples). Again, the general solutions assume utility estimates for different job–machine pairs.

For such problems, various Mixed-Integer Linear Programming Problem (MILP) formulations are given as in [17,19,20]. Some resemble our problem, but they are not an exact fit because of the specificities that all samples should be eventually gathered into the same node, a robot can carry one sample at a time, there are limited amounts of energy, and we do not know a priori a relationship between number of samples and number of robots. Furthermore, we provide a second MILP for the case of problems infeasible due to energy requirements. We further relax the complex MILP solutions into sub-optimal solutions as non-convex Quadratic Programming (QP) and iterative heuristics. Our goal is to draw rules of choosing the appropriate method for a given problem, based on extensive tests.

Some preliminary mathematical formulations that generalize traveling salesman problems are included in [21], with targeted application to exploring robots that must collect and analyze multiple heterogeneous samples from a planetary surface. Other works focus on specific applications as task allocation accomplished by agents with different dynamics [22], or allocation in scenarios with heterogenous robots that can perform different tasks [23]. Various works propose auction-based mechanisms for various assignments problems or develop and apply distributed algorithms for specific cases with equal number of agents and tasks [24]. Research [25] assumes precedence constraints on available tasks and builds solutions based on integer programming forms and auction mechanisms. However, we do not include auction-based methods here. The closest solution we propose may be our iterative heuristic algorithm from Section 4.2, which can be viewed as a specific greedy allocation method [17].

Our problem can be seen as a particular case of Capacity and Distance constrained Vehicle Routing Problem (CDVRP) [26] with capacity equal to one and the distance constrains related to the limited energy. For this problem, many algorithms have been provided, for the exact methods [27,28] and for heuristic methods [29–32]. However, our problem is different in multiple aspects. First, in the CDVRP problem, the capacity of each vehicle (robot in our case) should be greater than the demand of each vertex [26]. In our case this cannot hold since the robot capacity is one (we assume that each robot can transport maximum one good) and the number of goods at the vertices is, in many cases, greater than one. Second, up to our knowledge, the heuristics considered in this work, which include relaxing the optimal solution of the MILP to a QP problem, have not been considered for the CDVRP. Finally, most of the works on CDVRP try to characterize worst-case scenarios through cost differences between heuristic and optimal methods.

In this paper, we are also interested in the computational complexity and we evaluate the proposed solutions using numerical simulations. To the best of our knowledge, none of the mentioned works contains a directly applicable formulation that yields a solution for our specific problem. This work builds on solutions reported in [33–35]. In [33] we constructed a MILP problem that solves the minimum time sample gathering problem. Different than in [33], we also design a MILP solution that can be used when the initial problem is infeasible due to scarce energy limits. Besides the MILP formulations, one of the main goals of this research are to provide computationally efficient sub-optimal solutions for the targeted problems. The MILP solution was relaxed to a QP formulation in [34]. Here we also construct an iterative sub-optimal solution inspired by [35] as a QP alternative. Based on extensive simulations that involve the three proposed solutions, we conclude with a decision scheme that helps a user to choose the appropriate method for a specific problem instance.

The purpose of this work is to plan a team of mobile agents such that they gather the samples scattered throughout the environment into a storage facility. The problem's hypothesis consists in a team of mobile robots which must bring to a deposit region a set of samples that exist in an environment at known locations. As main contributions we claim the new mathematical models for the considered problems and the numerical evaluation of the obtained solutions.

The environment is modeled by a graph where an arc weight corresponds to consumed energy and time for moving between the linked nodes. Each robot has limited energy, and the goal is to collect all samples in minimum time. Because the robots are initially deployed in the storage (deposit) node and given the static nature of the environment, it is customary to build movement plans before the agents start to move. The problem reduces to a specific case of optimal assignment or task allocation [15,17,18]. The paper combines results from our previous research reported in [34,36,37]. We first build an optimal solution involving a MILP formulation.

Then, we design two solutions with lower complexities, one as a Quadratic Programming (QP) relaxation of the initial MILP, and the other as an Iterative Heuristic (IH) algorithm. A numerical evaluation between the formulated solutions yields criteria as time complexities for computing robotic plans and the difference of costs between these plans. Based on these criteria, a decision diagram is provided such that a user can easily choose the appropriate method to embed.

The remainder of the paper is structured as follows. Section 2 formulates the targeted problem, outlines the involved assumptions and introduces an example that will be solved throughout the subsequent sections. Section 3 details two optimal solutions, from which the first will be relaxed to a QP formulation, while the second can be used when robots have low energy supplies. The sub-optimal methods based on QP optimization or IH algorithm are presented in Section 4. The developed methods are numerically evaluated and compared in Section 5 and rules are given for choosing the proper method for a specific problem instance.

2. Problem Formulation

Consider a team of N_R identical robots that are labeled with elements of set $R = \{r_1, r_2, \ldots, r_{N_R}\}$. The robots "move" on a weighted graph $G = (V, E, c)$, where $V = \{v_1, v_2, \ldots, v_{|V|}\}$ is the finite number of nodes (also called locations or vertices), $E \subseteq V \times V$ is the adjacency relationship corresponding to graph edges, and $c : E \to \mathbb{R}_+$ is a cost (weight) function.

We mention that there are multiple approaches for creating such finite-state abstractions of robot control capabilities in a given environment [1,2]. A widely used idea is to partition the free space into a set of regions via cell decomposition methods, each of these regions corresponding to a node from V [6,11,38]. The graph edges correspond to possible robot movements between adjacent partition regions, i.e., $\forall v, v' \in V$, if an agent can move from location v to v' without visiting any other node from graph, then $(v, v') \in E$. Each edge corresponds to a continuous feedback control law for the robot such that the desired movement is produced, and various methods exist for designing such control laws based on agent dynamics and partition types [39,40]. Alternatives to cell decomposition methods, as visibility graphs or generalized Voronoi diagrams, can also produce discrete abstractions in form of graphs or transition systems [1].

We assume that the graph G is connected, and the adjacency relationship E is symmetric, i.e., if a robot from R can travel from location v to v', then it can also move from v' to v. For any $(v, v') \in E$, we consider that the cost $c(v, v')$ represents the amount of *energy* spent by the robot for performing the movement from v to v' and that $c(v', v) = c(v, v')$. By assuming identical agents with constant velocity and a homogenous environment (i.e., the energy for following an arc is proportional with the distance between linked nodes), we denote the *time* necessary for performing the movement from location v to v' by $\gamma \cdot c(v, v')$, where $\gamma \in \mathbb{R}_+$ is a fixed value.

Initially, all agents are deployed in a storage (deposit) node $v_{|V|}$ (labeled for simplicity as the last node in graph G), and each robot $r \in R$ has a limited amount of energy, given by map $\mathcal{E} : R \to \mathbb{R}_+$, for performing movements on abstraction G. For homogenous environments and constant moving speeds, energy $\mathcal{E}(r)$ can be easily linked with battery level of robot r, with the distance it can travel, or the sum of costs of followed edges.

There are N_S *samples* or valuable items scattered throughout the environment graph G. The samples are indexed (labeled) with elements of set $S = \{1, 2, \ldots, N_S\}$, while a map $\pi : S \to V$ shows to which node each sample belongs.

Problem 1. *For every robot $r \in R$ find a moving strategy on G such that:*

- *the team of robots gathers (collects) all samples from graph G in the storage node $v_{|V|}$ within minimum time;*

- *each robot can carry at most one sample at any moment;*

- *the total amount of energy spent by each robot is at most its initially available energy.*

Remark 1 (NP-hardness). *Our problem is related to the so-called Set Partitioning Problems (SPPs) [17], which are employed in various task allocation problems for mobile agents. A SPP formulation aims to find a partition of a given set such that a utility function defined over the set of acceptable partitions with real values is maximized. Various SPPs are solved by using Operations Research formulations that employ different standard optimization problems. In our case, the given set is S (samples to be collected), while the desired partition should have N_R disjoint subsets of S, each subset corresponding to the samples a robot should collect. The utility relates to the necessary time required for collecting all samples (being a maximum value over utilities of elements of obtained partition), while the partition is acceptable if each robot has enough energy to collects its samples. The maximization over individual utilities show that our problem is more complicated than standard SPPs. Since an SPP is NP-hard [41], we conclude that Problem 1 is also NP-hard. Therefore, we expect computationally intensive solutions for optimally solving Problem 1, while sub-optimal relaxations may be used when an optimal solution does not seem tractable.*

Since the sample deployment and robot energy limits are known, the searched solution is basically an off-line computed plan (sequence of nodes) for each robot such that the mission requirements are fulfilled. The first requirement from Problem 1 can be regarded as a global target for the whole team (properly assign robots to collect samples such that the overall time for accomplishing the task is minimized), while the last two requirements are related to robot capabilities. As in many robot planning approaches where global tasks are accomplished, we do not account for inter-robot collisions when developing movement plans. In real applications, such collisions can be avoided by using local rules during the actual movement, and the time (or energy) offset induced by such rules is negligible with respect to the total movement time (or required energy). Clearly, in some cases Problem 1 may not have a solution due to insufficient available energy for robots, these situations will be discussed during solution description.

Example 1. *For supporting the problem formulation and solution development, we introduce an example that will be discussed throughout the next sections. Thus, we assume an environment abstracted to the graph from Figure 1, composed by 10 nodes $(v_1, v_2, \ldots, v_{10})$, with the deposit v_{10}. The costs for moving between adjacent nodes are marked on the arcs from Figure 1, e.g., $c(v_{10}, v_8) = 2$. The team consists of 3 robots labeled with elements of set $R = \{r_1, r_2, r_3\}$. For simplicity of exposition, we consider $\gamma = 1$ (the constant that links the moving energy with necessary time) and equal amounts of energy for robots, $\mathcal{E}(r) = 100, \forall r \in R$. There are 14 samples scattered in this graph (labeled with numbers from 1 to 14), with locations given by map π, e.g., $\pi(9) = \pi(10) = v_3$.*

The problem requires a sequence of movements for each robot such that all the 14 samples are gathered into storage v_{10}, each robot being able to carry one sample at any time.

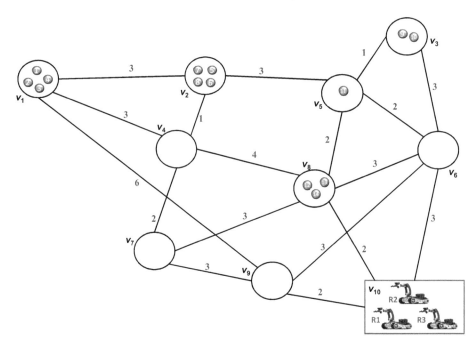

Figure 1. Example: environment graph with 10 nodes, 3 robots, and 14 samples. The robot moving costs are marked on graph edges, and the samples are represented by the blue discs.

3. Mathematical Model and Optimal Solution

To solve Problem 1, defined in the previous section, our approach consists from the following main steps:

(i)   Determine optimal paths in graph G from storage node to all nodes containing samples.
(ii)   Formulate linear constraints for correctly picking samples and for not exceeding robot's available energy based on a given allocation of each robot to pick specific samples.
(iii) Create a cost function based on robot-to-sample allocation and on necessary time for gathering all samples.
(iv)   Cast the above steps in a form suitable for applying existing optimization algorithms and thus find the desired robot-to-sample allocations.

Step (i) is accomplished by running a Dijkstra algorithm [42] on the weighted graph G, with source node $v_{|V|}$ and with multiple destination nodes: $v \in V$ for which $\exists s \in S$ such that $\pi(s) = v$. Please note that a single run of Dijkstra algorithm returns minimum cost paths to multiple destinations.

Let us denote with $path(v)$ the obtained path (sequence of nodes) from deposit $v_{|V|}$ to node $v \in V$ and with $\omega(v)/2$ its cost. Due to symmetrical adjacency relationship of G, the retour path from v to storage is immediately constructed by following $path(v)$ in inverse order. The retour path is denoted by $path^{-1}(v)$ and it has the same cost $\omega(v)/2$. (Because are interested in the round-trip cost between nodes $v_{|V|}$ and v, we denote the one-way cost by $\omega(v)/2$ and thus the cost of the full path is simply denoted by $\omega(v)$.). Therefore, for collecting and bringing to storage location the sample $s \in S$, a robot spends $\omega(\pi(s))$ for the round-trip given by $path(\pi(s))$, $path^{-1}(\pi(s))$.

For solving steps (ii)–(iv), let us first define a decision function as $x : N_R \times N_S \rightarrow \{0, 1\}$ by:

$$x(r,s) = \begin{cases} 1, & \text{if robot } r \text{ picks sample } s \\ 0, & \text{otherwise} \end{cases} , \forall r \in R, s \in S. \tag{1}$$

The actual values returned by map x for any pair $(r,s) \in R \times S$ give the most important part of solution to Problem 1, and these values are unknown, yet. The outcomes of x will constitute the decision variables in an optimization problem that is described next.

Step (ii) is formulated as the following set of linear constraints:

$$\sum_{r \in R} x(r,s) = 1, \qquad\qquad \forall s \in S$$

$$\sum_{s \in S} \left(\omega(\pi(s)) \cdot x(r,s) \right) \leq \mathcal{E}(r), \quad \forall r \in R \tag{2}$$

where the first set of equalities impose that exactly one robot is sent to collect each sample, while the subsequent inequalities guarantee that the energy spent by robot r for collecting all its assigned samples does not exceed its available energy.

The cost function from step (iii) corresponds to the first requirement of Problem 1. It means to minimize the maximum time among all robots required for collecting the assigned samples, and it formally translates to finding outcomes of x that minimize the objective function $J(x)$ from

$$J(x) = \min_{x} \max_{r \in R} \left(\gamma \cdot \sum_{s \in S} \left(\omega(\pi(s)) \cdot x(r,s) \right) \right). \tag{3}$$

The objective function from (3) and the constraints from (2) form a minimax optimization problem [43,44]. However, decision function x should take binary values. To use available software tools when solving for values of x, step (iv) transforms the minimax optimization into a MILP problem [44,45], by adding an auxiliary variable $z \in \mathbb{R}_+$ and additional constraints that replace the max term from (3). This results in the following MILP optimization:

$$
\begin{aligned}
\min_{x,z} \quad & z \\
\text{s.t.:} \quad & \sum_{r \in R} x(r,s) = 1 \, , \; \forall s \in S \\
& \sum_{s \in S} \left(\omega(\pi(s)) \cdot x(r,s) \right) \leq \mathcal{E}(r) \, , \; \forall r \in R \\
& \gamma \cdot \sum_{s \in S} \left(\omega(\pi(s)) \cdot x(r,s) \right) \leq z \, , \; \forall r \in R \\
& x(r,s) \in \{0,1\} \, , \; \forall (r,s) \in R \times S \\
& z \geq 0
\end{aligned}
\tag{4}
$$

The MILP (4) can be solved by using existing software tools [46–48]. The solution is guaranteed to be globally optimal because both the feasible set defined by the linear constraints from (4) and the objective function (z) are convex [45,49]. Thus, solution of (4) gives the optimal outcomes for x (unknown decision variables $x(r,s)$), as well as the time z in which the team solves Problem 1.

Please note that the obtained map x indicates the samples that must be collected by each robot, as in (1). However, it does not impose any order for collecting these samples. For imposing a specific sequencing, each robot is planned to collect its allocated samples in the ascending order of the necessary costs. This means that robot r first picks the sample s for which $x(r,s) = 1$ and $\omega(\pi(s)) \leq \omega(\pi(s'))$, $\forall s' \in S$ with $x(r,s') = 1$, and so on for the other samples. The optimum path for collecting sample s from node $\pi(s)$ was already determined in step (i). Under the above explanations, Algorithm 1 includes the steps for obtaining an optimal solution for Problem 1.

Algorithm 1: Optimal solution

    Input: $G, R, S, \mathcal{E}, \pi$

    Output: Robot movement plans

1   Find on graph G paths $path(v)$ and costs $\omega(v)$ for every $v = \pi(s), s \in S$

2   Solve MILP optimization (4)

3   **if** *solutions z and x are obtained* **then**

4   **for** $r \in R$ **do**

5          $plan_r = \varnothing$

6          $S_{r:collects} = \{s \in S \mid x(r,s) = 1\}$

7   Sort set $S_{r:collects}$ in ascending order based on costs $\omega(\pi(s)), s \in S_{r:collects}$

8   **for** $s \in S_{r:collects}$ **do**

9   Append $path(\pi(s))$ to plan of robot r, $plan_r$

10   Insert command to collect sample s in $plan_r$

11   Append $path^{-1}(\pi(s))$ to $plan_r$

12   Insert command to deposit sample s in $plan_r$

13   *Return* plans $plan_r, \forall r \in R$

14   **else**

15   Problem 1 is infeasible

16   *Return*

Example 2. *We apply the optimal solution from this section on the example introduced in Section 2. The Dijkstra algorithm (line 1 from Algorithm 1) returns paths and corresponding energy for collecting each sample, e.g., $path(\pi(1)) = path(v_1) = v_{10}, v_9, v_1$ and $\omega(\pi(1)) = 16$. The MILP (4) was solved in about 0.7 s and it returned an optimal solution with $z = 54$ (time for fulfilling Problem 1) and allocation map x. Based on robot-to-sample allocations x, lines 3–13 from Algorithm 1 yield the robotic plans for collecting samples from the following nodes (the sequences of nodes and the collect/deposit commands are omitted due to their length):*

$$
\begin{aligned}
&\text{Robot } r_1 \text{ collects samples from:} \\
&(v_8),\ (v_8),\ (v_2),\ (v_1),\ (v_1) \quad (\text{time} : 54) \\
&\text{Robot } r_2 \text{ collects samples from:} \\
&(v_3),\ (v_2),\ (v_2),\ (v_1) \quad (\text{time} : 54) \\
&\text{Robot } r_3 \text{ collects samples from:} \\
&(v_8),\ (v_5),\ (v_3),\ (v_2),\ (v_1) \quad (\text{time} : 52)
\end{aligned}
\tag{5}
$$

In the remainder of this section we focus on the situation in which the mobile robots cannot accomplish Problem 1 due to energy constraints.

Remark 2. Relaxing infeasible problems: *If MILP (4) is infeasible, this means that Problem 1 cannot be solved due to insufficient available energy of robots for collecting all samples.*

Intuitive argument. Whenever (4) has a non-empty feasible set (the set defined by the linear constraints), it returns an optimal solution from this set [45,49]. The feasible set can become empty only when the first two sets of constraints and the fourth ones from (4) are too stringent. The third set of constraints cannot imply the emptiness of feasible set, because there is no upper bound on z. It results that (4) has no solution whenever the first, second and fourth sets of its constraints cannot simultaneously hold. The fourth constraints cannot be relaxed, and therefore only the first two sets may imply the infeasibility of (4). This proves the remark, since the first constraints require all samples to be collected, while the second ones impose upper bounds on consumed energy.

MILP (4) has a non-empty feasible when the required energy for collecting all samples is small enough, or when the available energy limits $\mathcal{E}(r)$ are large enough. This is because the connectedness of G implies that the coefficients $\omega(\pi(s))$ are finite, $\forall s \in S$. Therefore, the first two sets of constraints from (4) could be satisfied even by an initial solution of form $x(r,s) = 1$ for a given $r \in R$, $\forall s \in S$, and $x(r',s) = 0$ for any $r' \in R \setminus \{r\}$.

This aspect yields the idea that one can relax the first constraints from (4) whenever there is no solution, i.e., collect as many samples as possible with the available robot energy. Such a formulation is given by the MILP problem (6), which allows that some samples are not collected (inequalities in first constraints) and imposes a penalty in the cost function for each uncollected sample. Basically, for a big enough value of $W > 0$ from (6), any uncollected sample would increase the value of the objective function more than the decrease resulted from saved energy. The constant W can be lower-bounded by:

$$W > \gamma \cdot \sum_{s \in S} \omega(\pi(s)).$$

For this lower bound, the cost function increases whenever a sample s is not collected, because the term z decreases with $\gamma \cdot \omega(\pi(s))$ and the second term increases with more than this value. Since MILP optimization returns a global optimum, minimizing the cost function under constraints from (6) guarantees that the largest possible number of samples are collected while minimizing the necessary time.

Observe that when (6) is employed, the returned value of the minimized function does not represent the time for collecting all samples, but this time is given by the returned z.

$$
\begin{aligned}
\min_{x,z} \quad & z - W \cdot \sum_{s \in S} \sum_{r \in R} x(r,s) \\
\text{s.t.:} \quad & \sum_{r \in R} x(r,s) \leq 1 \,, \ \forall s \in S \\
& \sum_{s \in S} \left(\omega(\pi(s)) \cdot x(r,s) \right) \leq \mathcal{E}(r) \,, \ \forall r \in R \\
& \gamma \cdot \sum_{s \in S} \left(\omega(\pi(s)) \cdot x(r,s) \right) \leq z \,, \ \forall r \in R \\
& x(r,s) \in \{0,1\} \,, \ \forall (r,s) \in R \times S \\
& z \geq 0
\end{aligned}
\tag{6}
$$

The optimization problem from Equation (6).

4. Sub-Optimal Planning Methods

In the general case, a MILP optimization is NP-hard [50]. The computational complexity increases with the number of integer variables and with the number of constraints, but exact complexity orders or upper bounds on computational time cannot be formulated [51]. These notes imply that for some cases the complexity of MILP (4) or (6) may render the solution from Section 3 as being computationally intractable, although the optimization is run off-line, i.e., before robot movement.

This section includes two approaches for overcoming this issue. Section 4.1 reformulates the MILP problem (4) as in [34] and obtains a QP formulation. Section 4.2 proposes an IH algorithm, inspired by allocation ideas from [35].

4.1. Quadratic Programming Relaxation

We aim to relax the binary constraints from MILP (4), and for accomplishing this we embed them into a new objective function. The idea starts from various penalty formulations defined in [52], some being related to the so-called big M method [49].

Let us replace the binary constraints $x(r,s) \in \{0,1\}$ from (4) with lower and upper bounds of 0 and respectively 1 for outcomes of map x. At the same time, add to the cost function of (4) a penalty term depending on $M > 0$, as shown in the objective

$$\min_{x,z} z + M \cdot \sum_{r \in R} \sum_{s \in S} \left(x(r,s) \cdot (1 - x(r,s)) \right). \tag{7}$$

For a large enough value of the penalty parameter M, the minimization of the new cost function from (7) tends to yield a binary value for each variable $x(r,s)$. This is because only binary outcomes of x imply that the second term from sum (7) vanishes, while otherwise this term has a big value due to the large M. The quadratic objective from (7) can be re-written in a standard form of an objective function of a QP problem, and together with the remaining constraints from (4) we obtain the QP formulation:

$$
\begin{aligned}
\min_{x,z} \quad & z + M \cdot \sum_{r \in R} \sum_{s \in S} x(r,s) - M \cdot \sum_{r \in R} \sum_{s \in S} \left(x(r,s) \right)^2 \\
\text{s.t.:} \quad & \sum_{r \in R} x(r,s) = 1 , \ \forall s \in S \\
& \sum_{s \in S} \left(\omega(\pi(s)) \cdot x(r,s) \right) \leq \mathcal{E}(r) , \ \forall r \in R \\
& \gamma \cdot \sum_{s \in S} \left(\omega(\pi(s)) \cdot x(r,s) \right) \leq z , \ \forall r \in R \\
& 0 \leq x(r,s) \leq 1 , \ \forall (r,s) \in R \times S \\
& z \geq 0
\end{aligned}
\tag{8}
$$

Under the above informal explanations and based on formal proofs from [52], the MILP (4) and QP (8) have the same global minimum for a sufficiently large value of parameter M (The actual value of M is usually chosen based on numerical ranges of other data from the optimization problem, as values returned by maps ω and $\mathcal{E}$.).

Remark 3 (Sub-optimality or failure). *Please note that the cost function from (8) is non-convex, because of the negative term in $x(r,s)^2$. Therefore, optimization (8) could return local minima, while in some cases the obtained values of x may even be non-binary. If the obtained outcomes of x are binary, the value of completion time z for collecting all samples is directly returned as the cost of QP (8), while otherwise a large cost is obtained due to the non-zero term in M from (7).*

The QP optimization (8) can be solved with existing software tools [46,48]. As noted, it may return a sub-optimal solution. Nevertheless, such a sub-optimal solution is preferable when the optimal solution from Section 3 is computationally intractable. If the QP returns a (local minimum) solution with non-integer values for outcomes of map x, then this result cannot be used for solving Problem 1.

Example 3. *Consider again the example from the end of Section 2. The paths in G and outcome values of map ω were already computed as in Section 3, where the optimal cost from MILP (4) was 54. QP (8) was solved in less than 0.4 s and it led to a sub-optimal total time of 60 for bringing all samples in v_{10}. The robots were allocated to collect samples as follows:*

$$
\begin{aligned}
&\textit{Robot } r_1 \textit{ collects samples from:} \\
&(v_8), \ (v_3), \ (v_2), \ (v_1), \ (v_1) \quad (\text{time}: 60) \\
&\textit{Robot } r_2 \textit{ collects samples from:} \\
&(v_8), \ (v_8), \ (v_3), \ (v_2), \ (v_1) \quad (\text{time}: 48) \\
&\textit{Robot } r_3 \textit{ collects samples from:} \\
&(v_5), \ (v_2), \ (v_2), \ (v_1) \quad\quad\quad (\text{time}: 52)
\end{aligned}
\tag{9}
$$

A similar QP relaxation may be constructed for MILP (6) by considering $M >> W$.

4.2. Iterative Solution

This subsection proposes an alternative sub-optimal allocation method, described in Algorithm 2. The method iteratively picks an uncollected sample whose transport to deposit requires minimum energy (line 7) and assigns it to a robot that has spent less energy (time) than other agents (lines 8–15). If a robot does not have enough energy to pick the current sample s, it is removed from further assignments (lines 16–17), because the remaining samples would require more energy than $\omega(\pi(s))$. If the current sample s cannot be allocated to any robot, the procedure is stopped (lines 18–19), and in this case some samples remain uncollected. When Algorithm 2 reaches line 20, the robot assignments constitute a solution to Problem 1 for collecting all samples from S. The robot-to-sample allocations returned by Algorithm 2 can be easily transformed to robot plans, as in lines 4–13 from Algorithm 1. The total time for completing the mission can be easily computed by maximizing over the times spent by each robot.

Algorithm 2: Iterative heuristic solution

Input: $R, S, w, \mathcal{E}$

Output: Robot-to-sample assignments

1 $R_{assign} = R$

2 $S_{uncollected} = S$

3 Set $x(r, s) = 0, \forall (r, s) \in R \times S$

4 Let $\mathcal{E}_{consumed}(r) = 0, \forall r \in R$

5 **while** $S_{uncollected} \neq \varnothing$ **do**

6 $\quad$ $S_0 = S_{uncollected}$

7 $\quad$ Pick $s \in S_{uncollected}$ s.t. $\omega(\pi(s)) = \min\limits_{s \in S_{uncollected}} \omega(\pi(s))$

8 $\quad$ Sort R_{assign} based on ascending order of consumed robot energy ($\mathcal{E}_{consumed}$)

9 $\quad$ **for** $r \in R_{assign}$ **do**

10 $\quad\quad$ **if** $w(s) \leq \mathcal{E}(r)$ **then**

11 $\quad\quad\quad$ $x(r, s) = 1$ (assign sample s to robot r)

12 $\quad\quad\quad$ $\mathcal{E}(r) := \mathcal{E}(r) - w(s)$

13 $\quad\quad\quad$ $\mathcal{E}_{consumed}(r) := \mathcal{E}_{consumed}(r) + w(s)$

14 $\quad\quad\quad$ $S_{uncollected} := S_{uncollected} \setminus \{s\}$

15 $\quad\quad\quad$ Break "for" loop

16 $\quad\quad$ **else**

17 $\quad\quad\quad$ $R_{assign} := R_{assign} \setminus \{r\}$

18 $\quad$ **if** $S_{uncollected} = S_0$ **then**

19 $\quad\quad$ *Return* current robot-to-sample allocations x

20 *Return* robot-to-sample allocations x

Under these ideas, Algorithm 2 can be seen as a greedy approach (first collect samples that require less energy/time), while the allocations to robots with less spent energy tries to reduce the overall time until the samples are collected.

Observe that this IH solution always returns a solution for collecting some (if not all) samples, whereas MILP from Section 3 may become computationally impracticable, while QP (8) may fail in providing a solution (Remark 3). Moreover, the software implementation of Algorithm 2 does not require additional tools, in contrast with specific optimization packages needed by MILP and QP solutions. A detailed analysis of the three methods is the goal of Section 5.

Example 4. *For illustrating Algorithm 2 on the example considered in the previous sections, we give here the sample picking costs: $\omega(\pi(s)) = 16$, $s = 1,\ldots,4$, $\omega(\pi(s)) = 14$, $s = 5,\ldots,8$, $\omega(\pi(s)) = 10$, $s = 9,10$, $\omega(\pi(11)) = 8$, $\omega(\pi(s)) = 4$, $s = 12,\ldots,14$. First, IH solution allocates sample 12 to r_1, then 13 and 14 to r_2, r_3, sample 11 to r_1 and so on. Algorithm 1 was run in 0.015 s and it yielded the following robotic plans:*

$$
\begin{aligned}
&\text{Robot } r_1 \text{ collects samples from:} \\
&(v_8), \ (v_5), \ (v_2), \ (v_2), \ (v_1) \quad (\text{time} : 56) \\
&\text{Robot } r_2 \text{ collects samples from:} \\
&(v_8), \ (v_3), \ (v_2), \ (v_1), \ (v_1) \quad (\text{time} : 60) \\
&\text{Robot } r_3 \text{ collects samples from:} \\
&(v_8), \ (v_3), \ (v_2), \ (v_1) \quad (\text{time} : 44)
\end{aligned}
\tag{10}
$$

5. Numerical Evaluation and Comparative Analysis

5.1. Additional Examples

Besides the remarks and examples from Sections 3 and 4, we present some slight modifications of the Example from Section 2 with the purpose of emphasizing the need for a comparative analysis of the three proposed solutions.

Example A: Let us add one more sample in node v_1 of the environment from Figure 1, leading to a total number of 15 samples. By running the MILP optimization (4), a solution was obtained in almost 40 s and it leads to an optimum time of 60. The QP relaxation (8) was run in almost 0.4 s (negligible increase from example from Section 4.1), and it implied a time cost of 62 for collecting all samples. The IH solution from Algorithm 2 yielded a cost of 60 in 0.016 s (practically no different to in Section 4.2). The actual robotic plans are omitted for this case.

Example B: If we assume a team of 4 robots for Example A, the MILP running time exhibits a significant decrease, being solved in less than 0.1 s. The QP optimization was solved in 0.5 s, and the IH in less than 0.02 s. The resulted time costs were 44 for MILP (optimum) and 50 for QP and IH (sub-optimum).

Example C: By adding one more robot to the team from Example B, the MILP optimization did not finish in 1 h, so it can be declared computationally unfeasible for this situation. The QP optimization finished in slightly more than 0.5 s, while the IH in around 0.02 s. Both QP and IH solutions returned a cost of 40.

Similar modifications of the above examples suggested the following empirical ideas:

- The running time of the MILP optimization may exhibit unpredictable behaviors with respect to the team size and to the number and position of samples, leading to impossibility of obtaining a solution in some cases;
- In contrast to MILP, the running times of the QP and IH solutions have insignificant variations when small changes are made in the environment;
- When MILP optimization finishes, the sub-optimal costs obtained by solutions from Section 4 are generally acceptable when compared to the optimal cost;
- In some cases the QP cost was better than the one obtained by IH, while in other cases the vice versa, but again the observed differences were fairly small.

The above ideas were formulated only based on a few variations of the same example. However, they motivate the more extensive comparison performed in the next subsections between the computation feasibility and outcomes of MILP, QP, and IH solutions.

Remark 4. *As mentioned, complexity orders of MILP and QP solutions cannot be formally given. However, as is customary in some studies, we here recall the number of unknowns and constraints of these optimizations. MILP (4) and QP (8) have each $N_R \times N_S + 1$ unknowns (from which $N_R \times N_S$ are binary in case of MILP (4))* *and $2N_R + N_S$ linear constraints. IH solution has complexity order $\mathcal{O}(N_R \times N_S)$, based on iterative loops from Algorithm 2 but, in all our studies the execution time of the algorithm is very small.*

Real-time example: Sample collecting experiments were implemented on a test-bed platform by using two Khepera robots equipped with plows for collecting items [53]. For exemplification, a movie is available at https://www.youtube.com/watch?v=2BQiWvquP7w. In the mentioned scenario, the graph environment is obtained from a cell decomposition [1,11] and a greedy method is employed for planning the robots. Although the collision avoidance problem is not treated in this paper, in the mentioned experiment the possible collisions are avoided by pausing the motion of one robot. In future work we intend to embed formal tools inspired by resource allocation techniques for collision and deadlock avoidance [54,55].

5.2. Numerical Experiments

All the simulations to be presented were implemented in MATLAB [48] and were performed on a computer with Intel i7 quad-core processor and 8 GB RAM.

The numerical experiments were run for almost 15 days, and they were organized by considering the following aspects:

(i)   Time complexity orders cannot be a priori given for MILP or for non-convex QP optimization problems. Thus, the time for obtaining a planning solution solving Problem 1 is to be recorded as an important comparison criterion.

(ii)   The complexity of MILP, QP, and IH solutions does not directly depend on the size of environment graph G, except for the initial computation of map ω that further embeds the necessary information from the environment structure. Therefore, the running time of either solution is influenced by two parameters: the number of robots (N_R) and the number of samples (N_S).

(iii)   Based on item (ii), we consider variation ranges $(N_R, N_S) \in [2, \ldots, 10] \times [2, \ldots, 50]$, with unit increment steps for N_R and N_S. Please note that the cases of 1 robot and/or 1 sample are trivial, and therefore are not included in the above parameter intervals.

(iv)   To obtain reliable results for item (i), for each pair (N_R, N_S) we have run a set of 50 trials. For each trial we generated a random distribution of samples in a 50-node graph. To maintain focus on time complexity, we assumed sufficiently large amounts of robot energy $\mathcal{E}$, such that Problem 1 is not infeasible due to these limitations.

(v)   For each trial, the MILP optimization was deemed *failed* whenever (4) did not return a solution in less than 1 min. This is because in multiple situations we observed that if no solution is obtained in less than 30–40 s, then the MILP (4) does not finish even after 2–3 h.

(vi)   For each trial from item (iv), the QP solution was deemed *failed* whenever it yielded non-binary outcomes of map x (see Remark 3). The IH solution is always *successful*.

(vii)   For each proposed solution, for each pair (N_R, N_S) from (iii) and based on trials from (iv), we computed the following comparison criteria:

- *success rate*, showing the percentage of trials when the solution succeeded in outputting feasible plans;
- *computation time*, averaged over the successful trials of a given (N_R, N_S) instance;
- *solution cost*, i.e., time for gathering all samples, for each successful situation.

5.3. Results

The results from item (vii) allow us to draw some rules that guide a user to choose MILP, QP, or IH solution when solving a specific instance of Problem 1. The following figures present and comment these results, leading to the decision diagram from the end of this section.

Figure 2 illustrates the success rates of the optimization problems from Sections 3 and 4.1, respectively. For a clearer understanding, Figure 3 presents pairs (N_R, N_S) when MILP (4) fails in more than 50% from each set of 50 trials (see item (v) from Section 5.2). It is noted that MILP generally returns optimal solutions for small values of N_R and N_S and fails (because of optimization time limit) for larger values. QP returns usually returns feasible solutions, excepting some cases with small values of N_R and N_S.

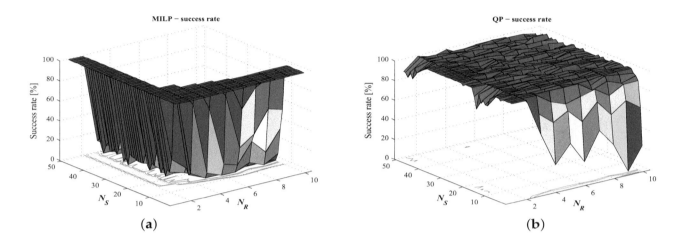

Figure 2. Success rates of MILP (**a**) and QP (**b**) vs. number of robots N_R and number of samples N_S.

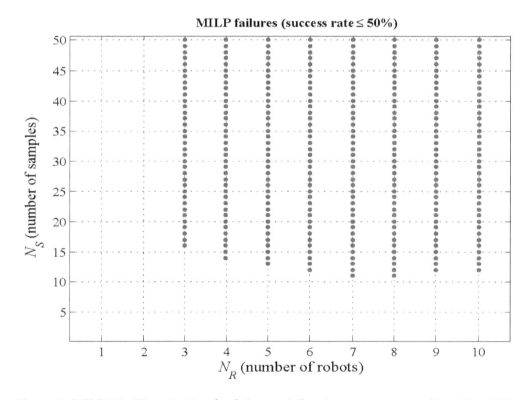

Figure 3. MILP (4): 2D projection for failures, defined as success rate of less than 50%.

Figure 4 presents the average computation time over the successful trials, for each solution we proposed. The representation is omitted for pairs (N_R, N_S) when there are less than 5 (from 50) successful trials—as it is often the case for MILP solver, when $N_R \geq 3$ and $N_S \geq 13$. MILP time may sudden variations, whereas the times for QP and IH indicates predictable behaviors. The IH time is very small (note axis limits in Figure 4) and exhibits negligible variations versus N_R and almost linear increases versus N_S, due to the main iteration loop from Algorithm 2.

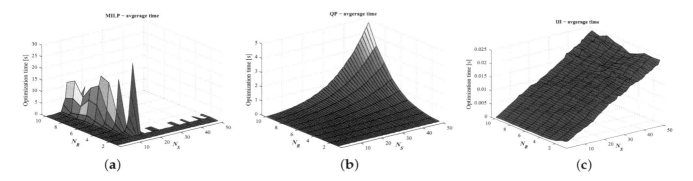

Figure 4. Average optimization times, for (N_R, N_S) pairs for which at least 5 feasible solutions from the 50 tests were obtained: (**a**) MILP (4), (**b**) QP (8), (**c**) IH from Algorithm 2.

To suggest the confidence intervals of values plotted in Figure 4, we mention that:

- For $N_R = 3$ and $N_S = 15$, when the success rate of each optimization exceeds 98%, the standard deviations of optimization times are: 15 for MILP, 0.01 for QP, 0.002 for IH;

- For $N_R = 10$ and $N_S = 50$, when QP has 96% success rate, the standard deviations of optimization times are: 0.23 for QP and 0.003 for IH.

Figure 5 presents the averaged relative differences of costs yielded by the three proposed solutions for solving Problem 1. As visible in Figure 5a, the QP (sub-optimal) cost is usually less than 120% . . . 130% of optimal MILP cost. More specifically, from all the 25,000 trials, in 8443 cases (about 33%) both optimizations succeeded. After averaging cost differences versus (N_R, N_S), we obtained 348 points for representing Figure 5a, and in 330 cases the cost difference was less than 20%. Figure 5b compares the costs yielded by the sub-optimal solutions from Section 4 by representing variations of the IH cost related to the QP one. It follows that usually IH yields a higher cost than QP, but the difference decreases below 5% . . . 10% with the increase in problem complexity. Further studies can be conducted towards formulating a conjecture that gives a formal tendency for the variation of cost difference based on problem size. However, one issue for such a study is mainly given by the necessity of obtaining the optimal cost even for large problems, i.e., solving large MILP optimizations.

For more complex problems ($N_R > 10$, $N_S > 50$), the time tendencies from Figure 4b,c and the cost differences from Figure 5b suggest that the IH solution is preferable as a good trade-off between planning complexity and resulted cost.

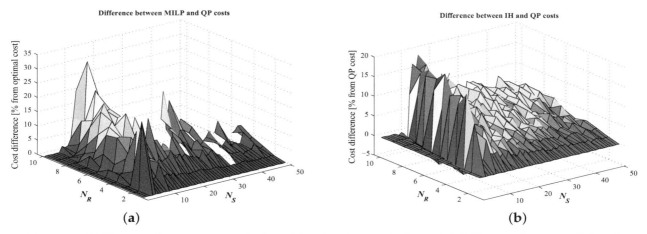

Figure 5. Differences between costs induced by the three solutions: (**a**) difference between QP and MILP costs, related to the optimal MILP cost; (**b**) difference between IH and QP sub-optimal costs, related to QP cost.

Extensive simulations yielded quantitative comparison criteria. Based on this information a decision scheme Figure 6 is given for indicating the proper method to be used in a specific problem instance when a fast computation scenario is considered.

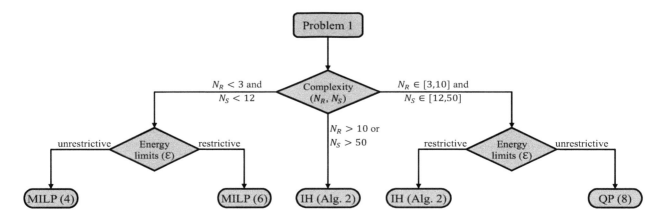

Figure 6. Decision diagram for choosing an appropriate solution for Problem 1 when a fast computational scenario is considered.

6. Conclusions

This paper details three methods for planning a team of robots such that multiple samples (items) from the environment are collected and deposited into a storage location. The environment is represented by a weighted graph, and the robots have limited amounts of energy for performing movements on this graph. The goal is to plan the agents such that the samples are collected in minimum time, under the assumption that each robot can carry at most one item at a time. The first solution is given by a MILP formulation that can be too complex to solve when there are many samples or many robots in the team. The second solution provides a QP relaxation that represents in some cases a good trade-off between the time for finding movement plans and the cost difference from the optimal one. The third solution is an IH algorithm that yields plans even when the QP fails due to low amounts of available energy for robots.

Based on the results reported in the previous subsection, the recommendations for a user that solves Problem 1 are the following: MILP (6) or IH algorithm are to be used when the robots have small amounts of energy in comparison with the energy required for moving to sample locations. Otherwise, MILP (4) and QP (8) may fail in providing any solution for such restrictive scenarios.

The usage of QP or IH solutions generally yields an acceptable loss in total time for accomplishing the mission whenever the MILP optimization becomes computationally intractable on a decently powerful computer. We emphasize that the above recommendations resulted from an intensive campaign of numerical simulations and they could not be drawn by only inspecting the formal solutions. Extensive simulations yielded quantitative comparison criteria. Based on this information a decision scheme is given for indicating the proper method to be used in a specific problem instance. Besides suggestions from Figure 6, we recall that MILP and QP solutions involve existing optimization tools.

A real-time experiment was performed for illustrating a sample gathering solution. Although the collision avoidance problem is not treated in this paper, in the mentioned experiment the possible collisions are avoided by pausing the motion of one robot. In future work we intend to embed formal tools inspired by resource allocation techniques for collision and deadlock avoidance, while considering the effects of acceleration and deceleration of the mobile robots and restricted energy.

Author Contributions: The authors contributed equally to this work, each of them being involved in all research aspects.

References

1. Choset, H.; Lynch, K.; Hutchinson, S.; Kantor, G.; Burgard, W.; Kavraki, L.; Thrun, S. *Principles of Robot Motion: Theory, Algorithms and Implementation*; MIT Press: Cambridge, MA, USA, 2005.
2. LaValle, S.M. *Planning Algorithms*; Cambridge University Press: Cambridge, UK, 2006.
3. Siciliano, B.; Khatib, O. *Springer Handbook of Robotics*; Springer: Berlin, Germany, 2008.
4. Belta, C.; Bicchi, A.; Egerstedt, M.; Frazzoli, E.; Klavins, E.; Pappas, G.J. Symbolic Planning and Control of Robot Motion. *IEEE Robot. Autom. Mag.* **2007**, *14*, 61–71. [CrossRef]
5. Imeson, F.; Smith, S.L. A Language For Robot Path Planning in Discrete Environments: The TSP with Boolean Satisfiability Constraints. In Proceedings of the IEEE Conference on Robotics and Automation, Hong Kong, China, 31 May–7 June 2014; pp. 5772–5777.
6. Mahulea, C.; Kloetzer, M. Robot Planning based on Boolean Specifications using Petri Net Models. *IEEE Trans. Autom. Control* **2018**, *63*, 2218–2225. [CrossRef]
7. Fainekos, G.; Girard, A.; Kress-Gazit, H.; Pappas, G. Temporal logic motion planning for dynamic robot. *Automatica* **2009**, *45*, 343–352. [CrossRef]
8. Ding, X.; Lazar, M.; Belta, C. {LTL} receding horizon control for finite deterministic systems. *Automatica* **2014**, *50*, 399–408. [CrossRef]
9. Schillinger, P.; Bürger, M.; Dimarogonas, D. Simultaneous task allocation and planning for temporal logic goals in heterogeneous multi-robot systems. *Int. J. Robot. Res.* **2018**, *37*, 818–838. [CrossRef]
10. Kloetzer, M.; Mahulea, C. LTL-Based Planning in Environments With Probabilistic Observations. *IEEE Trans. Autom. Sci. Eng.* **2015**, *12*, 1407–1420. [CrossRef]
11. Berg, M.D.; Cheong, O.; van Kreveld, M. *Computational Geometry: Algorithms and Applications*, 3rd ed.; Springer: Berlin, Germany, 2008.
12. Kloetzer, M.; Mahulea, C. A Petri net based approach for multi-robot path planning. *Discret. Event Dyn. Syst.* **2014**, *24*, 417–445. [CrossRef]
13. Cassandras, C.; Lafortune, S. *Introduction to Discrete Event Systems*; Springer: Berlin, Germany, 2008.
14. Silva, M. Introducing Petri nets. In *Practice of Petri Nets in Manufacturing*; Springer: Berlin, Germany, 1993; pp. 1–62.
15. Burkard, R.; Dell'Amico, M.; Martello, S. *Assignment Problems*; SIAM e-Books; Society for Industrial and Applied Mathematics (SIAM): Philadelphia, PA, USA, 2009.
16. Mosteo, A.; Montano, L. *A Survey of Multi-Robot Task Allocation*; Technical Report AMI-009-10-TEC; Instituto de Investigación en Ingeniería de Aragón, University of Zaragoza: Zaragoza, Spain, 2010.
17. Gerkey, B.; Matarić, M. A formal analysis and taxonomy of task allocation in multi-robot systems. *Int. J. Robot. Res.* **2004**, *23*, 939–954. [CrossRef]
18. Korsah, G.; Stentz, A.; Dias, M. A comprehensive taxonomy for multi-robot task allocation. *Int. J. Robot. Res.* **2013**, *32*, 1495–1512. [CrossRef]
19. Shmoys, D.; Tardos, É. An approximation algorithm for the generalized assignment problem. *Math. Program.* **1993**, *62*, 461–474. [CrossRef]
20. Atay, N.; Bayazit, B. Emergent task allocation for mobile robots. In Proceedings of the Robotics: Science and Systems, Atlanta, GA, USA, 27–30 June 2007.
21. Cardema, J.; Wang, P. Optimal Path Planning of Multiple Mobile Robots for Sample Collection on a Planetary Surface. In *Mobile Robots: Perception & Navigation*; Kolski, S., Ed.; IntechOpen: London, UK, 2007; pp. 605–636.
22. Chen, J.; Sun, D. Coalition-Based Approach to Task Allocation of Multiple Robots with Resource Constraints. *IEEE Trans. Autom. Sci. Eng.* **2012**, *9*, 516–528. [CrossRef]
23. Das, G.; McGinnity, T.; Coleman, S.; Behera, L. A Distributed Task Allocation Algorithm for a Multi-Robot System in Healthcare Facilities. *J. Intell. Robot. Syst.* **2015**, *80*, 33–58. [CrossRef]
24. Burger, M.; Notarstefano, G.; Allgower, F.; Bullo, F. A distributed simplex algorithm and the multi-agent assignment problem. In Proceedings of the American Control Conference (ACC), San Francisco, CA, USA, 29 June–1 July 2011; pp. 2639–2644.

25. Luo, L.; Chakraborty, N.; Sycara, K. Multi-robot assignment algorithm for tasks with set precedence constraints. In Proceedings of the IEEE International Conference on Robotics and Automation (ICRA), Shanghai, China, 9–13 May 2011; pp. 2526–2533.

26. Toth, P.; Vigo, D. (Eds.) *The Vehicle Routing Problem*; Society for Industrial and Applied Mathematics: Philadelphia, PA, USA, 2001.

27. Laporte, G.; Nobert, Y. Exact Algorithms for the Vehicle Routing Problem. In *Surveys in Combinatorial Optimization*; Martello, S., Minoux, M., Ribeiro, C., Laporte, G., Eds.; North-Holland Mathematics Studies; North-Holland: Amsterdam, The Netherlands, 1987; Volume 132, pp. 147–184.

28. Toth, P.; Vigo, D. Models, relaxations and exact approaches for the capacitated vehicle routing problem. *Discret. Appl. Math.* **2002**, *123*, 487–512. [CrossRef]

29. Raff, S. Routing and scheduling of vehicles and crews: The state of the art. *Comput. Oper. Res.* **1983**, *10*, 63–211. [CrossRef]

30. Christofides, N. Vehicle routing. In *The Traveling Salesman Problem: A guided Tour of Combinatorial Optimization*; Wiley: New York, NY, USA, 1985; pp. 410–431.

31. Laporte, G. The vehicle routing problem: An overview of exact and approximate algorithms. *Eur. J. Oper. Res.* **1992**, *59*, 345–358. [CrossRef]

32. Chen, J.F.; Wu, T.H. Vehicle routing problem with simultaneous deliveries and pickups. *J. Oper. Res. Soc.* **2006**, *57*, 579–587. [CrossRef]

33. Kloetzer, M.; Burlacu, A.; Panescu, D. Optimal multi-agent planning solution for a sample gathering problem. In Proceedings of the IEEE International Conference on Automation, Quality and Testing, Robotics (AQTR), Cluj-Napoca, Romania, 22–24 May 2014.

34. Kloetzer, M.; Ostafi, F.; Burlacu, A. Trading optimality for computational feasibility in a sample gathering problem. In Proceedings of the International Conference on System Theory, Control and Computing, Sinaia, Romania, 17–19 October 2014; pp. 151–156.

35. Kloetzer, M.; Mahulea, C. An Assembly Problem with Mobile Robots. In Proceedings of the IEEE 19th Conference on Emerging Technologies Factory Automation, Barcelona, Spain, 16–19 September 2014.

36. Panescu, D.; Kloetzer, M.; Burlacu, A.; Pascal, C. Artificial Intelligence based Solutions for Cooperative Mobile Robots. *J. Control Eng. Appl. Inform.* **2012**, *14*, 74–82.

37. Kloetzer, M.; Mahulea, C.; Burlacu, A. Sample gathering problem for different robots with limited capacity. In Proceedings of the International Conference on System Theory, Control and Computing, Sinaia, Romania, 13–15 October 2016; pp. 490–495.

38. Tumova, J.; Dimarogonas, D. Multi-agent planning under local LTL specifications and event-based synchronization. *Automatica* **2016**, *70*, 239–248. [CrossRef]

39. Belta, C.; Habets, L. Constructing decidable hybrid systems with velocity bounds. In Proceedings of the 43rd IEEE Conference on Decision and Control, Nassau, Bahamas, 14–17 December 2004; pp. 467–472.

40. Habets, L.C.G.J.M.; Collins, P.J.; van Schuppen, J.H. Reachability and control synthesis for piecewise-affine hybrid systems on simplices. *IEEE Trans. Autom. Control* **2006**, *51*, 938–948. [CrossRef]

41. Garey, M.; Johnson, D. "Strong" NP-Completeness Results: Motivation, Examples, and Implications. *J. ACM* **1978**, *25*, 499–508. [CrossRef]

42. Cormen, T.; Leiserson, C.; Rivest, R.; Stein, C. *Introduction to Algorithms*, 2nd ed.; MIT Press: Cambridge, MA, USA, 2001.

43. Ding-Zhu, D.; Pardolos, P. *Nonconvex Optimization and Applications: Minimax and Applications*; Spinger: New York, NY, USA, 1995.

44. Polak, E. *Optimization: Algorithms and Consistent Approximations*; Spinger: New York, NY, USA, 1997.

45. Floudas, C.; Pardolos, P. *Encyclopedia of Optimization*, 2nd ed.; Spinger: New York, NY, USA, 2009; Volume 2.

46. Makhorin, A. GNU Linear Programming Kit. 2012. Available online: http://www.gnu.org/software/glpk/ (accessed on 4 January 2019).

47. SAS Institute. The Mixed-Integer Linear Programming Solver. 2014. Available online: http://support.sas.com/rnd/app/or/mp/MILPsolver.html (accessed on 4 January 2019).

48. The MathWorks. *MATLAB®R2014a (v. 8.3)*; The MathWorks: Natick, MA, USA, 2006.

49. Wolsey, L.; Nemhauser, G. *Integer and Combinatorial Optimization*; Wiley: New York, NY, USA, 1999.

50. Vazirani, V.V. *Approximation Algorithms*; Springer: New York, NY, USA, 2001.

51. Earl, M.; D'Andrea, R. Iterative MILP Methods for Vehicle-Control Problems. *IEEE Trans. Robot.* **2005**, *21*, 1158–1167. [CrossRef]

52. Murray, W.; Ng, K. An Algorithm for Nonlinear Optimization Problems with Binary Variables. *Comput. Optim. Appl.* **2010**, *47*, 257–288. [CrossRef]

53. Tiganas, V.; Kloetzer, M.; Burlacu, A. Multi-Robot based Implementation for a Sample Gathering Problem. In Proceedings of the International Conference on System Theory, Control and Computing, Sinaia, Romania, 11–13 October 2013; pp. 545–550.

54. Kloetzer, M.; Mahulea, C.; Colom, J.M. Petri net approach for deadlock prevention in robot planning. In Proceedings of the IEEE Conference on Emerging Technologies Factory Automation (ETFA), Cagliari, Italy, 10–13 September 2013; pp. 1–4.

55. Roszkowska, E.; Reveliotis, S. A Distributed Protocol for Motion Coordination in Free-Range Vehicular Systems. *Automatica* **2013**, *49*, 1639–1653. [CrossRef]

7

Automated Enemy Avoidance of Unmanned Aerial Vehicles based on Reinforcement Learning

Qiao Cheng *, Xiangke Wang *, Jian Yang and Lincheng Shen

College of Intelligence Science and Technology, National University of Defense Technology, Changsha 410073, China; yj_ntx@163.com (J.Y.); lcshen@nudt.edu.cn (L.S.)
* Correspondence: qiao.cheng@nudt.edu.cn (Q.C.); xkwang@nudt.edu.cn (X.W.)

Abstract: This paper focuses on one of the collision avoidance scenarios for unmanned aerial vehicles (UAVs), where the UAV needs to avoid collision with the enemy UAV during its flying path to the goal point. Such a type of problem is defined as the enemy avoidance problem in this paper. To deal with this problem, a learning based framework is proposed. Under this framework, the enemy avoidance problem is formulated as a Markov Decision Process (MDP), and the maneuver policies for the UAV are learned based on a temporal-difference reinforcement learning method called Sarsa. To handle the enemy avoidance problem in continuous state space, the Cerebellar Model Arithmetic Computer (CMAC) function approximation technique is embodied in the proposed framework. Furthermore, a hardware-in-the-loop (HITL) simulation environment is established. Simulation results show that the UAV agent can learn a satisfying policy under the proposed framework. Comparing with the random policy and the fixed-rule policy, the learned policy can achieve a far higher possibility in reaching the goal point without colliding with the enemy UAV.

Keywords: enemy avoidance; reinforcement learning; decision making; hardware-in-the-loop simulation; unmanned aerial vehicles

---

1. Introduction

Unmanned Aerial Vehicles (UAVs) have received considerable attention in many areas [1], such as commercial, search and rescue, military, and so on. In the military area, there are applications such as the surveillance [2,3], target tracking [4,5], target following [6,7], and so on. Among these applications, collision avoidance is one of the most important concerns [8], especially in unsafe environment. In such cases, a UAV should keep safe separation with various kinds of objects, such as static obstacles [9,10], teammates [11], and moving enemies. The strategies toward different approaching objects are different due to specific requirements in dealing with those objects. This paper focuses on avoiding the collision with moving enemies. There are many researches on collision avoidance problems. However, relatively fewer works are on the avoidance of moving enemies, comparing with those on the avoidance of static obstacles and flying teammates. Furthermore, the uncertain motion of enemies and the necessity to attack enemies create more challenges on the avoidance of moving enemies than avoiding other objects. Besides, the mission of the UAV, such as reaching a specific goal destination, should also be considered. Therefore, the avoidance of moving enemies is a challenge problem, and such a problem is defined as the enemy avoidance problem in this paper.

There are many approaches to handle the collision avoidance problem in different stages [12]. Many of those approaches rely on models for the dynamic of the environment and UAVs. However, the accuracy of these models can sometimes greatly affect the performance of those methods. Moreover, building these models is not easy work, and is even impractical. On the other hand, a complex model means heavy computation load when making decisions. Therefore, learning methods are increasingly used in collision avoidance problems, which are based on collected data. However, most of these learning methods are used to predict the effect of the decision, not directly used for the decision making. Different from other learning methods, reinforcement learning is a very popular method for sequential decision making problems [13]. It can learn to make decisions incrementally based on feedback from the environment. Therefore, it can generate a good policy even if the models of the environment are unknown. Since a sequence of appropriate actions are required to avoid the enemy UAV, the enemy avoidance problem can be regarded as a sequential decision making problem. Therefore, this paper proposes a framework which incorporates the reinforcement learning to deal with the enemy avoidance problem.

Many methods for the collision avoidance problem discretize the state space to make decisions [14,15]. However, this paper studies the enemy avoidance problem in continuous state space. Therefore, the function approximation technique, which can handle continuous space, is also embodied in the proposed framework.

As for the UAVs, most researches [16–18] focus on quadrotors rather than fixed-wing UAVs in collision avoidance problems. However, the dynamics of quadrotors and fixed-wing UAVs are different. In addition, the UAV in the enemy avoidance problem has the mission to reach the goal point, and thereby needs to keep away from the enemy UAV and even attack enemy UAVs. In practical application, the fixed-wing UAVs are more suitable for such a problem scenario for their better mission fulfillment properties, such as higher endurance and greater speeds. Therefore, this paper focuses on the enemy avoidance problem of fixed-wing UAVs.

Since learning the policy on the real UAV platform would bring about great consumption, a hardware-in-the-loop (HITL) simulation system is constructed. With hardware-in-the-loop, the simulation system can provide very consistent properties to that of the real environment, which highly respects the kinecmatic and maneuver constraints of the UAVs. Furthermore, it saves the energy to build a model for the related hardware, which is usually very hard to build accurately. Comparing with the real UAV platform, the HITL simulation system can repeat the experiments as many times as needed without worrying about UAV costs.

The contributions of this paper are summarized as follows.

(i)   An interesting new problem called the enemy avoidance problem is defined, which can be a good adding up scenario to the collision avoidance problem. The newly defined problem is different from most of the existing collision avoidance problems, for it is to avoid the collision with the enemy UAV rather than static obstacles or moving teammates.

(ii)   A novel framework is proposed to learn the policy for the decision making UAV. The proposed framework formulates the enemy avoidance problem as a Markov Decision Process (MDP) problem, and solves the MDP problem by a temporal-difference reinforcement learning method called Sarsa. The Cerebellar Model Arithmetic Computer (CMAC, [19]) technique is also embodied in the proposed framework for the generalization of the continuous state space. With this framework, such a decision making problem is transformed from the usual computational problem to a learning problem. Besides, it can learn the policy with an unknown environment model, and can make decisions based on continuous state space rather than discrete ones like most existing works do.

(iii)   A hardware-in-the-loop (HITL) simulation environment for the enemy avoidance problem is constructed, which is used for the policy learning and policy testing experiments. Different from the simulation environment in most of the existing works, this HITL simulation system saves a lot of model designing trouble, and has better consistency to the real environment, such as the environment noise. When comparing with real environment platforms, the HITL simulation system has the advantage of saving experimental cost.

The remainder of this paper is outlined as follows. Section 2 gives some reviews on the related literature. The enemy avoidance problems are presented in Section 3. The proposed framework for the enemy avoidance problem is elaborated in Section 4. Section 5 details the construction of the hardware-in-the-loop simulation environment. Simulation experiments and results are illustrated in Section 6. Finally, Section 7 concludes the whole work and discusses future works.

2. Literature Review

There are many researches on collision avoidance problems, and different methods are used to solve different collision avoidance problems. Therefore, this section will give a summary about several widely used methods in collision avoidance problems, as well as a comparison between our work and these existing works.

One of the most widely used methods is to formulate the collision avoidance problem as an optimization problem, while considering all kinds of constrains. Therefore, to avoid collision is to solve the optimization problem with appropriate methods under different constrains. The work in [20] formulates the collision avoidance problem as a convex optimization problem, and seeks for a suitable control constraint set for participating UAV based on reachable sets and tubes for UAVs. This method is limited to linear systems. The collision avoidance in work [21] is formulated as a set of linear quadratic optimization problems, which are solved with an original geometric based formulation. To handle flocking control with obstacle avoidance, work [22] proposes a UAV distributed flocking control algorithm based on the modified multi-objective pigeon-inspired optimization (MPIO), which considers both the hard constraints and the soft ones. Our previous works [23,24] formulate the conflict avoidance problem as a nonlinear optimization problem, and then use different methods to solve such an optimization problem. The work in [23] proposes a two-layered mechanism to guarantee safe separation, which finds the optimal heading change solutions with the vectorized stochastic parallel gradient descent-based method, and finds the optimal speed change solutions with a mixed integer linear programming model. The work in [24] uses the stochastic parallel gradient descent (SPGD) method to find the feasible initial solutions, and then uses the Sequential quadratic programming (SQP) algorithm to compute the local optimal solution. Even for the obstacle avoidance problem in other areas, the optimization methods are also used. For example, two swarm based optimization techniques are used in work [25] to offer obstacle-avoidance path planning for mobility-assisted localization in wireless sensor networks (WSN), which are grey wolf optimizer and whale optimization algorithm. The main difference between the collision avoidance for UAVs and the obstacle-avoidance path planning in WSN lies in the constrains and objective of the optimization model. Usually, solving the optimization problem requires a lot of computation. Therefore, our work does not formulate the enemy avoidance problem as an optimization problem, but formulates it as an MDP problem and solves the MDP incrementally by interaction with the environment.

Another kind of method for solving collision avoidance problems is to predict the potential collision with certain techniques. The work in [26] proposes an approach based on radio signal strengths (RSS) measurements to obtain position estimation of the UAV, and to detect the potential collisions based on the position estimations, and then to distribute the UAVs at different altitudes to avoid collision. The work in [27] proposes a model-based learning algorithm that enables the agent to learn an uncertainty-aware collision prediction model through deep neural networks, so as to avoid the collision with unknown static obstacles. The work in [28] proposes a data-driven end-to-end motion planing approach which helps the robot navigate to a desired target point while avoiding collisions with static obstacles without the need of a global map. This approach is based on convolutional neural networks (CNNs), and the robot is provided with expert demonstrations about navigation in a given virtual training environment. One of the problems for such kind of methods is that high capacity learning algorithms like deep learning tend to overfit when little training data is available. Therefore, the work in [29] collects a lot of crash samples to build a dataset by crashing their drone 11,500 times. The used data driven approach demonstrates such negative data is also crucial for learning how to navigate without collision. However, to collect both positive data and negative data for prediction is very costly. Different from these works, our work aims to obtain a policy with the proposed framework.

The policy is a mapping from the state directly to the action, therefore, no prediction of the collision is needed.

As reinforcement learning gains its popularity in decision making problems, there are works that use different reinforcement learning to solve the collision avoidance problem. The work in [30] proposes to combine Model Predictive Control (MPC) with reinforcement learning to learn obstacle avoidance policies for the UAV in a simulation environment. In this method, the MPC is used to generate data at the training time, and the deep neural network policies are trained with an off-policy reinforcement learning method called guided policy search based on the generated data. The work in [31] proposes a geometric reinforcement learning algorithm for UAV path planning, which constructs a specific reward matrix to include the geometric distance and risk information. This algorithm considers the obstacles as risk and builds a risk model for the obstacles, which is used in constructing the reward matrix. The work in [32] models the UAV collision avoidance problem as a Partially Observable Markov Decision Process (POMDP) and uses Monte Carlo Value Iteration (MCVI) to solve the POMDPs, which can cope with high-dimensional continuous-state space in a collision avoidance problem. The work in [33] formulates the problem of collision avoidance as an MDP and a POMDP, and uses generic MDP/POMDP solvers to generate avoidance strategies. Though the framework proposed in our work is also based on reinforcement learning, many details are different from the these works. For example, this paper uses neither special data generating process, nor complex reward function designing. Besides, the environment transition model is unknown in this paper, and a different reinforcement learning method is adopted.

Another big difference between the existing works and our work is that the collision avoidance problem in this paper is not exactly the same as those in previous works. First, the UAV in this paper needs to avoid collision with a moving enemy UAV, not static obstacles [25] or teammates [11]. Besides, the actions the UAV uses to avoid collision include both heading angle change and velocity change. The work in [34] investigated strategies for multiple UAVs to avoid collision with moving obstacles, which is a little similar with collision with enemy UAV. However, their work assumes all UAVs and all obstacles have constant ground speeds, and the direction of the velocity vector of an obstacle is constant. Therefore, they only consider change in direction of the UAV for collision avoidance, and do not consider change in velocity of the UAV. Furthermore, this paper does not attempt to build an environment model, but approximates it by continuous interaction with the environment. Similarlly, work [35] also approximates the unmodeled dynamics of the environment, but it uses back propagation neural networks and proposes a tree search algorithm to find the near optimal conflict avoidance solutions. In addition, this paper considers continuous state space in the decision making process, not the discrete one like many other related works do.

3. Problem Definition

We call the problem posed in this paper the enemy avoidance problem, which is different from the usual collision avoidance problems or path planning problems. Before proposing methods to solve this problem, we first give a detailed description for the enemy avoidance problem, as well as the related assumptions and definitions.

3.1. Problem Description

For the convenience of the research, we define the enemy avoidance problem in a fixed region. There are two UAVs flying toward each other in the region, namely the decision making UAV f and the enemy UAV e. The fixed region is where the two UAVs may collide with each other. The decision making UAV f enters the region from the left side, while the enemy UAV e enters the region from the right side. Both UAVs are flying toward their own goal points. Let G_f and G_e denote the goal point of the decision making UAV f and that of the enemy UAV e, respectively. The goal point G_f for the decision making UAV f is located near the right edge of the region, while the goal point G_e for the enemy UAV e is located near the left edge of the region. Both goal points are on the middle line of the region, as shown in Figure 1.

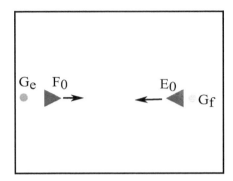

Figure 1. The enemy avoidance scenarios.

In such a region, there is no hovering requirement or taking off and landing requirements, therefore the fixed-wing UAV can be easily applied in such problem settings.

The mission of the decision making UAV f is to reach the goal point G_f safely and with as little cost as possible. However, the decision making UAV f and the enemy UAV e are flying in opposite directions along the middle line of the region toward their own goal points, which poses the decision making UAV f the danger of collision with the enemy UAV e. Therefore, the decision making UAV f needs to avoid the enemy UAV e during its flight towards the goal point G_f. Ways to avoid collision with the enemy UAV include changing the heading angle or the velocity, and attacking the enemy UAV. Though changing the heading angle can let the decision making UAV avoid flying directly into the enemy UAV if the the enemy UAV happens to be in the heading direction of the decision making UAV, inappropriate heading angle change may make the decision making UAV fly too far away from the goal point G_f. Similarly, changing the velocity at an appropriate time can also avoid collision with the enemy UAV, such as accelerating to pass the enemy UAV before the collision or decelerating to wait for the enemy UAV to pass. Successfully attacking the enemy UAV can also provide good insurance for the decision making UAV to fulfill its mission. However, the attacking action may fail in destroying the enemy UAV, and the decision making UAV suffers certain losses when using attacking action. Therefore, the decision making UAV should not use the attacking action too often. To achieve the mission requirements, the decision making UAV cannot use just one single avoiding action, but should arrange all the actions in an appropriate sequence.

On the other hand, the enemy UAV simply flies toward the goal point G_e with constant velocity and heading angle if the decision making UAV does not collide with it or attack it successfully. The scenario is supposed to end as soon as the decision making UAV has collided with or successfully attacked the enemy UAV, or the decision making UAV has reached the goal point G_f successfully or has been out of the region.

To arrange an appropriate action sequence is the process of decision making, which is also the main focus of this work. In decision making, such an action sequence is called the policy. Based on the action chosen by the on-board agent at each decision making step, the decision making UAV can adjust its flying attitude or attack the enemy UAV. The agent makes decisions based on both its own information from its on-board sensor system and the enemy UAV's information from the ground station. The action executed by the decision making UAV makes the UAV changes its state in the environment. On the other hand, the enemy UAV also updates its states in the environment and senses its own states from the environment with its on-board sensor system. The ground station captures all UAVs' information, and then transmits the information to the decision making UAV. Figure 2 presents the overall decision making process of the enemy avoidance problem. Therefore, how the decision making agent uses the gathered information to make decisions for avoiding collision with the enemy UAV is what this paper is going to solve. Furthermore, the time and the location at which the enemy

UAV enters the region are not fixed each time. Therefore, the decision making ability of the decision making UAV should be able to generalize to different enemy avoidance cases.

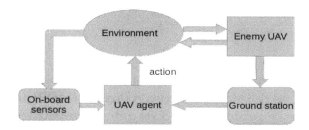

Figure 2. The decision making framework.

3.2. Assumptions and Definitions

At first, we need to make some assumptions about the posed enemy avoidance problem.

Assumption 1. *The enemy UAV has constant desired velocity and desired heading angle, while the actual velocity and heading angle of the enemy UAV oscillate a little around the desired ones.*

Assumption 2. *The decision making UAV can change its attitude and attack the enemy UAV during its flight according to the action decided by the UAV agent.*

Assumption 3. *The heights of the UAVs are not considered. Therefore, all the distances in the problem are simply computed by two dimensions.*

Assumption 4. *Each UAV obtains its own position and attitude (velocity and heading angle) with its on-board sensor system. The decision making UAV can obtain information about all the UAVs through the ground station.*

Assumption 5. *There are no other UAVs in the region except the decision making UAV and the enemy UAV, as well as no obstacles in the region.*

In these assumptions, Assumptions 1 and 3 are used to simplify the enemy avoidance problem, so that we can focus more on other more important factors in the enemy avoidance problem. The researched results then can be used as the basis of more practical problems. With Assumption 5, this paper can focus on the collision avoidance of the enemy UAV, and does not need to consider collision with other UAVs and obstacles.

Furthermore, we give some definitions that will be used in solving the enemy avoidance problem.

Definition 1. *(Distance). The distance between two points $a = (x_a, y_a)$ and $b = (x_b, y_b)$ is calculated with the following equation:*

$$d_{ab} = \sqrt{(x_a - x_b)^2 + (y_a - y_b)^2} \tag{1}$$

Definition 2. *(Reaching Goal). A UAV f is regarded as having reached a goal G when the following condition is satisfied:*

$$d_{fg} < r_g \tag{2}$$

where d_{fg} denotes the distance between the UAV f and the goal point G, and r_g is the specified goal radius.

Definition 3. *(Collision). A UAV f is regarded as having collided with the enemy UAV e when the following condition is satisfied:*

$$d_{fe} < r_c \tag{3}$$

where d_{fe} denotes the distance between the UAV f and the enemy UAV e, and r_c is the specified collision radius.

Definition 4. *(Attacking Probability). The success of an attacking action is defined by the attacking probability P, which is specified by the following equation:*

$$P = e^{1 - \frac{d_{fe}}{30}} \tag{4}$$

4. Problem Sovling

This paper proposes a new framework to solve the enemy avoidance problem, which formulates the enemy avoidance problem as the Markov Decision Process (MDP) and learns the decision making policy for the enemy avoidance problem based on reinforcement learning. Firstly, the detail of formulating the enemy avoidance problem as the Markov Decision Process (MDP) is presented, which is the basis of the reinforcement learning. Secondly, how to learn the policy based on a temporal-difference reinforcement learning method called Sarsa is elaborated, as well as the embodied function approximator called CMAC for the generalization of the continous state.

4.1. Formulate the Problem as the MDP

Reinforcement learning has been widely used in sequential decision making problems which are formulated as the Markov Decision Process (MDP). Typically, an MDP comprises of four elements: the state set $\mathcal{S}$, the action set $\mathcal{A}$, the transition function $\mathcal{T}$, and the reward function $\mathcal{R}$. When an agent is in a state $s \in \mathcal{S}$, it can choose an action $a \in \mathcal{A}$ according to a policy π. After executing the action a, the agent will enter into the next state $s' \in \mathcal{S}$ according to the transition function $\mathcal{T}$, and will receive an immediate reward r according to the reward function $\mathcal{R}$. In this enemy avoidance problem, the environment transition function $\mathcal{T}$ is unknown, and will be learned by interaction with the environment. To formulate the enemy avoidance problem as the MDP, the state space $\mathcal{S}$, the action space $\mathcal{A}$, and the reward function $\mathcal{R}$ are defined as follows.

4.1.1. State Space

In this enemy avoidance problem, the design of the state space mainly considers the positions and attitudes of the UAVs, as well as the position of the goal point G_f. However, these raw data are not used directly as the state variables. Instead, higher-level variables based on these data are defined. To be specific, the state space contains three sets of variables:

(i)   Variables about the status of the decision making UAV f:

- v_f: The velocity of the decision making UAV f.
- ψ_f: The heading angle of the decision making UAV f.

(ii)   Variables about the goal point G_f:

- d_{fg}: The distance from the decision making UAV f to the goal point G_f.
- ω_g: The angle between the north direction and the line from the decision making UAV f to the goal point G_f.

(iii)   Variables about the status of the enemy UAV e:

- v_e: The velocity of the enemy UAV e.
- ψ_e: The heading angle of the enemy UAV e.
- d_{fe}: The distance from the decision making UAV f to the enemy UAV e.
- ω_e: The angle between the north direction and the line from the decision making UAV f to the enemy UAV e.

As we can see, there are 8 state variables in total. Figure 3 illustrates these state variables. The value ranges for all the velocity variables are $[10, 20]$, while the value ranges for all the angle variables are $[0, 360)$. Suppose the length and width of the region are l and w, respectively. Thus, the value ranges of all the distance variables are $(0, d)$, where $d = \sqrt{l^2 + w^2}$. The reference frame is set in this way: the X axis points to the North, and the Y axis points to the East. Besides, the system neglects the rotation and acceleration of the earth, and the earth is assumed to be flat.

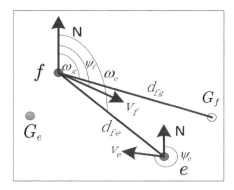

Figure 3. The denotations for the state variables.

4.1.2. Action Space

At each decision making step, the agent needs to decide an action for execution, so as to change the altitude of the decision making UAV. Six actions are defined in this paper: flying toward the goal, accelerating, decelerating, increasing the heading angle, decreasing the heading angle, and attacking the enemy. Denote these actions with $A = \{a_0, a_1, ..., a_5\}$. Each action corresponds to a way to change the desired attitude. The details are listed as follows.

- a_0: Fly toward the goal straightly, while keeping the desired velocity as the same in the previous step.
- a_1: Increase the desired velocity with ΔV and fly with the same desired heading angle as the previous step.
- a_2: Decrease the desired velocity with ΔV and fly with the same desired heading angle as the previous step.
- a_3: Increase the desired heading angle with $\Delta \phi$ and fly with the same desired velocity as the previous step.
- a_4: Decrease the desired heading angle with $\Delta \phi$ and fly with the same desired velocity as the previous step.
- a_5: Attack the enemy UAV, while the desired attitude changes as that of a_0.

The desired height is set with a fixed value h in all cases, for the flying height is not considered in this problem.

4.1.3. Reward Function

There are two types of rewards in this enemy avoidance problem, denoted by r_1 and r_2, respectively. The first type of reward r_1 is the usual reward given at each decision making step, which aims to encourage the agent to reach the goal point in as few steps as possible. The other reward r_2 is the reward given at the end of the episode for different ending reasons. There are four situations that will end an episode. Denote these four situations by $S_T = \{s_{T1}, s_{T2}, s_{T3}, s_{T4}\}$, which are defined as follows.

- s_{T1}: The decision making UAV has collided with the enemy UAV.
- s_{T2}: The decision making UAV has reached the goal point.
- s_{T3}: The decision making UAV has attacked the enemy UAV successfully.
- s_{T4}: The decision making UAV has been out of the region.

Since the mission of the decision making UAV is to reach the goal point successfully, the agent will be rewarded with a very big value when the mission is fulfilled. Cases that the decision making

UAV needs to avoid, such as colliding with the enemy UAV or out of the region, should be punished. Successfully attacking the enemy UAV can guarantee the fulfillment of the mission for the decision making UAV, therefore it should be rewarded. However, since the attacking action is costly and may fail, it will be better to limit the use of the attacking action. Considering this, a small punishment is given when attacking action is used. Therefore, we define the reward $r = r_1 + r_2$, where $r_1 = -1$, and r_2 is defined as below:

$$r_2 = \begin{cases} -20 & a = a_5; \\ -100 & s = s_{T1}; \\ -100 & s = s_{T4}; \\ 2 & s = s_{T3}; \\ 500 & s = s_{T2}. \end{cases} \tag{5}$$

4.2. Learning Policy with the Sarsa Method

In the first part of the new framework, the enemy avoidance problem has been formulated as an MDP. For the second part of the new framework, the agent of the decision making UAV is allowed to learn the policy based on the Sarsa reinforcement learning method.

There are mainly three classes of methods for solving the reinforcement learning problem [13]: dynamic programming, Monte Carlo methods, and temporal-difference (TD) learning. Dynamic programming methods require a complete and accurate model of the environment (the transition function $\mathcal{T}$), while the Monte Carlo methods are not suitable for step-by-step incremental computation. Only the temporal-difference methods require no model and are fully incremental. In the enemy avoidance problem, the environment transition function $\mathcal{T}$ is unknown, and the policies need to be learned by continuous interaction with the environment. Therefore, the temporal-difference methods are more suitable for the enemy avoidance problem. As one of the temporal-difference reinforcement learning methods, the Sarsa method is chosen to be used in the proposed framework to learn the policy for the enemy avoidance problem.

Since the state space in the enemy avoidance problem is continuous, it is impractical to visit each state with each action infinitely often. Therefore, certain function approximations are needed to generalize the state space from relatively sparse interaction samples and with fewer variables than there are states. In this paper, the CMAC (cerebellar model arithmetic computer, [19]) technique is also embodied in the proposed framework to approximate the Q-value function when learning with the Sarsa method.

The details of how the proposed framework embodies the Sarsa method and the CMAC function approximator to learn the policy for the enemy avoidance problem are elaborated as follows.

4.2.1. Sarsa Method

The Sarsa method is an on-policy temporal-difference learning method, where the agent attempts to update the policy that is used to make decisions for the decision making UAV f at the same time. Different from most reinforcement learning methods where the main goal is to estimate the optimal value function, the Sarsa agent learns an action-value function $Q(s, a)$ rather than a state-value function $V(s)$. For the enemy avoidance problem, the Sarsa agent updates its action-value function $Q(s, a)$ after every transition from a state $s \in S$, where s is not an episode ending situation. That is to say, $s \notin S_T$. If the state $s' \in S_T$, then $Q(s', a') = 0$.

The updating rule is given in Equation (6), where γ is the discount rate, and α is a step-size parameter. Every element of the quintuple of the enemy avoidance events, (s, a, r, s', a'), are used in this updating rule. Such a quintuple makes up a transition from one state-action pair of the enemy avoidance problem to the next, and therefore gives rise to the name Sarsa for the algorithm.

$$Q(s, a) \leftarrow Q(s, a) + \alpha(r + \gamma Q(s', a' - Q(s, a)) \tag{6}$$

In the proposed framework, the Sarsa algorithm [13] is adapted into the enemy avoidance problem for the agent to learn the policy, which is presented in Algorithm 1.

Algorithm 1 Sarsa Algorithm for the Enemy Avoidance Problem

1: Initialize $Q(s, a)$ arbitrarily
2: **for** each episode **do**

3:     Initialize s as all the UAVs having entered the problem region
4:     Choose $a \in A$ for s using policy derived from Q with ϵ−greedy
5:     **for** each step of episode **do**

6:         Take action a, observe r, s'
7:         Choose $a' \in A$ for s' using policy derived from Q with ϵ−greedy
8:         $Q(s, a) \leftarrow Q(s, a) + \alpha(r + \gamma Q(s', a') - Q(s, a))$
9:         $s \leftarrow s'; a \leftarrow a'$
10:     **end for**
11:     **until** $s \in S_T$
12: **end for**

4.2.2. CMAC Function Approximation

In the proposed framework for the enemy avoidance problem, the state space is continuous. Therefore, the learning agent for the decision making UAV needs to use function approximation to generalize from limited experienced states. With function approximation, the action-value function $Q(s, a)$ of the enemy avoidance problem is maintained in a parameterized functional form with parameter vector $\vec{\theta}$. In this framework, the linear function form is used, as presented by Equation (7), where $\vec{\phi}_{(s,a)}$ is the feature vector of the function approximation.

$$Q(s, a) = \vec{\theta}^T \vec{\phi}_{(s,a)} \tag{7}$$

The CMAC (cerebellar model arithmetic computer, [19]) is one of those linear function approximators, and thus is used to construct the feature vector $\vec{\phi}_{(s,a)}$ in the proposed framework. To update the parameter vector $\vec{\theta}$, the gradient-descent method is adopted in the proposed framework as well.

The CMAC discretizes the continuous state space of the enemy avoidance problem by laying infinite axis-parallel tilings over all the eight state variables and then generalizes them via multiple overlapping tilings with some offset. Each element of a tiling is called a tile, which is a binary feature, as shown by Equation (8).

$$\phi_{(s,a)}(i) = \begin{cases} 1 & \text{tile } i \text{ is activated.} \\ 0 & \text{otherwise.} \end{cases} \tag{8}$$

Therefore, the CMAC maintains $Q(s, a)$ of the enemy avoidance problem in the following form:

$$Q(s, a) = \sum_{i=1}^{n} \theta(i)\phi_{(s,a)}(i) = \sum_{i \in I(\vec{\phi}_{(s,a)})} \theta(i) \tag{9}$$

where $I(\vec{\phi}_{(s,a)})$ is the set of tiles that are activated by the pair (s, a) in the enemy avoidance problem, whose tile values are 1.

The parameter vector $\vec{\theta}$ is adjusted by the gradient-descent method, whose updating rule is as follows:

$$\vec{\theta}_{t+1} = \vec{\theta}_t + \alpha\delta_t \tag{10}$$

where δ_t is the usual TD error,

$$\delta_t = r_{t+1} + \gamma Q_t(s_{t+1}, a_{t+1}) - Q_t(s_t, a_t) \qquad (11)$$

5. Simulation Environment

Since it is hard to build an accurate environment model for the enemy avoidance problem, and the RL agent needs to approximate the environment model through continuous interaction with the environment, this paper builds a hardware-in-the-loop (HITL) simulation system for the enemy avoidance problem. In this HITL simulation system, the kinecmatic and dynamic of UAVs are modeled by the X-plane simulators, while the maneuver and control properties of the system are confined by the hardware controller called Pixhawk. With such a HITL simulation system, the simulation can be more consistent to real flying, and can reduce the energy of building a complex environment model and save the cost of a real flying test.

In this section, how the HITL simulation system is constructed will be detailed. After this, the simulation process for an episode will be given.

5.1. System Construction

The X-plane flight simulator is used in this paper to simulate the flying dynamics of both the decision making UAV and the enemy UAV, and each X-plane is controled by a Pixhawk (PX4) flight controller. The Pixhawk is a hardware which is also used in controlling the real UAVs. In the simulation system, there is a ground station which can broadcast the information of all the UAVs to every UAV. The PX4 can control the flying of the UAVs according to the desired attitude. The desired attitude is composed of the desired velocity, the desired heading angle, and the desired height. In each UAV, there is an on-board sensor system which is used to sense the position and the attitude information of the UAV. Besides, each UAV has an agent for the communication and the decision control. To be specific, the agent for the decision making UAV has three modules: communication module, decision making module, and translator module. The on-board sensor system sends the sensed information to the agent through the communication module, while the communication module also sends the received information to the ground station and receives the enemy UAV's information from the ground station. Based on all the information received by the communication module, the decision making module then decides which action the decision making UAV should take, while the translator module interprets the action into desired attitude and sends it to the PX4 for the UAV flying control. Since the enemy UAV in this paper is assumed to fly towards the goal point G_e directly all along the process, there is no decision making module in the enemy agent. Therefore, the enemy agent is composed of two modules: communication and control module. The communication module of the enemy agent has the same function as that of the agent for the decision making UAV. In the enemy agent, the control module sends the desired attitude to the PX4, where the desired velocity and the desired height are fixed at the initialization of the simulation, and the desired heading angle is calculated based on the relationship between the goal point and the position of the enemy UAV.

The structure of the simulation system is illustrated by Figure 4.

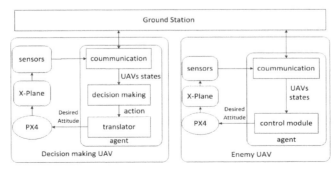

Figure 4. The structure of the simulation system.

Since another purpose of this paper is to explore the collision avoidance solution for the fixed-wing UAV, the X-plane simulator is set to use the fixed-wing simulation model. The simulation environment is shown as in Figure 5.

Figure 5. The simulation environment.

Beside receiving and sending the flying information of the UAVs, the ground control station also needs to send commands to all the UAVs. With these commands, the simulation process can be well controlled by the ground station.

5.2. Simulation Process

In order to collect as many samples as possible for the policy learning, the simulation should be carried for many episodes. Each episode is run with the same process, as shown by Figure 6. There are five steps for an episode, listed as follows.

(i) Both PX4s are set on the mission mode, so as to control the two UAVs to loiter around their own loiter points outside of the region. The loiter points are send to the PX4 from the ground station before the simulation begins.

(ii) The ground station sends a command asking all the UAVs to fly to their goal points. Both PX4s are set to the offboard mode, so that the UAV agents can guide their own UAVs to fly towards their goal points. In this stage, the agents use the simple-fly pattern to guide the flying of the UAVs. In the simple-fly pattern, the desired heading angle of the UAV is the angle toward the goal point directly, while the desired velocity and the desired height stay unchanged.

(iii) When both UAVs have entered the region for the enemy avoidance problem, the ground station sends a command to inform both UAV agents that the new episode begins. Upon receiving this command, the decision making agent changes into its decision making pattern from the simple-fly pattern, during which the agent can either learn the policy with reinforcement learning or make decisions with certain policy. On the other hand, the enemy agent still guide the enemy UAV fly straightly toward its goal point G_e with simple-fly pattern during this stage.

(iv) When the simulation meets one of those ending conditions, the ground station sends a command to both UAVs to inform the ending of the episode. Both PX4s are set back to the mission mode after receiving this command from the ground station.

(v) Under the mission mode, both PX4s control their own UAVs fly back to their loiter points again, since the loiter points remain as their next point in the mission mode. When both UAVs have returned to their loiter points, the ground station sends the command to make both UAVs fly into the region again for the start of the next episode.

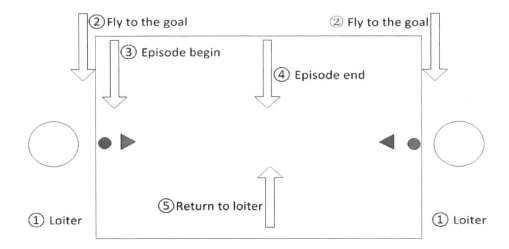

Figure 6. The simulation process.

6. Implementation and Results

Based on the HITL simulation environment built in Section 5, the decision policy for the enemy avoidance problem will be learned with the new framework constructed in Section 4. Furthermore, a fixed-rule policy and a random policy are designed to compare with the policy learned by the Sarsa based framework.

The experiment settings are as follows. The region for the enemy avoidance problem has a length of $l = 600$ m and a width of $w = 450$ m. The outside loiter center of each UAV is 100 meters away from the nearest region edge, and the loiter radium is 100 m. The goal point G_f is inside the region, which is 150 m away from the right edge of the region. The decision making UAV loiters around the goal point G_f with a loiter radius of 50 m when it has arrived the goal point G_f, and the goal radius is $r_g = 100$ m. The collision radius is $r_c = 40$ m. Set $\Delta V = 1.0$ m/s and $\Delta \phi = 5.0$ degree.

6.1. Learning with the Sarsa Based Framework

First, we use the newly constructed framework based on the Sarsa method and the CMAC function approximator to learn the policy for the enemy avoidance problem in the HITL simulation environment.

As in reinforcement learning, the goal is to maximize the expected accumulate rewards, thus the action-value function Q is an effective metric to measure the performance of the policy learning.

Considering the mission of the decision making UAV, it has to reach the goal point G_f successfully as soon as possible. On the other hand, the number of steps that an episode lasts can partially indicates how long it takes the decision making UAV to reach the goal point G_f. Therefore, the number of steps for an episode can be used as a metric to measure the performance to some extend.

However, there are several ending reasons. It can take very few steps to end the episode if the decision making UAV collides with or attacks the enemy UAV at a very early stage of an episode. Therefore, the number of steps alone is not enough to measure the performance of the learning in the enemy avoidance problem. To this end, the numbers of episodes that ending for different reasons are calculated, so as to measure more accurately how the learning changes the episode ending reasons.

The experiment for learning the policy with the proposed framework has been run for 14,362 episodes. The related parameter settings are: $\epsilon = 0.01$, $\alpha = 0.1$, $\gamma = 0$, and the number of tiles laid for each variable is $n = 32$. The results of the above three metrics are presented as follows.

Figure 7 illustrates how the Q-value changes with the increase of the episodes during the policy learning process. The figure is drawn with data filtered by a window of 1000 episodes. From Figure 7, it can be seen that the Q-value increases as the number of episodes increases, and becomes stable after around 10,000 episodes.

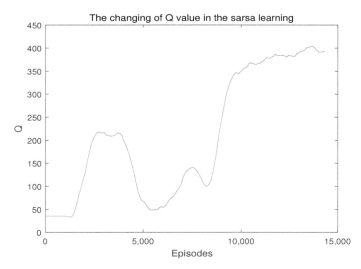

Figure 7. The *Q*-value for each episode in the policy learning.

Figure 8 shows how the number of steps for each episode changes as the number of policy learning episodes increases. The figure is also drawn with data filtered by a window of 1000 episodes. In Figure 8, the number of steps fluctuates very much at the beginning, but converges to a relatively small number as the number of episodes increases, and it comes to a plateau after around 10,000 episodes.

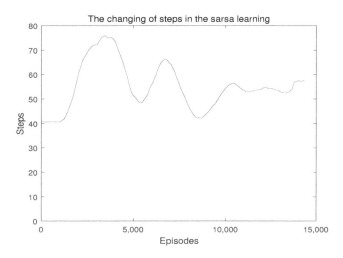

Figure 8. The number of steps for each episode in the policy learning.

Figure 9 presents how the accumulated numbers of different ending reasons change during the policy learning process. From Figure 9, it can be see that the episodes are mostly ended for collision with the enemy UAV at the initial period of the learning process. After about 1500 episodes, the number of episodes ending for reaching the goal begins to increase rapidly, which surpasses those of other ending reasons very soon. The number of episodes ending for successfully attacking the enemy UAV has a rise from about the 4000th episode to about the 8000th episode, and surpasses the number of episodes ending for collision with the enemy UAV during this period. Except for ending for reaching the goal point, the numbers of other ending reasons all stop to increase after about 9000 episodes. These results mean that the learning agent gradually learns to avoid the enemy UAV and attack the enemy UAV, and then it learns to limit the usage of the attacking action, and finally it has learned a very good policy to reach the goal point.

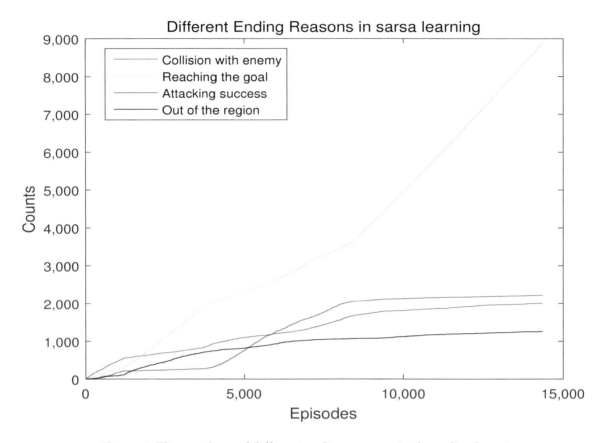

Figure 9. The numbers of different ending reasons in the policy learning.

From the above results for three metrics, it can be seen that the policy learning process under the proposed framework converges after about 10,000 episodes. Such episode number for converging is a reflection of the sample complexity of this learning based method. The proposed framework is to reduce the computation cost on the price of the sample complexity. However, such policy learning can be done before the real-time decision making, while decision making depending on some computation methods need to bear those computation costs at each real-time decision making step. Therefore, the proposed framework still has its advantages on these points.

6.2. Designing Comparison Policies

In order to show the effectiveness of the policy learned by the Sarsa based framework, we design a set of rules to guide the flying of the decision making UAV based on human experience, which is called the fixed-rule policy. As a more baseline comparison, a random policy is also used for the comparison. The detail of these two comparison policies are introduced in the following.

6.2.1. Fixed-Rule Policy

The fixed-rule policy is designed with human experience. It composes of several if-then rules. At each decision making step, the agent examines the current state and decides which rule can be used. The detail of the designed rules are listed as in Algorithm 2.

The designing of these rules aims at guiding the UAV flying toward the goal as straightforward as possible and trying to avoid the enemy UAV at the same time. However, these if-then rules are very simple ones due to the limitation of the human knowledge. Therefore, the hypothesis here is that the Sarsa based framework can help the agent discover more latent rules to guide the flying of the decision making UAV.

Algorithm 2 Fixed-rule Policy

1: **if** the distance to the nearest enemy<100 m **then**

2:     **if** decision steps from the last attacking action>10 decision steps **then**

3:         execute the attacking action

4:     **else**

5:         fly to the goal

6:     **end if**

7: **else**

8:     **if** the enemy UAV in the flying direction (within 2.5 degree) **then**

9:         **if** the enemy is on the right side **then**

10:           increase the flying angle

11:         **else**

12:           decrease the flying angle

13:         **end if**

14:     **else**

15:         fly to the goal

16:     **end if**

17: **end if**

6.2.2. Random Policy

The purpose of designing this random policy is to set a fundamental baseline policy for the effectiveness comparison of all other policies. The hypothesis is that those effective policies should all have better performance than this random policy.

In this random policy, the agent chooses action at each decision making step randomly. The probabilities of choosing each action are equal, so that no special favor is given to any action.

6.3. Policy Comparison Results

To test the effectiveness of the policy learned with the Sarsa based framework, another set of experiments are carried on the same HITL simulation platform. For comparison, the fixed-rule policy and the random policy are also tested with the same experimental settings. Each policy is run for 1200 episodes, and the results are summarized as follows.

Firstly, the numbers of steps taken in each episode for using different policies are compared in Figure 10. The figure is drawn with data filtered with a window of 100 episodes. It can be seen that the number of steps are stable in the testing process, no matter which policy is used. However, it takes different numbers of steps to end an episode when different policies are used. Overall, the random policy takes the least number of steps, while the policy learned with the Sarsa based framework takes the most. Though more number of steps means more cost, it is also a reflection of higher possibility to reach the goal point and better policy to avoid the enemy UAV. Such understanding is from the following two aspects. First, the episode ending for reaching the goal point takes more steps than the episode ending before reaching the goal point for other reasons, because of the longer distance. On the other hand, it needs more steps to fly away to avoid the collision with the enemy UAV than to fly straightly towards the goal. To more accurately show the performance of different policies, more results are presented below.

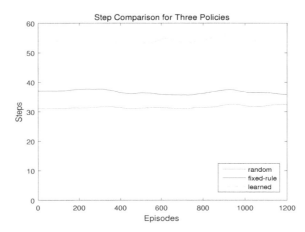

Figure 10. The steps of three policies.

6.3.1. Results of the Random Policy

Figure 11 illustrates how the numbers of episodes ending for different reasons change as the number of the total episode increases under the random policy. From Figure 11, it can be seen that there are mainly three ending reasons for these episodes, which are collision with the enemy UAV, reaching the goal point successfully, and attacking the enemy UAV successfully. Besides, there is nearly no case for flying out of the region. The number of the episodes ending for reaching the goal point increases slowest among the three main ending reasons, while the episode number for successfully attacking of the enemy UAV increases the quickest. This indicates that it is easier to end the episode by attacking the enemy UAV. The reason for this should be that ending the episode by successfully attacking takes only an attacking action as long as the attacking action is executed at a relatively near distance from the enemy UAV. On the contrast, to reach the goal point successfully is much harder, for it needs to avoid the moving enemy UAV by taking many decision steps in an appropriate sequence.

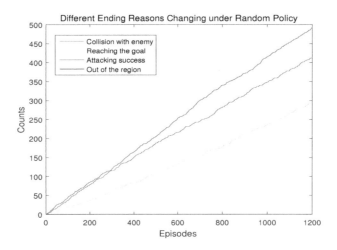

Figure 11. The numbers of different ending reasons when using the random policy.

Figure 12 presents the total episode numbers of different ending reasons when the random policy is used. It can be seen more clearly that the number of episodes ending for reaching the goal point is smaller than those for the other two ending reasons (collision with the enemy UAV and attacking the enemy UAV successfully). Therefore, the random policy is unable to guide the decision making UAV to fulfill its mission very successful, and the results of this policy only reflect different challenges for achieving different ending reasons.

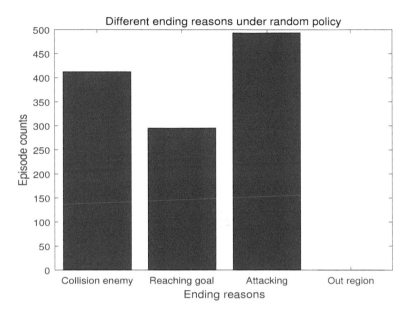

Figure 12. The total numbers of different ending reasons when using the random policy.

6.3.2. Results of the Fixed-Rule Policy

When the fixed-rule policy is used, the numbers of episodes ending with different reasons increase during the testing process, as shown in Figure 13. Still, there are three main ending reasons, which are collision with the enemy UAV, reaching the goal point successfully, and attacking the enemy UAV successfully. The number of episodes ending for attacking the enemy UAV is less than that of collision with the enemy UAV, while the number of the episodes ending for reaching the goal point successfully is nearly equal to that of collision with the enemy UAV. Compared with the random policy, these results show that the fixed-rule policy can reduce the unnecessary use of the attacking action, and can increase the use of some effective enemy avoidance actions so as to increase the probability of reaching the goal point successfully.

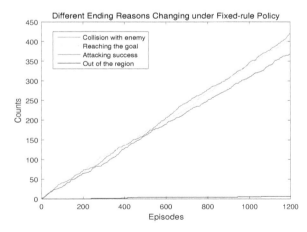

Figure 13. The numbers of different ending reasons when using the fixed-rule policy.

Figure 14 gives the total number of episodes for each ending reason in the fixed-rule policy experiment. Compared with that of the random policy, the fixed-rule policy can obviously reduce the number of episodes ending for attacking the enemy UAV, but is not effective in reducing the number of episodes for colliding with the enemy UAV.

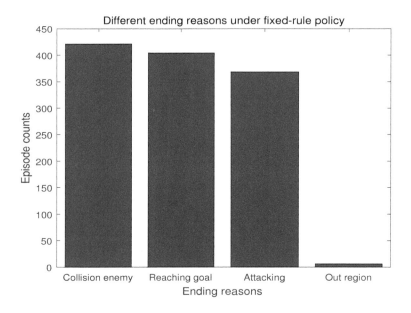

Figure 14. The total numbers of different ending reasons when using the fixed-rule policy.

6.3.3. Results of the Sarsa Learned Policy

Finally, the policy learned with the Sarsa based framework is also tested in the platform. In this experiment, the agent only takes action at each decision making step by following the learned policy, without any further learning.

Figure 15 presents how the numbers of episodes ending for different reasons change when the policy learned with the Sarsa based framework is adopted. Different from the other two policies, there is only one main ending reason: reaching the goal point. The numbers of episodes ending for both collision with the enemy UAV and attacking the enemy UAV are sharply reduced in the learned policy, and can be neglected when compared with that of the reaching goal case. Obviously, the policy learned with the Sarsa based framework can successfully guide the decision making UAV reaching the goal point.

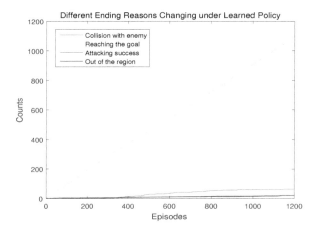

Figure 15. The numbers of different ending reasons when using the policy learned with the Sarsa based framework.

Figure 16 gives the total numbers of episodes ending for different reasons when the policy learned with the Sarsa based framework is used. It can be seen that the number of episodes for collision with the enemy UAV and the number of episodes for attacking the enemy UAV are both reduced to quite small numbers, which indicates the effectiveness of the policy learned with the Sarsa based framework.

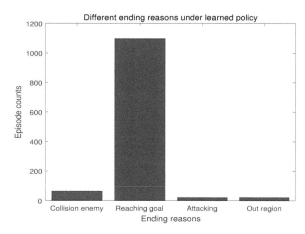

Figure 16. The total numbers of different ending reasons when using the learned policy.

7. Conclusions and Future Work

In this paper, we have defined the enemy avoidance problem, which is quite different from most collision avoidance problems. To solve the enemy avoidance problem, we have proposed a new framework, which formulates the enemy avoidance problem as the Markov Decision Process, and learns policy for the decision making UAV based on the Sarsa reinforcement learning method, and generalizes the continuous state space with the CMAC function approximation technique. Furthermore, we have constructed a HITL simulation system to learn the policy with the proposed framework and to verify the effectiveness of different policies for the enemy avoidance problem. The carried experiments show that the HITL simulation system is valid in presenting the enemy avoidance problem, and also suitable for the RL based framework. The Sarsa based framework can learn a satisfying policy for the enemy avoidance problem. Further experiments also show that the policy learned with the Sarsa based framework outperforms both the random policy and the fixed-rule policy in solving the enemy avoidance problem.

There are several advantages in our proposed method. Firstly, it does not require a mathematically built environment model to be computed when making decisions. Secondly, it learns the policy off-line, and the learned policy is a simple mapping from the state to the action. Therefore, little computation is needed for the on-line decision making. Thirdly, the policy is learned in the continuous space, and has good generalization ability. However, there are still some weaknesses in our proposed method. Firstly, samples are needed for the learning, which could also be costly. Secondly, there are a lot of simplifications in the researched problem, so inconsistency may occur if the method is applied directly into some more practical real problems.

Therefore, some future works can be explored. Firstly, a more practical problem setting could be studied. For example, equip the enemy UAV with more practical properties like changeable desired attitude and the attacking ability, or extend the two dimensional problem into three dimensions, or add noise like wind disturbance to the environment. Secondly, more efficient reinforcement learning methods and even some other techniques could be applied, so as to shorten the learning process. Thirdly, ways to combine the learning method with human experience could be explored, so as to make use of human experience to facilitate the learning and to discover latent knowledge by agent learning. Fourthly, methods which consider the collision with both obstacles and other UAVs would be more useful in improving the autonomy of the UAVs, and therefore could be studied in the future. Lastly, extending the problem to scenarios with more UAVs can be a very worth research area.

Author Contributions: Conceptualization, Q.C., X.W. and J.Y.; Funding acquisition, X.W. and L.S.; Investigation, Q.C. and J.Y.; Methodology, Q.C., X.W. and J.Y.; Project administration, L.S.; Supervision, X.W. and L.S.; Validation, Q.C.; Visualization, Q.C. and X.W.; Writing—original draft, Q.C.; Writing—review & editing, Q.C., X.W., J.Y. and L.S.

References

1. Yu, X.; Zhang, Y. Sense and avoid technologies with applications to unmanned aircraft systems: Review and prospects. *Prog. Aerosp. Sci.* **2015**, *74*, 152–166. [CrossRef]

2. Motlagh, N.H.; Bagaa, M.; Taleb, T. UAV-based IoT platform: A crowd surveillance use case. *IEEE Commun. Mag.* **2017**, *55*, 128–134. [CrossRef]

3. Gu, J.; Su, T.; Wang, Q.; Du, X.; Guizani, M. Multiple moving targets surveillance based on a cooperative network for multi-UAV. *IEEE Commun. Mag.* **2018**, *56*, 82–89. [CrossRef]

4. Liu, Y.; Wang, Q.; Hu, H.; He, Y. A Novel Real-Time Moving Target Tracking and Path Planning System for a Quadrotor UAV in Unknown Unstructured Outdoor Scenes. *IEEE Trans. Syst. Man Cybern. Syst.* **2018**. [CrossRef]

5. Yadav, I.; Eckenhoff, K.; Huang, G.; Tanner, H.G. Visual-Inertial Target Tracking and Motion Planning for UAV-based Radiation Detection. *arXiv* **2018**, arXiv:1805.09061.

6. Vanegas, F.; Campbell, D.; Roy, N.; Gaston, K.J.; Gonzalez, F. UAV tracking and following a ground target under motion and localisation uncertainty. In Proceedings of the IEEE Aerospace Conference, Big Sky, MT, USA, 4–11 March 2017; pp. 1–10.

7. Li, S.; Liu, T.; Zhang, C.; Yeung, D.Y.; Shen, S. Learning Unmanned Aerial Vehicle Control for Autonomous Target Following. *arXiv* **2017**, arXiv:1709.08233.

8. Mahjri, I.; Dhraief, A.; Belghith, A. A review on collision avoidance systems for unmanned aerial vehicles. In *International Workshop on Communication Technologies for Vehicles*; Springer: Berlin, Germany, 2015; pp. 203–214.

9. Gottlieb, Y.; Shima, T. UAVs task and motion planning in the presence of obstacles and prioritized targets. *Sensors* **2015**, *15*, 29734–29764. [CrossRef]

10. Ramasamy, S.; Sabatini, R.; Gardi, A.; Liu, J. LIDAR obstacle warning and avoidance system for unmanned aerial vehicle sense-and-avoid. *Aerosp. Sci. Technol.* **2016**, *55*, 344–358. [CrossRef]

11. Liu, Z.; Yu, X.; Yuan, C.; Zhang, Y. Leader-follower formation control of unmanned aerial vehicles with fault tolerant and collision avoidance capabilities. In Proceedings of the International Conference on Unmanned Aircraft Systems (ICUAS), Denver, CO, USA, 9–12 June 2015; pp. 1025–1030.

12. Jenie, Y.I.; van Kampen, E.J.; Ellerbroek, J.; Hoekstra, J.M. Taxonomy of conflict detection and resolution approaches for unmanned aerial vehicle in an integrated airspace. *IEEE Trans. Intell. Transp. Syst.* **2017**, *18*, 558–567. [CrossRef]

13. Sutton, R.S.; Barto, A.G. *Reinforcement Learning: An Introduction*; MIT Press: Cambridge, MA, UK, 1998.

14. Radmanesh, M.; Kumar, M.; Nemati, A.; Sarim, M. Dynamic optimal UAV trajectory planning in the national airspace system via mixed integer linear programming. *Proc. Inst. Mech. Eng. G J. Aerosp. Eng.* **2016**, *230*, 1668–1682. [CrossRef]

15. Ong, H.Y.; Kochenderfer, M.J. Short-term conflict resolution for unmanned aircraft traffic management. In Proceedings of the IEEE/AIAA 34th Digital Avionics Systems Conference (DASC), Prague, Czech Republic, 13–17 September 2015.

16. Alonso-Mora, J.; Naegeli, T.; Siegwart, R.; Beardsley, P. Collision avoidance for aerial vehicles in multi-agent scenarios. *Auton. Robot.* **2015**, *39*, 101–121. [CrossRef]

17. Dentler, J.; Kannan, S.; Mendez, M.A.O.; Voos, H. A real-time model predictive position control with collision avoidance for commercial low-cost quadrotors. In Proceedings of the IEEE Conference on Control Applications (CCA), Buenos Aires, Argentina, 19–22 September 2016; pp. 519–525.

18. Alvarez, H.; Paz, L.M.; Sturm, J.; Cremers, D. Collision avoidance for quadrotors with a monocular camera. In *Experimental Robotics*; Springer: Cham, Switzerland, 2016; pp. 195–209.

19. Albus, J.S. *Brains, Behaviour, and Robotics*; Byte Books: Perterborough, NH, USA, 1981.

20. Zhou, Y.; Baras, J.S. Reachable set approach to collision avoidance for UAVs. In Proceedings of the IEEE 54th Annual Conference on Decision and Control (CDC), Osaka, Japan, 15–18 December 2015; pp. 5947–5952.

21. D'Amato, E.; Mattei, M.; Notaro, I. Bi-level Flight Path Planning of UAV Formations with Collision Avoidance. *J. Intell. Robot. Syst.* **2018**, *93*, 193–211. [CrossRef]

22. Qiu, H.; Duan, H. A multi-objective pigeon-inspired optimization approach to UAV distributed flocking among obstacles. *Inf. Sci.* **2018**, in press. [CrossRef]

23. Yang, J.; Yin, D.; Cheng, Q.; Shen, L. Two-Layered Mechanism of Online Unmanned Aerial Vehicles Conflict Detection and Resolution. *IEEE Trans. Intell. Transp. Syst.* **2017**. [CrossRef]

24. Yang, J.; Yin, D.; Shen, L.; Cheng, Q.; Xie, X. Cooperative Deconflicting Heading Maneuvers Applied to Unmanned Aerial Vehicles in Non-Segregated Airspace. *J. Intell. Robot. Syst.* **2018**, *92*, 187–201. [CrossRef]

25. Alomari, A.; Phillips, W.; Aslam, N.; Comeau, F. Swarm Intelligence Optimization Techniques for Obstacle-Avoidance Mobility-Assisted Localization in Wireless Sensor Networks. *IEEE Access* **2017**, *6*, 22368–22385. [CrossRef]

26. Masiero, A.; Fissore, F.; Guarnieri, A.; Pirotti, F.; Vettore, A. UAV positioning and collision avoidance based on RSS measurements. *Int. Arch. Photogramm. Remote Sens. Spat. Inf. Sci.* **2015**, *40*, 219. [CrossRef]

27. Kahn, G.; Villaflor, A.; Pong, V.; Abbeel, P.; Levine, S. Uncertainty-aware reinforcement learning for collision avoidance. *arXiv* **2017**, arXiv:1702.01182.

28. Pfeiffer, M.; Schaeuble, M.; Nieto, J.; Siegwart, R.; Cadena, C. From perception to decision: A data-driven approach to end-to-end motion planning for autonomous ground robots. In Proceedings of the IEEE International Conference on Robotics and Automation (ICRA), Singapore, 29 May–3 June 2017; pp. 1527–1533.

29. Gandhi, D.; Pinto, L.; Gupta, A. Learning to fly by crashing. In Proceedings of the IEEE/RSJ International Conference on Intelligent Robots and Systems (IROS), Vancouver, BC, Canada, 24–28 September 2017; pp. 3948–3955.

30. Zhang, T.; Kahn, G.; Levine, S.; Abbeel, P. Learning deep control policies for autonomous aerial vehicles with mpc-guided policy search. In Proceedings of the IEEE International Conference on Robotics and Automation (ICRA), Stockholm, Sweden, 16–21 May 2016; pp. 528–535.

31. Zhang, B.; Mao, Z.; Liu, W.; Liu, J. Geometric reinforcement learning for path planning of UAVs. *J. Intell. Robot. Syst.* **2015**, *77*, 391–409. [CrossRef]

32. Bai, H.; Hsu, D.; Kochenderfer, M.J.; Lee, W.S. Unmanned aircraft collision avoidance using continuous-state POMDPs. *Robot. Sci. Syst. VII* **2012**, *1*, 1–8.

33. Temizer, S.; Kochenderfer, M.; Kaelbling, L.; Lozano-Pérez, T.; Kuchar, J. Collision avoidance for unmanned aircraft using Markov decision processes. In Proceedings of the AIAA Guidance, Navigation, and Control Conference, Toronto, ON, Canada, 2–5 August 2010; p. 8040.

34. Seo, J.; Kim, Y.; Kim, S.; Tsourdos, A. Collision avoidance strategies for unmanned aerial vehicles in formation flight. *IEEE Trans. Aerosp. Electron. Syst.* **2017**, *53*, 2718–2734. [CrossRef]

35. Yang, J.; Yin, D.; Cheng, Q.; Shen, L.; Tan, Z. Decentralized cooperative unmanned aerial vehicles conflict resolution by neural network-based tree search method. *Int. J. Adv. Robot. Syst.* **2016**, *13*. [CrossRef]

8

Loop Closure Detection based on Multi-Scale Deep Feature Fusion

Baifan Chen, Dian Yuan *, Chunfa Liu and Qian Wu

School of Automation, Central South University, Changsha 410083, China; chenbaifan@csu.edu.cn (B.C.); 15084843844@163.com (C.L.); wuqian945@csu.edu.cn (Q.W.)
* Correspondence: dianyuan@csu.edu.cn.

Abstract: Loop closure detection plays a very important role in the mobile robot navigation field. It is useful in achieving accurate navigation in complex environments and reducing the cumulative error of the robot's pose estimation. The current mainstream methods are based on the visual bag of word model, but traditional image features are sensitive to illumination changes. This paper proposes a loop closure detection algorithm based on multi-scale deep feature fusion, which uses a Convolutional Neural Network (CNN) to extract more advanced and more abstract features. In order to deal with the different sizes of input images and enrich receptive fields of the feature extractor, this paper uses the spatial pyramid pooling (SPP) of multi-scale to fuse the features. In addition, considering the different contributions of each feature to loop closure detection, the paper defines the distinguishability weight of features and uses it in similarity measurement. It reduces the probability of false positives in loop closure detection. The experimental results show that the loop closure detection algorithm based on multi-scale deep feature fusion has higher precision and recall rates and is more robust to illumination changes than the mainstream methods.

Keywords: loop closure detection; convolutional neural network; spatial pyramid pooling

1. Introduction

Loop closure detection has become a key problem and research hotspot in the field of mobile robot navigation, particularly in simultaneous localization and mapping (SLAM), because it can reduce the cumulative error of robot pose estimation and achieve accurate navigation in large-scale complex environments. Vision-based loop closure detection, also called visual place recognition, is when the robot identifies the places that have been visited before with images provided by the vision sensor during the navigation. For example, assume there are two images captured at the current time and at an earlier time, the problem of loop closure detection is to judge whether the places at the two moments are the same according to the similarity of these two images. Correct loop closure detection can add an edge constraint in the pose map to help optimize robot motion estimation further and build a consistent map. Wrong loop closure detection will lead to the failure of map building. Therefore, a good loop closure detection algorithm is crucial for consistent mapping and even for the entire SLAM system.

At present, the mainstream methods of visual loop closure detection are based on the Bag of Words (BoW), which cluster the visual features into some "words" and then describe an image in the form of a "words" vector. Thus, the visual loop closure detection problem is transformed into a similarity measure problem with the word vectors of the two images. However, the visual features in the BoW are all artificially designed by researchers in the field of computer vision, and they all belong to the low-level features and are sensitive to illumination changes. With the advent of various visual sensors, different visual features are designed based on the different characteristics of the sensors. However, the design of a new visual feature is often very difficult. In recent years, deep learning methods have developed rapidly. They start from the raw data of the sensor and automatically extract the abstract information of the data through a multi-layer neural network. Compared with traditional image processing, deep learning networks use multiple convolutional layers to extract features and use pooling layers to select features. The extracted image features are more advanced and abstract than traditional artificial visual features. Convolutional Neural Network (CNN) has been widely applied in image retrieval and image classification.

Considering the similarity between visual loop closure detection and image classification (they both need to extract the features of the image, and then complete the related tasks based on the extracted features), this paper applies CNN to loop closure detection and proposes a loop closure detection algorithm based on multi-scale deep feature fusion. The algorithm includes three modules: feature extraction layer, feature fusion layer and decision layer (as shown in Figure 1). We selected the first five convolutional layers of the pre-trained AlexNet network on the ImageNet dataset as the feature extraction layer, which can extract more advanced and more abstract features. In the feature fusion layer, we designed a multi-scale fusion operator with spatial pyramid pooling (SPP) [1] to fuse the deep features with different receptive fields and create a fixed length representation of an image. Finally, in the decision layer, we developed a similarity measurement method by calculating the distinguishability weight of features, which helps reduce the probability of false positives in loop closure detection. The results show that the loop closure detection algorithm has a high precision and recall rate. They also verify the algorithm's robustness to illumination changes.

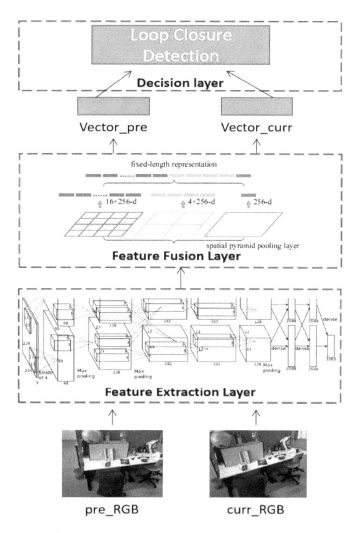

Figure 1. The framework of visual loop closure detection algorithm based on multi-scale deep feature fusion.

2. Related work

In recent years, scholars have done much research in the direction of loop closure detection algorithms based on vision. The classical algorithms can be roughly divided into two categories: the method based on the BoW (Bag of Word) [2] and the method based on the global descriptor. The first method extracts local features from the scene image and clusters them into multiple "words". Then the whole image is described in the form of vectors based on these "words". Thus, the visual loop closure detection problem is transformed into a similarity measure problem of the description vectors of the two images. BoW is the mainstream method for loop closure detection. A key problem of the BoW method is how to select local features of the image. The common feature points are SIFT [3], SURF [4] and ORB [5]. For example, Mei et al. [6] used the FAST [7] operator to extract the key points and then used the SIFT [3] as feature descriptor. Newman et al. [8] extracted FAST [7] key points and then calculated the descriptors using BRIEFF [9]. For the general case, the images described by the BoW can be compared one-to-one by the histogram or Hamming distance, and the closed loop is detected when the distance is less than a certain threshold. However, in a large-scale scene, search speed is very important, and some researchers have begun to apply the word tree to do efficient loop closure detection. Cummins et al. [10,11] applied Chow-Liu tree approximation to describe the correlation between words and words, and then proposed the classic FAB-MAP method. Glover et al. [12] made public the FAB-MAP development kit based on the work of Cummins et al., which provided convenience for researchers. Maddern et al. [13] proposed the CAT-SLAM method based on FAB-MAP, which combines loop closure detection with a local metric pose filter. Compared with FAB-MAP, the loop closure detection of CAT-SLAM is better. For the second method, the main idea is to describe the entire image with a global descriptor. Ulrich et al. [14] proposed that color histograms provide a compact representation of an image, which results in a system that requires little memory and performs in real-time. But it is very sensitive to changes in illumination. Dalai et al. [15] used histograms of oriented gradients (HOG) as the feature descriptor of the image, which gave very good results for person detection in cluttered backgrounds. GIST [16] had been demonstrated to be a very effective conventional image descriptor, capturing the basic structure of different types of scenes in a very compact way. Based on this, Murillo et al. [17] utilized global gist descriptor computed for portions of panoramic images and a simple similarity measure between two panoramas, which is robust to changes in vehicle orientation, while traversing the same areas in different directions.

Both of the two methods have their own advantages and disadvantages. Furgale et al. [18] proved that the global descriptor method is more sensitive to the camera pose than the BOW method. Milfold [19] and Naseer [20] proposed that the global descriptor method is more robust in the case of illumination changes. Therefore, some researchers have considered combining the two methods and proposed a method of using scene signatures. For example, McManus et al. [21] presented an unsupervised system that produces broad-region detectors for distinctive visual elements, which improved the accuracy of detection. However, the features used in these methods are low-level features and designed artificially in the field of computer vision. They are sensitive to the influence of light, weather and other factors, so these algorithms lack the necessary robustness.

With the disclosure of large-scale datasets (such as ImageNet [22]) and the upgrading of various hardware (such as GPU), deep learning has developed rapidly in recent years. Deep learning can extract abstract and high-level features of the input image through multi-layer neural networks, which is more robust to changes in environmental factors [23,24]. Therefore, it has been widely used in image classification [25] and image retrieval [26]. Considering that visual loop closure detection is similar to image classification and image retrieval, researchers have tried to apply deep learning to loop closure detection. Gao et al. [27,28] took advantage of Autoencoder to extract image features and used the similarity measurement matrix to detect closed loops, which got high accuracy on public datasets. He et al. [29] applied FLCNN (fast and lightweight convolutional neural networks) to extract image features and calculate the similarity matrix, which further improved the real-time and accuracy of loop closure detection. Xia et al. [30] extracted image features by PCANet, and proved that these features are superior to traditional manual design features. Hou et al. [31] used PlaceCNN to extract image features for loop closure detection, which got high accuracy even when the light changed. However, these methods relied on local deep features and ignored the scale problem.

3. Loop Closure Detection Algorithm Based on Multi-Scale Deep Feature Fusion

Different from traditional image classification, visual loop closure detection needs to determine whether the two moments are at the same location according to the similarity between the picture collected at the current time and the one taken earlier. Therefore, the algorithm in this paper has paired input, which corresponds to two branches in the algorithm as shown in Figure 1. The algorithm is divided into three layers: feature extraction, feature fusion and decision. The feature extraction layer extracts the deep feature of the input images. The feature fusion layer does multi-scale fusion and normalization of extracted features. The decision layer uses the fusion feature to detect the loop closure. The two branches of the algorithm are identical in the feature extraction layer and the feature fusion layer structure.

3.1. Feature Extraction Layer

The feature extraction layer is composed of two identical CNNs to extract features for two inputs separately. Compared to the early CNN, AlexNet [27] uses a much deeper network model to acquire features. Additionally, it adds modules such as the ReLU activation function, local response normalization (LRN), Dropout, etc., which can reduce the risk of over fitting. Moreover, it takes advantage of multi-GPU to improve the training speed of the network model. In view of these advantages, this paper refers to AlexNet.

The network model has a total of eight layers, consisting of five convolution layers and three fully connected layers. Only the first five layers of AlexNet are needed, as shown in Figure 2. Here, we set the input of the feature extraction layer to be an RGB image of $227 \times 227 \times 3$, and obtained 256 feature maps with a size of 6×6 through five convolution layers. (The output data size of each layer is indicated in Figure 2). The convolution layer contains the ReLU activation function and LRN processing, as well as max-pooling. Among them, LRN, which draws on the idea of "lateral inhibition" in neurobiology, is used to locally suppress neurons. When the activation function is ReLU, this "lateral inhibition" is very useful. It can prevent the model from over-fitting prematurely and speed up the training of the model. Its calculation formula is

$$b_{x,y}^i = a_{x,y}^i / \left(k + \alpha \sum_{j=\max(0,i-n/2)}^{\min(N-1,i+n/2)} \left(a_{x,y}^i\right)^2\right)^\beta,$$ (1)

where $a_{x,y}^i$ denotes the value of the i-th convolution kernel after applying the ReLU activation function at position (x, y); n is the number of convolution kernels adjacent in the same position, and N represents the total number of convolution kernels. k, n, α and β are tunable parameters; we set $k = 2, n = 5, \alpha = 0.0001, \beta = 0.75$ according to the empirical value.

3.2. Feature Fusion Layer

An RGB image can obtain several feature maps describing different deep features from features extraction layer. For general image classification tasks, a fully connected layer is usually added to weight and sum all the feature maps to obtain the classification result. However, each feature map only corresponds to a small area in the original input image, and its receptive field is small. The deep feature is a local feature. For the loop closure detection problem, we prefer to have different scales of features because this helps to accurately determine whether two images belong to the same scene. In addition, AlexNet [27] requires the fixed-size of the input image to be $227 \times 227 \times 3$. When the size of the input does not meet the requirements, cutting or compressing must happen, which causes the loss or distortion of some image information.

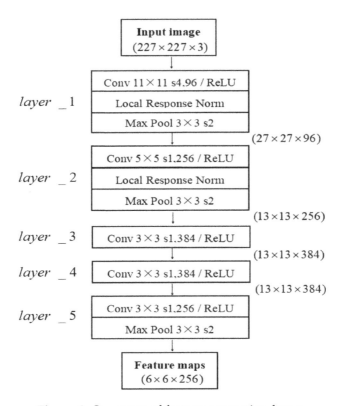

Figure 2. Structure of feature extraction layer.

To overcome this problem, we use spatial pyramid pooling (SPP) [1] to fuse multi-scale deep features. SPP divides the feature map into small patches of different sizes by different scales, and then gets each patch's feature. The receptive fields to the input images of these small patches are different, which means the patches correspond to different areas of the input image. Finally, the features extracted from the small patch of different sizes are combined to achieve the fusion of multi-scale features. In addition, the SPP layer can get a fixed-size output and eliminate the restriction that the input image size must be fixed.

Taking a single branch as an example, the feature fusion layer is shown in Figure 3. Here, the input of feature fusion layer is the output from feature extraction layer. After the SPP layer, we can obtain the fixed-length representation of the features, which is finally sent to softmax.

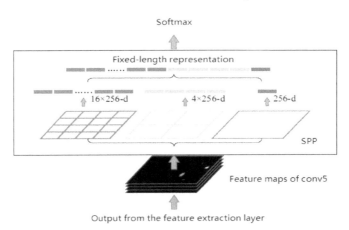

Figure 3. Feature fusion layer.

As seen in Figure 3, SPP divides each feature map by using three different scales 4×4, 2×2, 1×1. Each scale corresponds to one layer in the SPP; three scales indicate that the SPP has three layers colored as blue, green and gray. For example, the 4×4 scale divides a feature map into 4×4

small patches, and then extracts a feature from each small patch, so each feature map can extract 16 features. There are 256 feature maps as output from the feature extraction layer, and altogether there are 16 × 256 features (blue parts in Figure 3). The features extracted by the different scales of the SPP are concatenated together to obtain a fixed-length feature vector. In order to normalize the results of the SPP, we add a SoftMax layer. Assuming that the feature vector output by the SPP layer is $Z \in R^{n \times 1}$, the output of the SoftMax layer is:

$$Y = \left[\frac{e^{Z_1}}{\sum_{j=1}^n e^{Z_j}}, \cdots, \frac{e^{Z_n}}{\sum_{j=1}^n e^{Z_j}} \right]^T, Y \in R^{n \times 1}, \tag{2}$$

The number of SPP layers affects the output size of the SoftMax layer. If the number of layers is different, the performance of the corresponding algorithm will be different. In the experiment part, the effects of SPP in different layers on the performance of the algorithm are discussed.

3.3. Decision Layer

The decision layer is in charge of loop closure detection. Suppose the inputs of the two branches are Image_1 and Image_2, respectively, and their feature vectors of the SoftMax layer are f_1 and f_2. The vector dimension is determined by the number of SPP layers and the scale of each layer. Assuming that the dimensions of f_1 and f_2 are N, the similarity of two inputs can be calculated by Equation (3).

$$S_1(f_1, f_2) = 1 - \sum_{i=1}^N (f_{1i} - f_{2i}), \tag{3}$$

where f_1 and f_2 represent the i-th dimension of the feature vector. Setting a threshold T, when $S_1(f_1, f_2) > T$, it means that Image_1 and Image_2 correspond to the same place and a closed loop is detected. Equation (3) treats each feature node fairly, but the distinguishability of each feature node is different. Features such as walls and ground are more common in scenes and their distinguishability is relatively small, while the traffic signs are more distinguishable. If the feature nodes with different distinguishability are treated fairly, more false positives (different places in similar scenes) will occur when detecting the loop closure. Considering this character, we add a weight to each feature node and modify Equation (3) as Equation (4). The weight's value indicates the distinguishability of the feature to the scene.

$$S_1(f_1, f_2) = 1 - \sum_{i=1}^N \delta_i (f_{1i} - f_{2i}), \tag{4}$$

where δ_i represents the weight of the i-th feature node, and the larger the value of δ_i, the greater the distinguishability of its corresponding feature in the scene. The weights can be learned by training and accord with Gaussian distribution. The value of δ_i is calculated as follows:

$$\delta_i = \exp(-\frac{(\overline{h}_i - u)^2}{2\delta^2}), \tag{5}$$

where $\overline{h}_i$ represents the average response of the i-th feature node. If the average response of a feature node is larger, such features are more common (such as ground and sky), and the δ_i is smaller. If the average response of a feature node is smaller and lower than the mean value of u, such features are not common (such as noise), and δ_i is also smaller. When $\overline{h}_i$ is near the mean value of u, the distinguishability is relatively large, and the corresponding weight value is also relatively large. After the network model is trained, the value of $\overline{h}_i$ can be calculated and retained by the test. u and δ are tunable parameters and set according to experience.

4. Parameter Training

4.1. Training Method

Since SPP in the feature fusion layer is a special maxing pooling layer, it has no parameters to be trained. Here we only need to train the parameters of the convolution network in the feature extraction layer. The algorithm in this paper consists of two branches with the same structure in the feature extraction layer and the feature fusion layer. For the loop closure detection problem, its positive sample indicates the same location, and the negative sample indicates non-loop closure. Loop closure always happens at different locations and with very few times, which means a large number of classes and small labeled samples. Taking these into consideration, this paper uses the Siamese [32] model to train.

The Siamese network is mainly used in the field of face recognition can solve the classification problem with small sample data well. Figure 4 shows the training model.

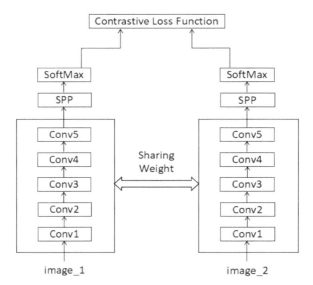

Figure 4. Parameter training model for feature extraction layer.

Contrastive loss function is used here. For the i-th pair of samples, we assume that the feature vectors of the SoftMax layer are f_{i1}, f_{i2} respectively, and the contrastive loss function is:

$$L = \frac{1}{2M}\sum_{i=1}^{M}\left[y_i d_i^2 + (1 - y_i)\max\left(m\arg in - d_i, 0\right)^2\right],\tag{6}$$

where M is the number of sample pairs; $d_i = \| f_{i1} - f_{i2} \|_2$ is the Euclidean distance of the two samples in the feature space, and y_i is the label of the i-th pair of samples. $y = 1$ is the positive sample, which means that the two images are similar and form a loop closure; $y = 0$ represents a negative sample, which means that the similarity of the two pictures is small and does not form a loop closure. The threshold $margin$ is set to 1 in the experiment. When $y = 1$ and the loss function is $L = \frac{1}{2M}\sum_{i=1}^{M}y_i d_i^2$, if their Euclidean distance d in the feature space is large, the current training model is not good and its loss value increases. When $y = 0$ and the loss function is $L = \frac{1}{2M}\sum_{i=1}^{M}[\max\left(m\arg in - d_i, 0\right)^2]$, if their Euclidean distance in the feature space is small, the loss value of the model will become large. The contrastive loss function can express the matching degree of paired samples, and it can be used for the model training.

4.2. Model Training

We use the AlexNet model pre-trained by the ImageNet dataset on Caffe [33], and set the parameters of the first five convolutional layers as the initial values of the parameters in the feature extraction layer. Meanwhile, we use the Matterport3D dataset to fine tune the model. The Matterport3D dataset is the world's largest public 3D dataset from 3D scanning solution provider Matterport. It is a large-scale RGB-D dataset containing 10,800 panoramic views from 194,400 RGB-D images of 90 building-scale scenes. The dataset was acquired by a Pro 3D camera. The camera rotates around the center of gravity at each sampling point, and samples the images at six rotation positions, each of which corresponds to 18 sets of pictures. This special way of data acquisition makes the camera cover a wider range of angles, and it provides many loop closure data. Figure 5a–f are some examples of RGB images in the MatterPort3D dataset.

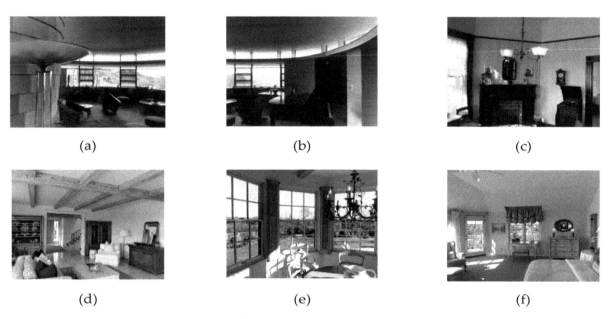

Figure 5. Examples of an RGB graph in the Matterport3D dataset.

We selected 30,000 pairs of positive samples (loop closure) and 30,000 pairs of negative samples (non-loop closure) in the Matterport3D data set. Then by some data enhancement means such as rotating and translating, the positive and negative samples were each doubled to 60,000. Of these, 10,000 pairs of positive samples and 10,000 pairs of negative samples were selected as test sets, and the other 50,000 pairs of positive samples and 50,000 pairs of negative samples were used for training. Some researchers have found that multi-scale training can improve the accuracy of the network containing SPP layers in the tasks of image classification and object detection. In view of this, this paper uses three different scales of data to train the model: 1280×1024, 227×227 and 180×180. The latter two data are cropped from the original image. In this way, the number of our training set is 300,000, including 150,000 pairs of positive samples and 150,000 pairs of negative samples.

The deep learning framework Caffe requires a fixed size of input images; therefore, we used a combination of single-scale and multi-scale to train the model. In one epoch of model training, single-scale images are input, while in each different epoch of model training, the input batches are of different scales.

In addition, in order to analyze the influence of different SPP layers in the feature fusion layer on the performance of the algorithm, three different SPP are used: a one-layer SPP with a scale of 1×1, a two-layer SPP with a scale of 1×1, 2×2, and a three-layer SPP with a scale of 1×1, 2×2, 4×4. For these three cases, we build three models and train them separately.

5. Experiment

In order to verify the performance and effectiveness of the algorithm, we tested the effects of different layers of SPP, the influence of the similarity measurement method in visual loop closure detection and the robustness to illumination changes. An Intel Core i7 processor of 2.8 GHz frequency and an NVIDIA GeForce GTX 1060 with MAX-Q Graphics card were used.

5.1. Dataset and Labeling

The dataset was provided by the computer vision group of the Technical University of Munich (TUM) [34]. We used the dataset's RGB-D image sequence, which includes Fr2/rpy, Fr2/large_with_loop, Fr2/pioneer_slam, Fr2/pioneer_slam2 and Fr3/long_office_household. However, the loop closure is not labeled in the TUM dataset and must be manually marked. Because of the 30 Hz/s sample frequency, there are too many images in the same place. Therefore, we selected the key frames with the ground truth of trajectory.

Keyframe selection includes the following steps.

For each image sequence, a key frame list $F = \{\}$ is set, and the first frame is added to F.

Sequentially compare the image f_j in the image sequence with the last frame

$$\boxed{f_i}$$

in the key frame list. Assuming that their corresponding poses are T_j, T_i which can be found in the ground truth, the relative translation of the two frames is $M_{j,i} = trans(T_j^{-1}T_i)$, where $trans(\bullet)$ represents the 2 norm of the translational part of the transformation matrix. If $0.1 < M_{j,i} < 1$, the frame f_j is added to the end of the key frame list.

Finally, the loop closure is manually labeled with the selected key frame sequence.

Figure 6a shows the trajectory ground truth of the Fr3/long_office_household sequence, and in Figure 6b red dots are the selected key frames and the black line segments are the loop closure. Figure 6c,d are one pair of loop closure images.

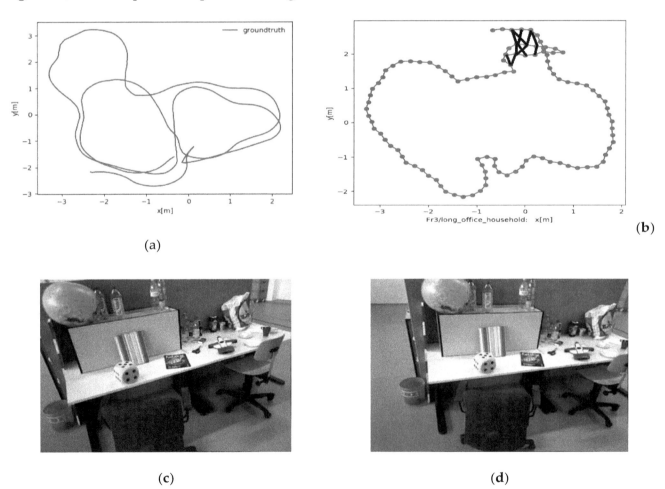

Figure 6. Keyframe selection and loop closure example in FR3/long_office_household.

5.2. Different layers of SPP

SPP is used in the feature fusion layer to fuse the extracted depth features of different scales. In order to verify how different layers of SPP influence the algorithm results, an SPP with 1×1, a two-layer SPP with 1×1, 2×2, and a three-layer SPP with 1×1, 2×2, 4×4 were used for experiments. The loop closure detection algorithm with these three different SPP layers, noted as spp1, spp12 and spp124, respectively, was compared with FabMap [10].

We selected 689 key frames from the Fr2/rpy, Fr2/large_with_loop and Fr3/long_office_household and labeled 45 loop closure places, which were noted as data_1. The experimental results of the data_1 are shown in Figure 7. It can be seen from Figure 7 that the P-R curves of spp1, spp12 and spp124 are basically on the upper right of the coordinate system, which means they have higher precision and recall rates. Our method can reach 100% precision at 50% recall. In addition, spp124 is better than spp12 and spp1, and spp12 is better than spp1. When the recall rate is 100%, spp124 reaches as high as 78% precision; spp12 is about 62% precision, and spp1 is less than 50% precision. It shows that the greater the layers of SPP, the higher the precision and recall rate of the algorithm.

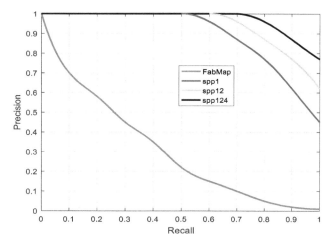

Figure 7. P-R curves on data_1.

The above algorithm was also tested in the Fr2/pioneer_slam and Fr2/pioneer_slam2 image sequences which are sampled in a very empty indoor environment and have many similar scenes, such as walls, boards and ground. We selected 435 key frames from the two sequences and labelled 20 loop closure places, which were noted as data_2. The test results on data_2 are shown in Figure 8.

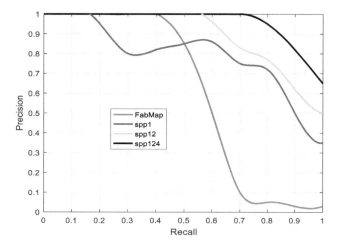

Figure 8. P-R curves on data_2.

As can be seen from Figure 8, spp124 has the highest precision-recall rate, followed by spp12. At the 50% recall rate, both spp124 and spp12 can reach 100% precision. For spp1, when the recall rate is less than 50%, the accuracy is higher than the accuracy of FabMap, but when the recall rate is greater than 50%, the accuracy of spp1 is lower than that of FabMap. It is worth mentioning that the performance of all algorithms on data_2 is worse than on data_1. Because there are many similar scenarios in data_2, the algorithms got some false positive results. Overall, the depth features extracted by CNN are more suitable for loop closure detection compared with the traditional artificial design visual features, and increasing the number of SPP layers in the feature fusion layer can improve the accuracy and recall rate.

5.3. Similarity Measurement

In order to verify the effect of weight adjustment in the similarity measurement, we compared the spp1, spp12, and spp124 methods with these added weight adjustments which were denoted as spp1+, spp12+, spp124+. The former directly uses Equation (3) to calculate similarity, and the latter uses Equation (4). In order to calculate δ_i from Equation (5), we selected 100,000 RGB images from the Matterport3D data set and calculated the average response of each node of the SoftMax layer as h_i. Since two branches are identical and share their parameters, only one branch needed to be calculated. The tunable parameter was set as $u = 0.5, \delta = 0.1$. The results of data_1 and data_2 are shown in Figures 9 and 10, respectively.

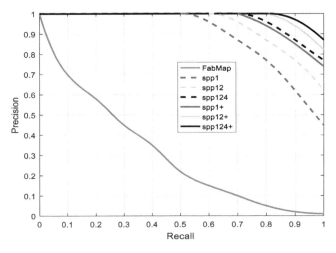

Figure 9. P-R curves of seven algorithms on data_1.

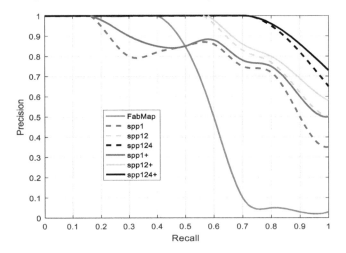

Figure 10. P-R curves of seven algorithms on data_2.

At a high recall rate, spp1+, spp12+ and spp124+ have higher precision than spp1, spp12 and spp124. When considering feature distinguishability, the visual loop closure detection algorithm reduced the probability of false positives and improved precision.

5.4. Illumination Changes

Illumination changes are the most critical factors affecting the visual loop closure detection. In order to test the robustness of the algorithm to illumination changes, we collected image sequences with different illuminations. We fixed a trajectory and then sampled image sequences at 12:30, 15:00, 17:30 and 19:00. Finally, four image sequences were obtained and named as VS1230, VS1500, VS1730 and VS1900 respectively, as shown in Figure 11. Each image sequence contained 200 frames and 20 closed loops.

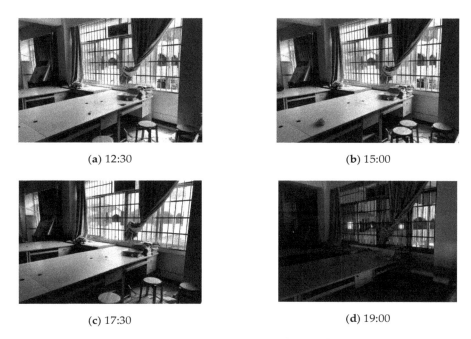

(a) 12:30   (b) 15:00

(c) 17:30   (d) 19:00

Figure 11. Sample images taken at four different times.

The tests were performed on these four image sequences, and the results are shown in Figures 12–15. Since the algorithms of spp1+, spp12+, spp124+ have been proven to be outstanding in 5.3, here we only compared them with FabMap.

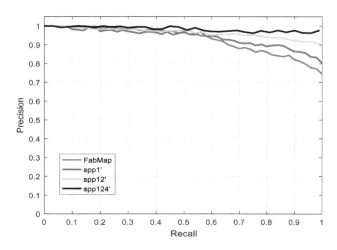

Figure 12. P-R curves of four algorithms on VS1230.

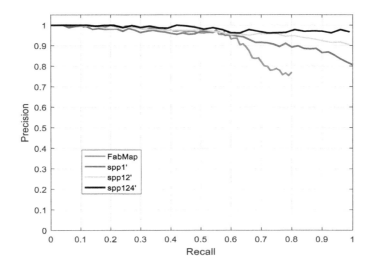

Figure 13. P-R curves of four algorithms on VS1500.

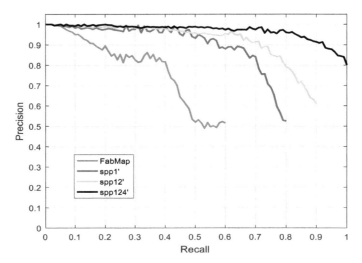

Figure 14. P-R curves of four algorithms on VS1730.

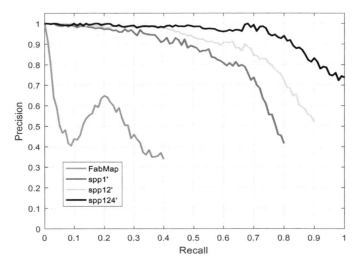

Figure 15. P-R curves of four algorithms on VS1900.

At 12:30, the P-R curves of the four algorithms are on the upper right of the coordinate system. However, as the time went by, all P-R values of the algorithms decreased, and the FabMap decreased more obviously because the illumination turned to dim. Especially in the VS1730 and VS1900 image sequences, the accuracy of FabMap was drastically lower at high recall rates. In the VS1900 image sequences, at 100% recall rate, the precision of spp124+ was still as high as 70% or more. This shows that the proposed algorithm is more robust to illumination changes than the traditional BoW method.

In general, for RGB-D SLAM, researchers prefer higher precision because the wrong closed loop leads to a completely wrong pose graph. Therefore, we also did statistical analysis about the average precision of spp1+, spp12+, spp124+, and FabMap on the four image sequences VS1230, VS1500, VS1730, and VS1900.

As can be seen from Figure 16, the average precision of the four algorithms was above 90% at 12:30, and at 17:30, only spp124+ and spp12+ were still above 90% precision. At 17:30, the average precision of FabMap is only 70%. At 19:00, the differences are more obvious. At this time, the average precision of spp124+ was still as high as 90%, and the other three were below 90%, especially FabMap which had an average accuracy of only 42%. It can be concluded that the algorithm of this paper is more robust than the traditional BoW method when the illumination changes.

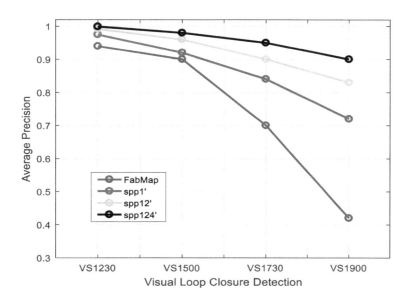

Figure 16. Average precision of spp1+, spp12+, spp124+, and FabMap on four image sequences.

6. Conclusions

This paper has proposed a loop closure detection algorithm based on multi-scale deep feature fusion. Our proposed algorithm has a higher accuracy and recall rate than the traditional BoW method, and it has better robustness to illumination changes. The key ideas that allowed us to achieve this efficiency are as follows. First, we used CNN to extract features. The features extracted in this way are more advanced and abstract, and have better robustness to illumination changes. Second, we used SPP to fuse the extracted features. By setting the multi-layer SPP with different receptive field sizes to fuse the different scale features in the image, it is more conducive to detecting loop closure. Last but not least, in the similarity calculation process of the decision layer, we added a weight to each feature node according to the distinguishability of the feature nodes to the scene. We have verified the benefits of the above points with experiments. In fact, the neural network used only plays the role of extracting more abstract features. Later, we will consider introducing the difference between the two image features in the middle layer of the deep network, and using the difference feature to detect loop closure.

Author Contributions: Conceptualization, B.C. and C.L.; Data curation, D.Y.; Formal analysis, B.C. and D.Y.; Investigation, Q.W.; Methodology, B.C. and C.L.; Project administration, B.C.; Resources, Q.W.; Validation, D.Y.; Writing—original draft, D.Y.; Writing—review & editing, B.C. and D.Y.

Key Research and Development Plan (2018YFB1201602) and Science and Technology Major Project of Hunan Province of China under Grant No. 2017GK1010.

References

1. He, K.; Zhang, X.; Ren, S.; Sun, J. Spatial pyramid pooling in deep convolutional networks for visual recognition. In *European Conference on Computer Vision*; Springer: Cham, Switzerland, 2014; pp. 346–361.

2. Sivic, J. Video Google: A text retrieval approach to object matching in videos. In Proceedings of the Ninth IEEE International Conference on Computer Vision, Nice, France, 13–16 October 2003.

3. Lowe, D.G. Distinctive image features from scale-invariant keypoints. *Int. J. Comput. Vis.* **2004**, *60*, 91–110. [CrossRef]

4. Bay, H. SURF: Speeded up robust features. In *European Conference on Computer Vision*; Springer: Berlin/Heidelberg, Germany, 2006.

5. Mur-Artal, R.; Tardos, J.D. ORB-SLAM2: An Open-Source SLAM System for Monocular, Stereo, and RGB-D Cameras. *IEEE Trans. Robot.* **2017**, *33*, 1255–1262. [CrossRef]

6. Mei, C.; Sibley, G.; Newman, P.; Reid, I. A Constant-Time Efficient Stereo SLAM System. In Proceedings of the 20th British Machine Vision Conference (BMVC), London, UK, 7–10 September 2009; pp. 1–11.

7. Rosten, E.; Drummond, T. Machine learning for high-speed corner detection. In *European Conference on Computer Vision*; Springer: Berlin/Heidelberg, Germany, 2006; pp. 430–443.

8. Churchill, W.; Newman, P. Experience-based navigation for long-term localisation. *Int. J. Robot. Res.* **2013**, *32*, 1645–1661. [CrossRef]

9. Calonder, M.; Lepetit, V.; Ozuysal, M.; Strecha, C.; Fua, P. BRIEF: Computing a local binary descriptor very fast. *IEEE Trans. Pattern Anal. Mach. Intell.* **2012**, *34*, 1281–1298. [CrossRef] [PubMed]

10. Schindler, G.; Brown, M.; Szeliski, R. City-Scale Location Recognition. In Proceedings of the IEEE Conference on Computer Vision and Pattern Recognition (CVPR), Minneapolis, MN, USA, 17–22 June 2007; pp. 1–7.

11. Cummins, M.; Newman, P. Probabilistic appearance-based navigation and loop closing. In Proceedings of the IEEE International Conference on Robotics and Automation, Roma, Italy, 10–14 April 2007; pp. 2042–2048.

12. Glover, A.; Maddern, W.; Warren, M.; Stephanie, R.; Milford, M.; Wyeth, G. Openfabmap: An open source toolbox for appearance-based loop closure detection. In Proceedings of the IEEE International Conference on Robotics and Automation (ICRA), Saint Paul, MN, USA, 14–18 May 2007; pp. 4730–4735.

13. Maddern, W.; Milford, M.; Wyeth, G. CAT-SLAM: Probabilistic localisation and mapping using a continuous appearance-based trajectory. *Int. J. Robot. Res.* **2012**, *31*, 429–451. [CrossRef]

14. Dalai, N.; Triggs, B. Histograms of oriented gradients for human detection. In Proceedings of the IEEE Conference Computer Vision and Pattern Recognition, San Diego, CA, USA, 20–25 June 2005; pp. 886–893.

15. Ulrich, I.; Nourbakhsh, I. Appearance-based place recognition for topological localization. In Proceedings of the IEEE International Conference on Robotics and Automation (ICRA), San Francisco, CA, USA, 24–28 April 2000; Volume 2, pp. 1023–1029.

16. Oliva and, A. Torralba. Building the gist of a scene: The role of global image features in recognition. *Vis. Percept. Prog. Brain Res.* **2006**, *155*, 23–36.

17. Murillo, A.C.; Kosecka, J. Experiments in place recognition using gist panoramas. In Proceedings of the IEEE International Conference on Computer Vision Workshops (ICCV), Kyoto, Japan, 27 September–4 October 2009; pp. 2196–2203.

18. Furgale, P.; Barfoot, T.D. Visual teach and repeat for long-range rover autonomy. *J. Field Robot.* **2010**, *27*, 534–560. [CrossRef]

19. Milford, M.J.; Wyeth, G.F. SeqSLAM: Visual route-based navigation for sunny summer days and stormy winter nights. In Proceedings of the IEEE International Conference on Robotics and Automation (ICRA), Saint Paul, MN, USA, 14–18 May 2012; pp. 1643–1649.

20. Naseer, T.; Spinello, L.; Burgard, W.; Stachniss, C. Robust visual robot localization across seasons using network flows. In Proceedings of the Twenty-Eighth AAAI Conference on Artificial Intelligence, Québec City, QC, Canada, 27–31 July 2014; pp. 2564–2570.

21. Mcmanus, C.; Upcroft, B.; Newmann, P. Scene Signatures: Localised and Point-less Features for Localisation. *Image Proc.* **2014**. [CrossRef]

22. Deng, J.; Dong, W.; Socher, R.; Li, L.J.; Li, K.; Fei-Fei, L. ImageNet: A large-scale hierarchical image database. In Proceedings of the IEEE Conference on Computer Vision and Pattern Recognition (CVPR), Miami, FL, USA, 20–25 June 2009; pp. 248–255.
23. Chatfield, K.; Simonyan, K.; Vedaldi, A.; Zisserman, A. Return of the devil in the details: Delving deep into convolutional nets. *Comput. Sci.* **2014**, 1–11.
24. Wan, J.; Wang, D.Y. Deep learning for content-based image retrieval: A comprehensive study. In Proceedings of the 22nd ACM International Conference on Multimedia, Orlando, FL, USA, 3–7 November 2014; pp. 157–166.
25. Krizhevsky, A.; Sutskever, I.; Hinton, G.E. ImageNet classification with deep convolutional neural networks. In Proceedings of the 25th International Conference on Neural Information Processing Systems, Lake Tahoe, NV, USA, 3–6 December 2012; pp. 1097–1105.
26. Babenko, A.; Slesarev, A.; Chigorin, A.; Lempitsky, V. Neural codes for image retrieval. In *European Conference on Computer Vision*; Springer: Cham, Switzerland, 2014; pp. 584–599.
27. Gao, X.; Zhang, T. Unsupervised learning to detect loops using deep neural networks for visual SLAM system. *Auton. Robot.* **2017**, *41*, 1–18. [CrossRef]
28. Gao, X.; Zhang, T. Loop closure detection for visual slam systems using deep neural networks. In Proceedings of the Chinese Control Conference, Hangzhou: Chinese Association of Automation, Hangzhou, China, 28–30 July 2015; pp. 5851–5856.
29. He, Y.L.; Chen, J.T.; Zeng, B. A Fast loop closure detection method based on lightweight convolutional neural network. *Comput. Eng.* **2018**, *44*, 182–187.
30. Xia, Y.; Li, J.; Qi, L.; Fan, H. Loop closure detection for visual SLAM using PCANet features. In Proceedings of the IEEE International Joint Conference, Vancouver, BC, Canada, 24–29 July 2016; pp. 2274–2281.
31. Hou, Y.; Zhang, H.; Zhou, S. Convolutional neural network-based image representation for visual loop closure detection. In Proceedings of the IEEE International Conference, Lijiang, China, 20–25 April 2015; pp. 2238–2245.
32. Chopra, S.; Hadsell, R.; Lecun, Y. Learning a similarity metric discriminatively, with application to face verification. In Proceedings of the IEEE Computer Society Conference on Computer Vision and Pattern Recognition, San Diego, CA, USA, 20–25 June 2005; pp. 539–546.
33. Jia, Y.; Shelhamer, E.; Donahue, J.; Donahue, J.; Karayev, S.; Long, J.; Girshick, R. Caffe: Convolutional Architecture for Fast Feature Embedding. In Proceedings of the 22nd ACM International Conference on Multimedia, Orlando, FL, USA, 3–7 November 2014.
34. Sturm, J.; Engelhard, N.; Endres, F.; Burgard, W.; Cremers, D. A benchmark for the evaluation of RGB-D SLAM systems. In Proceedings of the IEEE/RSJ International Conference on Intelligent Robots and Systems, Vilamoura, Portugal, 7–12 October 2012; pp. 573–580.

9

Comparison of Spray Deposition, Control Efficacy on Wheat Aphids and Working Efficiency in the Wheat Field of the Unmanned Aerial Vehicle with Boom Sprayer and Two Conventional Knapsack Sprayers

Guobin Wang [1], Yubin Lan [1,*], Huizhu Yuan [2], Haixia Qi [1], Pengchao Chen [1], Fan Ouyang [1] and Yuxing Han [1,*]

[1] National Center for International Collaboration Research on Precision Agricultural Aviation Pesticides Spraying Technology (NPAAC), South China Agricultural University, Guangzhou 510642, China; guobinwang@stu.scau.edu.cn (G.W.); qihaixia@scau.edu.cn (H.Q.); pengchao@stu.scau.edu.cn (P.C.); ouyangfan@scau.edu.cn (F.O.)

[2] State Key Laboratory for Biology of Plant Disease and Insect Pests, Institute of Plant Protection, Chinese Academy of Agricultural Sciences, Beijing 100193, China; hzhyuan@ippcaas.cn

* Correspondence: ylan@scau.edu.cn (Y.L.); yuxinghan@scau.edu.cn (Y.H.);

Abstract: As a new low volume application technology, unmanned aerial vehicle (UAV) application is developing quickly in China. The aim of this study was to compare the droplet deposition, control efficacy and working efficiency of a six-rotor UAV with a self-propelled boom sprayer and two conventional knapsack sprayers on the wheat crop. The total deposition of UAV and other sprayers were not statistically significant, but significantly lower for run-off. The deposition uniformity and droplets penetrability of the UAV were poor. The deposition variation coefficient of the UAV was 87.2%, which was higher than the boom sprayer of 31.2%. The deposition on the third top leaf was only 50.0% compared to the boom sprayer. The area of coverage of the UAV was 2.2% under the spray volume of 10 L/ha. The control efficacy on wheat aphids of UAV was 70.9%, which was comparable to other sprayers. The working efficiency of UAV was 4.11 ha/h, which was roughly 1.7–20.0 times higher than the three other sprayers. Comparable control efficacy results suggest that UAV application could be a viable strategy to control pests with higher efficiency. Further improvement on deposition uniformity and penetrability are needed.

Keywords: unmanned aerial vehicle; pesticide application; deposition uniformity; droplets penetrability; control efficacy; working efficiency

1. Introduction

Wheat is one of the major food crops in China; the planting area and yield account for 21.4% and 20.9%, respectively, in 2017 (Data from National Bureau of Statistics of China). However, there are many kinds of pests and diseases which harm the production of wheat. These pests and diseases are controlled basically by the application of chemical products.

The selection of the equipment to be used is a critical factor for chemical pest control. In China, more than 88% of sprayers are manually operated [1], which include electric or manual air-pressure knapsack sprayer and knapsack mist-blower sprayer. The quality of the application depends mainly on the skill of the operators. These types of equipment are of low cost, easily maintained and adequate to control periodic and localized problems. However, applications with knapsack sprayer generally lead to high chemical exposure of the operators [2,3] and postural discomfort [4]. The operational farm size is increasing with the growth of agricultural co-operatives, land leasing and contract farming, while the labor force is declining by urbanization and rural–urban migration [1], leading to these low-efficiency and labor-intensive equipment types no longer being suitable for crop protection. As one of the alternative equipment types, self-propelled boom sprayer appeared on the market, equipped with horizontal spray boom. These machines have relatively higher working efficiency, lower chemical exposure and higher deposition [5]. However, the complicated terrain and small farm size with separated plots limit the use of boom sprayer in China. In recent decades, to adapt to this unique operating environment and meet the shortage supply of the crop protection equipment, unmanned aerial vehicles (UAV) for pesticide application have been developed quickly in China. Comparing with the manned agricultural aircraft, UAVs do not require navigation station or airport, and the edge of field can be its landing site [6]. The low rate of no-load flight and less flight crew reduce the expenditure of operations and administration [6]. Meanwhile, UAVs have short turning radius due to hover and turn around flexibly in the air, which are suitable for working in rough terrain and small plots with high efficiency [1,7,8]. Comparing with the conventional ground crop protection machinery, UAVs operate with lower labor intensity, operator exposure and have a higher working efficiency, especially in rough terrain and small plots [1,6,9]. According to the statistics data by the Chinese Ministry of Agriculture, nearly 14,000 crop protection UAVs are used in the country. The spraying area approached 5.5 million hectares in 2017.

Because of the broad prospect of application, UAVs have attracted plenty of scholar's attention. In the aspect of optimizing operational altitudes and speeds, Qin et al. [7] optimized the flying parameters for preventing plant hoppers, showing that a flight height of 1.5 m and a flying velocity of 5 m/s achieved the maximum lower layer deposition and the most uniform distribution for HyB-15L UAV sprayer (Gao Ke Xin Nong Co. Ltd., Shenzhen, Guangdong, China).

The optimal parameters change with the type of UAV and the crop. In a spraying test of different shape (open center shape and round head shape) of circus trees, the 3W-LWS-Q60S UAV (Zhuhai Crop Guardian Aerial Plant Protection Co., Zhuhai, Guangdong, China) performs better when the working height is 1.0 m compared with 0.5 and 2 m [10]. In the aspect of deposition uniformity, a multi-spraying swath test is conducted with different UAV sprayers [9].

There is an obviously inconsistent amount of deposition in the longitudinal and lateral direction and this phenomenon has been reported in many studies [7,10,11]. The control efficacy on pests and diseases is one of the most important evaluation indices of chemical application. Quite different from the conventional large volume application, the UAV sprayer belongs to low volume (LV, 4.7–46.7 L/ha) or ultra-low volume (ULV, 0–4.7 L/ha) [12] spraying equipment with the spray volume in the range of 1–40 L/ha [6,11]. Meanwhile, with the same active ingredient applied per acre, the chemical concentration of UAV is particularly high. Qin et al. [7] studied the control efficacy of HyB-15L UAV with spraying Chlorpyrifos·Regent EC against plant hoppers and found that the insecticidal efficacy is 92% and 74% at 3 and 10 days after application, respectively. On the premise of guaranteed control efficacy, the working efficiency is another important evaluation index of application. Currently, UAVs mostly rely on semi-autonomous control with the flight altitude belonging to autonomous control. The control range is 200–300 m in visible distance with manual control [6]. The payload capacity of the aerial application UAVs is generally 5–25 kg [6,13,14]. Considering the limited payload and the flight range, the effective spray work rates of 2–5 ha/h can be achieved in a vineyard with a gasoline-powered helicopter (RMAX, Yamaha motor Co., Cypress, CA, USA) [11]. The working efficiency of different UAVs in the grain-filling stage of wheat was studied by Wang et al. [9], with the daily working area ranging from 13.4 to 18.0 ha in 8 h. Pesticide spray drift is an important environmental problem for aerial application. Compared with the manned agricultural aircraft [6,15,16], the droplet drift of UAV is effectively reduced with the lower flight height [6,7] and the downwash wind [17,18]. According to Xue et al. [19], under the wind speed of 3 m/s, 90% of drift droplets of

Z-3 UAV are located within a range of 8 m of the target area. Similar to Xue et al., Wang et al. [17] measured that 90% of drift droplets of a fuel powered single-rotor UAV are within 9.3–14.5 m under the wind speed of 0.76–5.5 m/s. Compared with the conventional boom sprayer [20,21], the droplets drift distance of UAV sprayer would be further.

Despite these preceding studies, research is focused mainly on parameter optimization, droplet deposition and biological efficacy of one equipment. Few studies compare different kinds of crop protection equipment, especially including UAV sprayer. Under the same working condition, the comparison of different crop protection equipment on deposition, control efficacy and working efficiency is very important for equipment selection and application quality analysis. The main objective of this research was to compare the application quality of a battery motive 3WTXC8-5 six-rotor UAV with a 3WX-280H self-propelled boom sprayer and two conventional knapsack sprayers (3WBS-16A2 electric air-pressure knapsack sprayer and WFB-18 knapsack mist-blower sprayer). The comparison items included the spray deposition on the plants and run-off, uniformity and penetrability of the deposition, deposition characterization (including droplet size, number of spray deposits and the area of coverage), pesticide efficacy on wheat aphid and working efficiency. The experimental results show that the control efficacy of the UAV on wheat aphid was comparable to other spraying equipment with the working efficiency significantly higher than others. Unfortunately, UAV still have many problems on deposition uniformity and droplets penetrability.

2. Materials and Methods

To compare the advantage and shortcoming of the UAV with other spraying equipment, we selected three typical types of crop protection equipment including a self-propelled boom sprayer and two conventional knapsack sprayers for field spray deposition, control efficacy on wheat aphid and working efficiency tests. The spray deposition was compared from four aspects: the total amount of deposition and losses to the ground, deposition uniformity, droplets penetration in the canopy and characterization of the deposition (including droplet size, number of spray deposits and the area of coverage).

2.1. Spray Equipment

A battery motive 3WTXC8-5 six-rotor UAV (Henan Tianxiucai Aviation protection machinery Co., Ltd., Kaifeng, China) (Figure 1A) was used in this study. The UAV was powered by two 12,000 mAh Li-Po batteries (Shenzhen Grepow battery Co., Ltd., Shenzhen, China). The flying time was 15–20 min with full tank. The flight speed was 12.6–14.4 km/h with two rotary cup atomizer arranged on both sides. The interval of nozzles was 0.85 m and the installation angle was vertically downward. The rotate speed of the disk was 10,000 rotations per minute. The chemicals were transferred from the tank to the nozzles by a HXB600 micro liquid pump (Shanghai Hallya Electric Co., Ltd., Shanghai, China) and the flow rate was 1.24 L/min. The accuracy of the flight height and flight velocity were controlled by the well-trained operator. The flight height was 1.0 m and the effective spraying width was 4.0 m. The spray volume of one sortie was close to 10 L/ha, which was equal to two times the tank capacity.

One of the comparison systems was the 3WX-280H self-propelled boom (SPB) sprayer produced by Sino-Agri Fengmao Plant Protection Machinery Company (Beijing, China) with the tank capacity of 280 L (Figure 1B). There are 12 ISO 04 nozzles (Spraying system Co.) installed vertically on the spray boom with the same interval of 0.5 m on the 6-m boom. The spraying height of the boom was 0.5 m from the top of the canopy. The spray pressure was 4 bar and the flow rate was 18.2 L/min. The operation speed was 6.0–6.5 km/h. Under these application conditions, the spray volume was close to 300 L/ha.

Figure 1. Four test sprayers: (**A**) 3WTXC8-5 six-rotor unmanned aerial vehicle (UAV) sprayer; (**B**) 3WX-280H self-propelled boom (SPB) sprayer; (**C**) WFB-18 knapsack mist-blower (KMB) sprayer; and (**D**) 3WBS-16A2 electric air-pressure knapsack (EAP) sprayer.

The two other conventional sprayers were WFB-18 knapsack mist-blower (KMB) sprayer (Sino-agri Fengmao, China) (Figure 1C) and 3WBS-16A2 electric air-pressure knapsack (EAP) sprayer (Chuangxing sprayer factory, Xinxiang, China) (Figure 1D). EAP sprayer was equipped with twin hollow cone nozzles and a pressure pump provided a maximum pressure of 4 bar and a flow rate of 1.6 L/min. The tank capacity was 16 L and the length of the lance of the EAP sprayer was 81 cm. The spray swath width was close to 2.5 m. The traveling speed in the test was approximately 1.1–1.3 km/h and the spray volume under these application conditions was close to 300 L/ha. The KMB sprayer was developed to improve the spraying efficiency of air-pressure knapsack sprayer, which was equipped with a tank, a spray pipe, a nozzle and a gasoline engine. The droplets were sprayed from the nozzle, which were further atomized by the high speed air flow. The high speed air flow was produced by high speed whirling impeller driven by the gasoline engine. The flow rate in this test was 2.0 L/min with the tank capacity of 18 L. The spraying swath width was close to 6.0 m. The traveling speed was approximately 2.7–3.0 km/h and the spray volume was close to 75 L/ha. The working height of the nozzle of two conventional sprayers was 0.5 m from the top of the canopy. The spraying patterns of two knapsack were both swinging spraying.

All working parameters and spray volumes for each sprayer were established taking into account local farmers practices. Before tests, all spray equipment performed a preliminary test to calibrate the equipment to ascertain the flow rate of the nozzles. After the flow rate was ascertained, the traveling speed was also calculated to obtain the stated application rate. To achieve the velocity, the operator needed to repeat several times before undertaking each trial until the desired traveling speed was reached. Each spraying treatment was done by a well-trained applicator.

2.2. Experiment Design

2.2.1. Field Plots

The tests were conducted at the agricultural experiment station of the Chinese Academy of Agricultural Science located at Xinxiang, Henan Province, China (latitude 35°8'8″, longitude 113°46'58″) (Figure 2). The experimental farm is a trapezoidal field nearly 50 ha in area and it consists of many square fields approximately 200 m on a side (Figure 2). The tested material was "Bainong AK58" wheat in the booting-filling stage on 27 April, which was sown on 10 October, 2014. The plant spacing, plant height, leaf area of the flag leaf, and planting density were 12 cm, 84.8 ± 4.1 cm, 1927.1 cm^2 and 4.1 × 10^6 plant/ha, respectively. The wheat in the whole test area grew well and consistently.

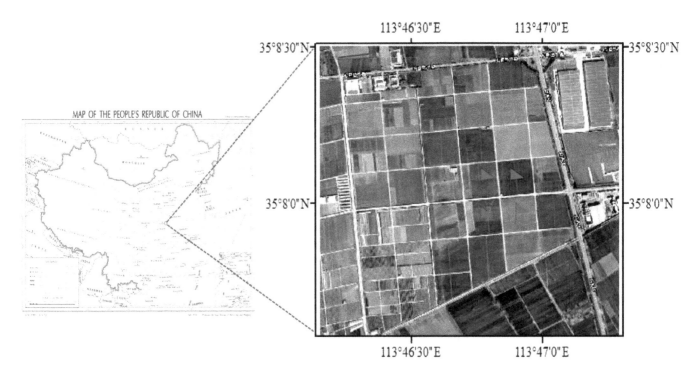

Figure 2. The test location and the brief overview of the studied wheat field. The experimental fields are marked with red flags.

2.2.2. Spray-Deposition Measurements

The experiment consists of five treatments: four kinds of spray equipment treatment and a blank control. The spray deposition, the control efficacy on wheat aphids and working efficiency were tested. The spray deposition and the control efficacy were tested in a 170 m × 190 m area (Figure 3A). In the test field, treatments were arranged within the location as a randomized complete block design with three replications, resulting in three blocks each with five plots corresponding to different treatments. Each plot was a 30 m × 50 m area. Ten-meter buffer zones [17,19] between plots were set to avoid the drift pollution (Figure 3A).

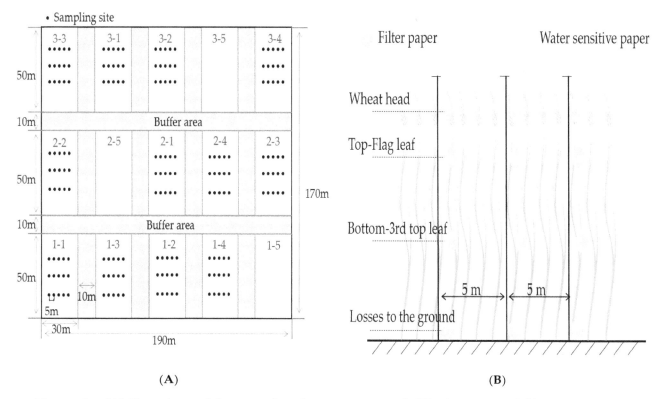

Figure 3. (A) Experimental layout of each treatment; and (B) placement of filter papers and water-sensitive paper at each sampling position within the wheat canopy.

Prior to the application in each treatment, sample collectors were placed at each plot in five equally-spaced sample sites. Each sample site was 5 m apart and spanned a total of 20 m. The sample sites were repeated three times with an interval of 15 m between each repetition. To avoid cross contamination between plots, sampling was arranged at the center of the plots, as shown in Figure 3A.

Sample collectors at each sample site consisted of one water-sensitive paper (WSP) (25 mm × 75 mm) and four filter papers (90 mm in diameter) (Figure 3B). The WSP was used to evaluate the characteristic of deposition such as an area of coverage, number of spray deposits and droplet size. The filter papers were used to measure the deposition distribution in the canopy. The WSPs were fixed horizontally on plastic rods through double-headed clamps. The heights of these clamps were adjusted to assist in positioning at a height equivalent to the head of the wheat canopy. The filter papers were used to simulate leaves to collect foliar deposition at different heights and losses to the ground. On each sampling site, three filter papers were attached on the wheat head, flag leaf and third top leaf of one wheat plant. To evaluate the losses to ground, a filter paper was placed on the ground.

In all tests, 70% Imidacloprid·Regent Water dispersible granule (WDG) at 85.7 g a.i./ha (active ingredient) (Zhejiang Sega Science and Technology Co., Ltd.) and allure red (80% purity, purchased from Beijing Oriental Care Trading Ltd.) were added into the tank. The pesticides were prepared according to the recommended dose. Allure red was used as the tracer according to the spraying area at 450 g/ha. Allure red, a water-soluble colorant, is frequently used in these types of studies [3,9,15]. It presents high recovery rate, high photostability and low acute toxicity in different species of animals according to the Joint FAO/WHO Expert Committee on Food Additives as well as the European Union's Scientific Committee for Food.

Nearly 30 s after spraying, five randomly selected wheat plants of over ground part, WSPs and filter papers of different canopy positions at each sampling site were collected and placed in labeled plastic zip-lock bags. Five random wheat plants were combined and bagged as one sample. The sample of wheat plants was used to measure the total deposition per plant. For each plot, there were 15 wheat plant samples, 60 filter papers (4 from each sampling site) and 15 WSPs. All samples were placed in zip-lock bags along with a label describing the treatment, replication, and location information. Samples were placed into light-proof seal box immediately after collection and transported to the laboratory for analysis.

Each filter paper and the wheat sample were washed in 0.02 L and 0.2 L of distilled water in the collection bags, respectively. Samples were agitated and vibrated for 10 min to allow the dye to dissolve into the water solution. Previous tests have shown that this methodology results in near total recovery of dye deposited on samples [22]. After vibration and elution, a sample portion of the wash effluent was filtered through a 0.22 μm membrane and the filtered wash effluent was poured into a cuvette to measure the absorbance value by a UV2100 ultraviolet and visible spectrophotometer (LabTech, Co. Ltd., Beijing, China) at an absorption wavelength of 514 nm. Spray deposits were quantified by comparison with similarly determined dye concentrations from spray tank samples and the area of the respective samples. The data were expressed as a quantity of dye (μg) deposited per unit area of the sample (cm^2) or quantity of dye (μg) deposited per plant. Note that throughout the text whenever the term "deposition" is used, it refers to the mass of dye (which would correspond to amount of active ingredient), not the mass of total spray mix, deposited on specified sampling surface. The distribution uniformity of the spray deposition in the canopy was analyzed using the value corresponding to the coefficient of variation (CV), calculated as the quotient between the standard deviation and the average of the spray deposits on the crop. In addition, for each application, the recovery rate was calculated using Equation (1):

$$R = (D \times P/A) \times 10^6 \tag{1}$$

where R is the recovery rate (%), D is the average deposited tracer solution per plant (μg/plant), P is the plant density (4.1×10^6 plant/ha), A is the additive amount of allure red (450 g/ha), and 10^6 is the unit conversion factor.

In the laboratory, the WSPs were scanned at a resolution of 600 dpi with a scanner. After that, imagery software DepositScan [23] (USDA, USA) was utilized to extract droplet deposits in the digital image and analyzed the droplet size, number of spray deposits and the area of coverage. The Volume Median Diameter (VMD) is a key index for reflecting the droplet size, which was used in this study.

The climatic conditions were recorded using a Kestrel 5500 digital meteorograph (Loftopia, LLC, USA), which recorded temperatures of 18.3–22.4 °C, a relative humidity of 46.3–53.7% and wind velocities of 3.6–10.8 km/h.

2.2.3. Investigation of Control Efficacy

To analyze the field efficacy of different spraying equipment with different spray deposition characteristics, we selected wheat aphid as the field test target according to the occurrence of diseases and pests. The wheat aphid was investigated and recorded four times according to pesticide field efficacy test criteria. Before the pesticide application on 27 April, the base number of the wheat aphids per hundred plants was more than 500, which meet the control criteria. The assessment was made by sampling five locations per plot on wheat aphid. The aphid number of 10 strains of wheat per location was investigated before spraying and the wheats were marked with a red string. After application on Days 1, 3, and 7, the number of aphid in the same location and plant was investigated again. The overall control effect against wheat aphid was calculated without regard to the types or instars of the wheat aphid. The mortality and control effects were obtained based on the population numbers of live insects in each zone before and after spraying. The control effect was calculated according to Equations (2) and (3):

$$\text{Mortality (\%)} = (\text{The number of pests before application} - \text{The number of pests after application})/\text{The number of pests before application} \times 100 \tag{2}$$

$$\text{Control effect (\%)} = [\text{Observed mortality (\%)} - \text{Control mortality (\%)}]/[100 - \text{Control mortality (\%)}] \times 100 \tag{3}$$

2.2.4. Working Efficiency Test

To better reflect the work efficiency (ha/h) of different equipment, the spraying area and the total spraying time of five filled tanks of chemical for each sprayer were recorded. The spray equipment was operated by experienced operators. The preparation time of mixing chemicals and other preparations, such as replacing the batteries or adding chemicals, did not count in the spraying time because they are greatly influenced by the cooperation and organization of the large-scale application. The spraying parameters are described in Section 2.2.1.

2.2.5. Statistical Analysis

Before the significant difference analysis, the percentage of the area of coverage on WSPs and the mortality of wheat aphids were transformed using $y = \arcsin\sqrt{X}/100$. The total deposition and losses to the ground, deposition on different canopies, number of spray deposits and droplet size were $\log (x + 1)$ transformed to stabilize wide variances and meet normality assumptions. After transformation, the data were analyzed for normality using the Kolmogorov–Smirnov test and for equal variances across the treatments and repeats using Levene's test ($p < 0.05$). The significant difference for the transformed data was conducted using analysis of variance (ANOVA) by Duncan's test at a significance level of 95% with SPSS v22.0 (SPSS Inc, an IBM Company, Chicago, IL, USA).

3. Results and Discussion

3.1. Spray Deposition

3.1.1. Total Deposition on the Crops and Losses to the Ground

The samples of the wheat plant were used to measure the total spray deposition per plant. The filter papers were used to measure the spray deposition in different wheat canopy and losses to the ground, and the recovery rate was calculated from the amount of the total deposition and additive. The total deposition of the UAV was not significantly different from the other sprayers (Table 1). Among the four sprayers, the EAP sprayer achieved the highest total deposition and the KMB sprayer achieved the lowest (Table 1). Compared with the SPB and KMB sprayer, the UAV and EAP sprayer had significantly higher recovery rates. The dose transfer process studied showed the transfer of pesticide from the spray tank to the target organism and in this process, losses of drift and run-off had great influence on deposition [24]. In this study, the SPB and KMB sprayer had relatively lower depositions and, correspondingly, those two sprayers had the higher run-offs, which were 0.39 and 0.50 $\mu g/cm^2$ (Table 1). Similar to other studies [25,26], it could be that high-volume spraying easily leads to run-off. In Table 1, Columns 4 and 5, compared with other sprayers, UAV had the lowest losses to the ground.

Table 1. The average (Avg.) and coefficient of variation (CV) of the total deposition on the plants, losses to the ground and the recovery rate of four sprayers.

Spray Equipment	Total Deposition		Losses to the Ground		Recovery Rate
	Avg. (µg/Plant)	CV (%)	Avg. (µg/cm²)	CV (%)	Avg. (%)
UAV sprayer	76.8 a	87.2	0.13 b	78.2	70.0 a
SPB sprayer	68.7 a	32.1	0.39 a	39.0	62.7 a
EAP Sprayer	84.8 a	84.4	0.14 b	118.2	77.3 a
KMB Sprayer	61.9 a	81.2	0.50 a	97.1	56.5 a

Note: Values followed by the same letter in the column do not differ statistically ($p < 0.05$; Duncan's Test).

3.1.2. The Uniformity of the Deposition

In addition to the total deposition, the uniformity of the deposition is also very important for controlling pests and diseases. The uniformity of the SPB sprayer was better than the others. The CV of the deposition was only 32.1%, which was significantly lower than the others (Table 1). This means it had a better deposition uniformity. However, the CVs of the deposition in this study was much greater than the value of Chinese National Standard requirement (10%) [27]. This may be due to the different measuring methods. In the study by Yang et al. [28], the CV of the deposition distribution ranged from 5% to 10% with different spray pressures and nozzle heights, which are far lower than the result of our study. This is because the Teejet pattern check was used by Yang et al., which is not influenced by the environment and the crop canopy. By comparison, the results in our study were a reflection of the actual uniformity of the deposition distribution on the crops.

Compared with the SPB sprayer, the UAV sprayer had a significant lower uniformity of the deposition distribution with the CV of 87.2% (Table 1). This result was larger than the Civil Aviation of China General Aviation Operation Quality and Technology Standard for ultra-low volume spraying, which is 60% [8]. The uniformity of the deposition distribution of the UAV was influenced by many factors, such as the types [9], the flight accuracy, the flight parameters [7], the spraying system, the biased downwash wind [18] and the meteorological condition. Precise flight route control and auto navigation are essential for chemicals application with improving the deposition distribution uniformity [14]. Xue et al. [8] developed an automatic navigation unmanned spraying system for the N-3 unmanned helicopter, which can significantly improve the deposition uniformity. The flight parameters also have a great influence on the deposition distribution uniformity. Various UAV sprayers under different flight parameters have been tested to find optimal spraying heights and speeds [7,10,18]. Qin et al. [7] found that the flight parameters affect not only the distribution uniformity on rice canopy but also the control efficacy on wheat hoppers. Bae and Koo [29] pointed out that most agricultural helicopters exhibit biased downwash, resulting in an uneven spray pattern. To address this problem, a roll-balanced agricultural helicopters with an elevated-pylon tail rotor system was developed. In their study, the uniformity of the spray patterns and area of coverage was improved.

The CVs of depositions of the EAP and KMB sprayer were 84.4% and 81.2%, respectively, which indicate a nonuniform deposition (Table 1), mainly due to the manual operation of the sprayer. The deposition uniformity was worse because it depended on the stability of the traveling speed along the spraying route and on the regularity of the arm movement of the operators.

3.1.3. Droplets Penetrability

In the test, filter papers were used to measure the tracer deposition on the different canopies of wheat. An analysis of the deposition on the wheat heads showed that the deposition of the low volume sprayer of the UAV and KMB sprayer were significantly higher than the SPB and EAP sprayer (Figure 4). The greatest deposition was achieved with the conventional KMB sprayer with 1.46 $\mu g/cm^2$, which was 18.7%, 80.2% and 111.6% more than the UAV, EAP and SPB sprayer, respectively (Figure 4). From the droplet size test results presented in Section 3.1.4, the VMD of droplets of KMB and UAV sprayer were lower than the other two sprayers. The main reason for higher wheat head deposition of UAV and KMB sprayer may be that fine droplets were better retained in the upper canopy [30].

The deposition of four sprayers on the flag leaf (top canopy) ranged from 0.57 to 1.02 $\mu g/cm^2$ (Figure 4). The UAV sprayer had the highest deposition on the flag leaf, which was 41.7–78.9% higher than the other sprayers. However, influenced by the great variability, the depositions on the flag leaf were not significantly different. In the study by Zhu et al. [31], the spray deposits decreased dramatically from the top to the bottom of the canopies and also tended to linearly decrease as the leaf area index increase. Due to the overlap and the block of the blades at the later growth stage of the wheat, the depositions on the lower parts were far less than top parts. The depositions of different sprayers were influenced by many factors, such as spraying pressure, spraying height and spraying pattern. The deposition of the UAV sprayer on the third top leaf (bottom canopy) was only 0.26 $\mu g/cm^2$, which was not significantly different from the other two conventional sprayers. However, it was 50.0% compared to the SPB sprayer and the deposition of SPB sprayer on the bottom was 0.52 $\mu g/cm^2$.

To improve the adhesive rate of solution and control efficacy, it is crucial to enhance the droplets penetrability and obtain a homogeneous deposition distribution [31], especially since many pests and diseases occur on the bottom of plants. Although many studies [18,19] have proven that the downwash airstream generated by the UAV is conducive to the disturbance of leaves and droplets penetration, the results from our study proved that the penetration of the droplets of UAV is still worse than boom sprayer. This result is similar to the results of Wang et al. [9] and Qin et al. [7]. There are many factors accounting for the poor droplet penetrability. One of the most important reasons is that the droplets were sprayed by the rotary cup atomizers, which lack the downward kinetic energy in comparison with hydraulic nozzles. Wang et al. [9] compared four kinds of UAV sprayer and found the flight height and the flight speed had a pronounced impact on droplet penetrability. In their study, the penetrability of the droplets was inversely proportional to the flight speed and the flight height. The flight parameter had an effect on the downwash flow, which further affected the droplets penetrability. Chen et al. [18] testified that the vertically downward wind had a significant effect on the penetrability. To improve the droplet penetrability of the UAV sprayer, the fly height and speed should be lower, thus ensuring operating efficiency, and suitable nozzles should be chosen to improve the droplets penetrability.

The deposition uniformity on different parts of the canopy was consistent with the total deposition. The SPB sprayer had the best deposition uniformity (CVs < 56%) on different canopies and other sprayers had lower uniformity (CVs > 70%).

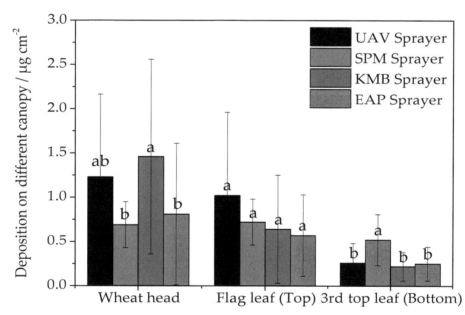

Figure 4. Deposition of four sprayers on the different canopies of wheat. Bars with different letters are significant different, Duncan's test, $p < 0.05$.

3.1.4. Characterization of Deposition

In the test, the WSPs were used to evaluate the characteristics of the deposition, which included droplet size, number of spray deposits and the area of coverage. Influenced by the spray volume and spraying system, the deposition characteristics of different sprayers were quite different (Figure 5). The spray volume of SPB sprayer approached 300 L/ha, which was identical to EAP sprayer belonging to the large volume application. Furthermore, the nozzles of SPB sprayer are similar to EAP sprayer, which are hydraulic nozzles. From the test results of WSPs, the VMD of the droplets of the SPB and EAP sprayer were 272.3 and 254.1 μm, respectively. They were not significantly different from each other (Figure 5). Although the VMD of droplets and the spray volume of SPB sprayer were similar to EAP sprayer, the area of coverage and the number of spray deposits of SPB sprayer were 75.9% and 73.9% greater than the EAP sprayer with no significant difference (Figure 5). From the CVs results of the total deposition, the greater area of coverage and number of spray deposits of SPB sprayer may be due to the better deposition uniformity. As a conventional knapsack sprayer, the spray volume of KMB sprayer was quite lower than the EAP sprayer of 75 L/ha. With the air assistance, the VMD of droplets of KMB sprayer was significantly lower than the EAP sprayer, which was only 154.7 μm. Results from previous studies had proved that the area of coverage is proportional to the spray volume [32]. The area of coverage of KMB sprayer was lower than the EAP sprayer with no significant difference (Figure 5). However, with finer droplets, the number of spray deposits of the KMB sprayer was quite similar to the EAP sprayer (Figure 5).

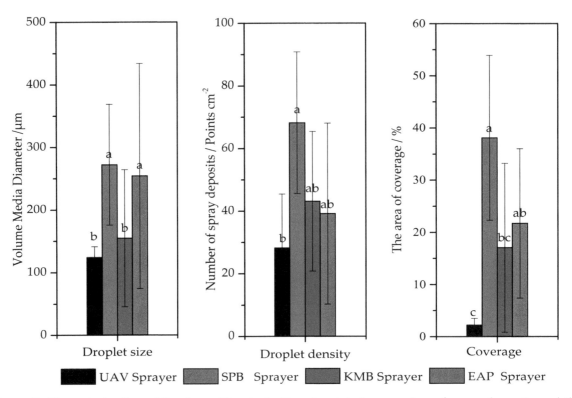

Figure 5. Characterization of the deposition including droplet size, number of spray deposits and the area of coverage.

The spray volume of UAV sprayer was 5 L/ha, which was lower than the other three sprayers. With the unique atomization method, the droplets were smashed by the high rotational speed of the rotary cup (10,000 rpm). The VMD of the droplets was 124.0 μm, which was finer than the three other sprayers (Figure 5). Another characteristic of this nozzle, which differed from hydraulic nozzle, is that the width of the droplet spectrum of this nozzle was narrower and the droplet size was more uniform with high rotational speed [33]. The result in this study verified that the CV of the VMD of the droplets was only 17.1%, which was far less than the other sprayers (CV > 35%) (Figure 5). Because of the lower spray volume, the area of coverage of the UAV was 2.2%, which were quite lower than the other sprayers and was only 5.8% to 12.8% of the other three sprayers. The number of spray deposits of the UAV was 28.2 points/cm^2, which was also significantly lower than others (Figure 5). This may go against the control of pests or diseases, especially when the contact pesticides were applied. Therefore, the focus of further studies will be improving the spreading coefficient of the droplets and increasing the area of the coverage, such as adding adjuvant in the tank [34,35], or changing into electrostatic-charged sprays [36,37].

Spraying systems had a great effect on the characteristics of deposition, including the area of coverage, number of spray deposits and droplet size. It is difficult to judge the application quality by a single index. Although the lower spray volume of UAV could lead to lower area of coverage and fewer deposits, the dose applied per area was not significant lower than other sprayers, because of the higher concentration of each droplets. Syngenta Crop Protection AG (Basel, Switzerland) recommends that the satisfactory results can be obtained by a number of spray deposits of at least 20–30 points/cm^2 for insecticide or pre-emergence herbicide applications, 30–40 points/cm^2 for contact post-emergence herbicide applications and 50–70 points/cm^2 for fungicide applications [23]. The number of spray deposits only needs to reach a certain threshold to achieve a good control efficacy [38].

3.2. Control of Wheat Aphids

The control efficacy of the four sprayers applying 70% Imidacloprid·Regent WDG on wheat aphids are indicated in Figure 6. Although the spray deposition characteristics of the four sprayers were significantly different from each other, the differences of control efficacy were not remarkable. The control efficacies on Day 7 after application were all beyond 70%. From the comparison of four sprayers, it was found that the SPB and KMB sprayers achieved the best control efficacy, while the efficacies were intermediate for EAP and UAV sprayer. Deposit structure plays a major role in toxin efficacy [24,38]. According to the results of the deposition, the SPB sprayer had the best deposition uniformity and penetrability, which benefit the control of wheat aphids, especially since wheat aphids tend to favor the lower portions of plants. The larger area of coverage and a larger number of spray deposits increased the odds of interaction between active ingredients and pests.

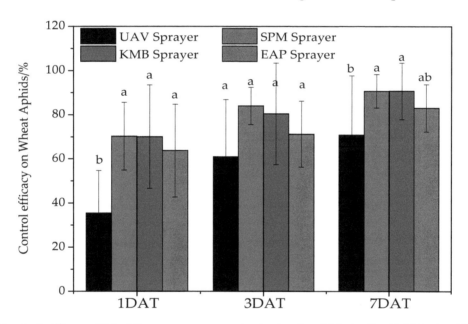

Figure 6. Control efficacy (%) of four sprayers against wheat aphids in the field tests at Days 1, 3, 7 after treatment. DAT, days after treatment.

Because UAV sprayers are still in the initial development stage in China, there are still many problems on the spraying system and working parameter. These problems lead to unevenness deposition and poor penetrability. The lower area of coverage and fewer spray deposits with the nonuniform deposition reduced the chance of aphids to be exposed to the insecticides. The control efficacy of UAV sprayer on Day 1 after application was significantly lower than other sprayers, which was only equal to 50.5% of the SPB sprayer. However, in time, the control efficacy increased. The reason could be that, with the movement of the aphids and the systemic action of the active ingredients, the chance of wheat aphids being exposed to the active ingredients increased. On Day 3 after application, the control efficacy increased to 60.9% and, on Day 7 after application, the control efficacy increased to 70.9%. Although it was also significantly lower than the SPB and KMB sprayers, it was still an acceptable result for farmers, especially in complex small plots or rice fields, where the SPB sprayer is hard to work and the KMB and EAP sprayer have a low work efficiency. Although the UAV sprayer, especially for the electrical multi-rotor UAV, used in the field, is an innovation, the low volume application is not a new technology. Many studies had been conducted to evaluate the feasibility of low volume application for pests and diseases control. In the laboratory, Maczuga and Mierzejewski [39] found that several spray deposits of 5–10 points/cm^2 on foliage after spraying were effective (90% mortality) against second and third instars. Washington et al. [40] investigated the effect of fungicide spray number of spray deposits (points/cm^2), droplet size and proximity of the spray deposits to fungal spores on the banana leaf surface. The test results suggest that the inhibition zones of two contact fungicides on the leaf surface extend beyond the visible edge of the spray droplet deposit and a mean droplet deposit density of 30 points/cm^2 can inhibit the germination of the ascospore below 1%. Latheef et al. [36] evaluated the efficacy of aerial electrostatic-charged sprays for season-long control of sweet potato whiteflies, and they concluded that the control efficacy of electrostatic-charged sprays with spray volume of 4.68 L/ha was comparable with those on cotton treated with conventional applications of 46.8 L/ha.

3.3. Working Efficiency

Working efficiency is also an important evaluation index for equipment selection. The UAV sprayer used in the test belongs to semi-autonomous control, which is the most common in China. By recording the spraying time of five sorties with the filled tank, the average spraying time of each sortie of the UAV sprayer was calculated as 0.095 h (Table 2). The average spraying area of each sortie was 0.39 ha with the spray volume of 5 L (Table 2). Calculated from the spraying time and area, the work efficiency of the UAV sprayer was 4.11 ha/h (Table 2). The work efficiency test results were consistent with Wang et al. [9], who reported that the working efficiency of UAV was at 13.4–18.0 hectare per 8 h, i.e., 1.68–2.25 ha/h. In his test, the operation items included the time of preparation, route planning, failure maintenance, and ground service, which accounted for 50% of the whole process. With micro-electronic technology development, the multi-sensor data fusion and real-time kinematic positioning technology and product will be applied in agriculture UAV, with which the UAV can achieve fully autonomous flight, significantly improving the working efficiency [41].

Table 2. Working efficiency (ha/h) of four sprayers.

Sprayer	Tank Capacity (L)	Spray Area (Means ± Standard Error, ha)	Spray Time (Means ± Standard Error, h)	Working Efficiency (ha/h)
UAV sprayer	5	0.39 ± 0.04	0.095 ± 0.01	4.11
SPB sprayer	280	0.93 ± 0.06	0.39 ± 0.03	2.38
KMB Sprayer	18	0.22 ± 0.02	0.14 ± 0.01	1.57
EAP Sprayer	16	0.039 ± 0.004	0.19 ± 0.01	0.21

Note: Spray area in the table means the area per spray with full tank. Spray time in the table means the time per spray with full tank.

Compared with the UAV sprayer, the SPB sprayer had a relatively lower working efficiency. The average spraying time was 0.39 h and the spraying area was 0.93 ha with the tank capacity of 280 L (Table 2). Although there are many other kinds of boom sprayers with a much longer boom, the complexity of the terrain and the farm size limit the usage of larger boom sprayers in China. The average farm size in China is amongst the smallest in the world of 0.67 ha and more than half (57.5%) of farms are small (<2 ha) [1]. Thus, this kind of small boom sprayer is widespread with the working efficacy of 2.38 ha/h (Table 2).

From the test results, the working efficiency of two conventional knapsack sprayers were 1.57 and 0.21 ha/h, respectively (Table 2). Conventional knapsack sprayers need to be carried on the back, which easily leads to exhausting and lower work efficiency. Especially for the EAP sprayer, the working efficacy was only 5.1% of the UAV sprayer. Although these knapsack sprayers also hold very large market share (88%, from China Agriculture Yearbook Editorial Committee 2016) in China, with the development of technology and the growth of farming size, these lower working efficiency sprayers will be tapered and the studies on high efficiency sprayers will be necessary.

4. Conclusions

In this study, four typical sprayers were used for pesticide application in the wheat field. The total deposition on the plants and losses to the ground, the uniformity of the deposition, droplets penetrability, characteristics of the deposition, control efficacy on wheat aphids and working efficiency were compared in this research. The conclusions are shown as follows:

1) The total deposition of the UAV sprayer was not significantly different from the other three sprayers, but the losses to the ground were the lowest.

2) The UAV sprayer had a poor deposition uniformity (CV = 87.2%) and droplets penetrability (0.26 µg/cm^2 on the third top leaf), which need to further improve in the future. By comparison, the SPB sprayer has the best deposition uniformity (CV = 32.1%) and droplets penetrability (0.52 µg/cm^2 on the third top leaf).

3) The area of coverage, number of spray deposits and droplet size varied with spray volume and the nozzle type of the different sprayers. Compared with other sprayers, the deposition characterizations of the UAV sprayer were a lower area of coverage (2.2%), a lfewer spray deposits (28.2 points/cm^2), finer droplet size (VMD = 124.0 µm), higher concentration and lower spray volume.

4) Although the spray deposition characterizations of the UAV sprayer were different from the three other sprayers, the control efficacy of applying 70% Imidacloprid·Regent WDG on wheat aphids was comparable with other sprayers. The control efficacy of UAV sprayer on Day 7 after the application was 70.9%.

5) The working efficiency of the UAV sprayer was 4.11 ha/h, which was 1.7, 2.6, and 20.0 times those of SPB, KMB and EAP sprayer, respectively. This is the greatest advantage of the UAV sprayer.

The experiment demonstrated the feasibility and high efficiency of the UAV sprayer. The deposition uniformity and the droplets penetrability of the UAV also need to be improved. Due to the lower coverage and poor deposition uniformity, the effective measures, such as optimizing the spraying system or adding adjuvant in the tank to improve deposition uniformity and penetrability would be needed in the future.

Author Contributions: G.W., Y.L., H.Y., H.Q., P.C., F.O. and Y.H. conceived the idea of the experiment. G.W. performed the experiments and analyzed the data. G.W. and Y.H. wrote and revised the paper.

References

1. Yang, S.; Yang, X.; Mo, J. The application of unmanned aircraft systems to plant protection in China. *Precis. Agric.* **2018**, *19*, 278–292. [CrossRef]
2. Zhang, X.; Zhao, W.; Jing, R.; Wheeler, K.; Smith, G.A.; Stallones, L.; Xiang, H. Work-related pesticide poisoning among farmers in two villages of Southern China: A cross-sectional survey. *BMC Public Health* **2011**, *11*, 429. [CrossRef] [PubMed]
3. Cao, L.; Cao, C.; Wang, Y.; Li, X.; Zhou, Z.; Li, F.; Yan, X.; Huang, Q. Visual determination of potential dermal and inhalation exposure using allura red as an environmentally friendly pesticide surrogate. *ACS Sustain. Chem. Eng.* **2017**, *5*, 3882–3889. [CrossRef]
4. Ghugare, B.D.; Adhaoo, S.H.; Gite, L.P.; Pandya, A.C.; Patel, S.L. Ergonomics evaluation of a lever-operated knapsack sprayer. *Appl. Ergon.* **1991**, *22*, 241–250. [CrossRef]
5. Sánchez-Hermosilla, J.; Rincón, V.J.; Páez, F.; Fernández, M. Comparative spray deposits by manually pulled trolley sprayer and a spray gun in greenhouse tomato crops. *Crop Prot.* **2012**, *31*, 119–124. [CrossRef]

6.   Xiongkui, H.; Bonds, J.; Herbst, A.; Langenakens, J. Recent development of unmanned aerial vehicle for plant protection in East Asia. *Int. J. Agric. Biol. Eng.* **2017**, *10*, 18–30. [CrossRef]

7.   Qin, W.C.; Qiu, B.J.; Xue, X.Y.; Chen, C.; Xu, Z.F.; Zhou, Q.Q. Droplet deposition and control effect of insecticides sprayed with an unmanned aerial vehicle against plant hoppers. *Crop Prot.* **2016**, *85*, 79–88. [CrossRef]

8.   Xue, X.; Lan, Y.; Sun, Z.; Chang, C.; Hoffmann, W.C. Develop an unmanned aerial vehicle based automatic aerial spraying system. *Comput. Electron. Agric.* **2016**, *128*, 58–66. [CrossRef]

9.   Wang, S.; Song, J.; He, X.; Song, L.; Wang, X.; Wang, C.; Wang, Z.; Ling, Y. Performances evaluation of four typical unmanned aerial vehicles used for pesticide application in China. *Int. J. Agric. Biol. Eng.* **2017**, *10*, 22–31. [CrossRef]

10.   Zhang, P.; Deng, L.; Lyu, Q.; He, S.; Yi, S.; Liu, Y.; Yu, Y.; Pan, H. Effects of citrus tree-shape and spraying height of small unmanned aerial vehicle on droplet distribution. *Int. J. Agric. Biol. Eng.* **2016**, *9*, 45–52. [CrossRef]

11.   Giles, D.; Billing, R. Deployment and performance of an unmanned aerial vehicle for spraying of specialty crops. In Proceedings of the International Conference of Agricultural Engineering, Zurich, Switzerland, 6–10 July 2014.

12.   Law, S.E. Embedded-electrode electrostatic-induction spray-charging nozzle: Theoretical and engineering design. *Trans. ASAE* **1978**, *21*, 1096–1104. [CrossRef]

13.   Zhang, C.; Kovacs, J.M. The application of small unmanned aerial systems for precision agriculture: A review. *Precis. Agric.* **2012**, *13*, 693–712. [CrossRef]

14.   Huang, Y.; Hoffmann, W.C.; Lan, Y.; Wu, W.; Fritz, B.K. Development of a spray system for an unmanned aerial vehicle platform. *Appl. Eng. Agric.* **2009**, *25*, 803–809. [CrossRef]

15.   Fritz, B. Meteorological effects on deposition and drift of aerially applied sprays. *Trans. ASAE* **2006**, *49*, 1295–1301. [CrossRef]

16.   Bird, S.L.; Perry, S.G.; Ray, S.L.; Teske, M.E. Evaluation of the AgDISP aerial spray algorithms in the AgDRIFT model. *Environ. Toxicol. Chem.* **2002**, *21*, 672–681. [CrossRef]

17.   Wang, X.; He, X.; Wang, C.; Wang, Z.; Li, L.; Wang, S.; Jane, B.; Andreas, H.; Wang, Z. Spray drift characteristics of fuel powered single-rotor UAV for plant protection. *Trans. ASAE* **2017**, *33*, 117–123, (In Chinese with English abstract). [CrossRef]

18.   Chen, S.; Lan, Y.; Li, J.; Zhou, Z.; Liu, A.; Mao, Y. Effect of wind field below unmanned helicopter on droplet deposition distribution of aerial spraying. *Int. J. Agric. Biol. Eng.* **2017**, *10*, 67–77. [CrossRef]

19.   Xue, X.; Tu, K.; Qin, W.; Lan, Y.; Zhang, H. Drift and deposition of ultra-low altitude and low volume application in paddy field. *Int. J. Agric. Biol. Eng.* **2014**, *7*, 23–28. [CrossRef]

20.   Zhao, H.; Xie, C.; Liu, F.; He, X.; Zhang, J.; Song, J. Effects of sprayers and nozzles on spray drift and terminal residues of imidacloprid on wheat. *Crop Prot.* **2014**, *60*, 78–82. [CrossRef]

21.   Wolters, A.; Linnemann, V.; Zande, J.C.v.d.; Vereecken, H. Field experiment on spray drift: Deposition and airborne drift during application to a winter wheat crop. *Sci. Total Environ.* **2008**, *405*, 269–277. [CrossRef] [PubMed]

22.   Qiu, Z.; Yuan, H.; Lou, S.; Ji, M.; Yu, J.; Song, X. The research of water soluble dyes of Allura Red and Ponceau-G as tracers for determing pesticide spray distribution. *Agrochemicals* **2007**, *46*, 323–325. (In Chinese with English abstract) [CrossRef]

23.   Zhu, H.; Salyani, M.; Fox, R.D. A portable scanning system for evaluation of spray deposit distribution. *Comput. Electron. Agric.* **2011**, *76*, 38–43. [CrossRef]

24.   Ebert, T.A.; Taylor, R.A.J.; Downer, R.A.; Hall, F.R. Deposit structure and efficacy of pesticide application. 2: Trichoplusia ni control on cabbage with fipronil. *Pestic. Sci.* **1999**, *55*, 793–798. [CrossRef]

25.   Rincón, V.J.; Sánchez-Hermosilla, J.; Páez, F.; Pérez-Alonso, J.; Callejón, Á.J. Assessment of the influence of working pressure and application rate on pesticide spray application with a hand-held spray gun on greenhouse pepper crops. *Crop Prot.* **2017**, *96*, 7–13. [CrossRef]

26.   Sánchez-Hermosilla, J.; Rincón, V.J.; Páez, F.; Agüera, F.; Carvajal, F. Field evaluation of a self-propelled sprayer and effects of the application rate on spray deposition and losses to the ground in greenhouse tomato crops. *Pest Manag. Sci.* **2011**, *67*, 942–947. [CrossRef] [PubMed]

27.   *Agricultural and Forestry Machinery–Inspection of Sprayers in Use–Part 2: Horizontal Boom Sprayers*; ISO/TC 23/SC 6. ISO 16122-2:2015; International Organization for Standardization (ISO): Geneva, Switzerland, 2015.

28. Yang, D.B.; Zhang, L.N.; Yan, X.J.; Wang, Z.Y.; Yuan, H.Z. Effects of droplet distribution on insecticide toxicity to asian corn borers (Ostrinia furnaealis) and spiders (Xysticus ephippiatus). *J. Integr. Agric.* **2014**, *13*, 124–133. [CrossRef]

29. Bae, Y.; Koo, Y.M. Flight attitudes and spray patterns of a roll-balanced agricultural unmanned helicopter. *Appl. Eng. Agric.* **2013**, *29*, 675–682. [CrossRef]

30. Hislop, E.C.; Western, N.M.; Butler, R. Experimental air-assisted spraying of a maturing cereal crop under controlled conditions. *Crop Prot.* **1995**, *14*, 19–26. [CrossRef]

31. Zhu, H.; Dorner, J.W.; Rowland, D.L.; Derksen, R.C.; Ozkan, H.E. Spray penetration into peanut canopies with hydraulic nozzle tips. *Biosyst. Eng.* **2004**, *87*, 275–283. [CrossRef]

32. Ferguson, J.C.; Chechetto, R.G.; Hewitt, A.J.; Chauhan, B.S.; Adkins, S.W.; Kruger, G.R.; O'Donnell, C.C. Assessing the deposition and canopy penetration of nozzles with different spray qualities in an oat (*Avena sativa* L.) canopy. *Crop Prot.* **2016**, *81*, 14–19. [CrossRef]

33. Craig, I.P.; Hewitt, A.; Terry, H. Rotary atomiser design requirements for optimum pesticide application efficiency. *Crop Prot.* **2014**, *66*, 34–39. [CrossRef]

34. Xu, L.; Zhu, H.; Ozkan, H.E.; Bagley, W.E.; Krause, C.R. Droplet evaporation and spread on waxy and hairy leaves associated with type and concentration of adjuvants. *Pest Manag. Sci.* **2011**, *67*, 842–851. [CrossRef] [PubMed]

35. Van Zyl, S.A.; Brink, J.-C.; Calitz, F.J.; Fourie, P.H. Effects of adjuvants on deposition efficiency of fenhexamid sprays applied to Chardonnay grapevine foliage. *Crop Prot.* **2010**, *29*, 843–852. [CrossRef]

36. Latheef, M.A.; Carlton, J.B.; Kirk, I.W.; Hoffmann, W.C. Aerial electrostatic-charged sprays for deposition and efficacy against sweet potato whitefly (*Bemisia tabaci*) on cotton. *Pest Manag. Sci.* **2009**, *65*, 744–752. [CrossRef] [PubMed]

37. Yanliang, Z.; Qi, L.; Wei, Z. Design and test of a six-rotor unmanned aerial vehicle (UAV) electrostatic spraying system for crop protection. *Int. J. Agric. Biol. Eng.* **2017**, *10*, 68–76. [CrossRef]

38. Ebert, T.A.; Taylor, R.A.J.; Downer, R.A.; Hall, F.R. Deposit structure and efficacy of pesticide application. 1: Interactions between deposit size, toxicant concentration and deposit number. *Pestic. Sci.* **1999**, *55*, 783–792. [CrossRef]

39. Maczuga, S.A.; Mierzejewski, K.J. Droplet size and density effects of bacillus thuringiensis kurstaki on gypsy moth (Lepidoptera: Lymantriidae) Larvae. *J. Econ. Entomol.* **1995**, *88*, 1376–1379. [CrossRef]

40. Washington, J.R. Relationship between the spray droplet density of two protectant fungicides and the germination of mycosphaerella fijiensis ascospores on banana leaf surfaces. *Pest Manag. Sci.* **1997**, *50*, 233–239. [CrossRef]

41. Lan, Y.; Chen, S.; Fritz, B.K. Current status and future trends of precision agricultural aviation technologies. *Int. J. Agric. Biol. Eng.* **2017**, *10*, 1–17. [CrossRef]

10

Modal Planning for Cooperative Non-Prehensile Manipulation by Mobile Robots

Changxiang Fan [1,*]**, Shouhei Shirafuji** [2] **and Jun Ota** [2]

[1] Department of Precision Engineering, Graduate School of Engineering, The University of Tokyo, 7-3-1 Hongo, Bunkyo-ku, Tokyo 113-8656, Japan
[2] Research into Artifacts, Center for Engineering, The University of Tokyo, 5-1-5 Kashiwanoha, Kashiwa-shi, Chiba 277-8568, Japan; shirafuji@race.u-tokyo.ac.jp (S.S.); ota@race.u-tokyo.ac.jp (J.O.)
* Correspondence: fan@race.u-tokyo.ac.jp.

Abstract: If we define a mode as a set of specific configurations that hold the same constraint, and if we investigate their transitions beforehand, we can efficiently probe the configuration space by using a manipulation planner. However, when multiple mobile robots together manipulate an object by using the non-prehensile method, the candidates for the modes and their transitions become enormous because of the numerous contacts among the object, the environment, and the robots. In some cases, the constraints on the object, which include a combination of robot contacts and environmental contacts, are incapable of guaranteeing the object's stability. Furthermore, some transitions cannot appear because of geometrical and functional restrictions of the robots. Therefore, in this paper, we propose a method to narrow down the possible modes and transitions between modes by excluding the impossible modes and transitions from the viewpoint of statics, kinematics, and geometry. We first generated modes that described an object's contact set from the robots and the environment while ignoring their exact configurations. Each multi-contact set exerted by the robots and the environment satisfied the condition necessary for the force closure on the object along with gravity. Second, we listed every possible transition between the modes by determining whether or not the given robot could actively change the contacts with geometrical feasibility. Finally, we performed two simulations to validate our method on specific manipulation tasks. Our method can be used in various cases of non-prehensile manipulations by using mobile robots. The mode transition graph generated by our method was used to efficiently sequence the manipulation actions before deciding the detailed configuration planning.

Keywords: non-prehensile manipulation; manipulation planning; contact planning; manipulation action sequences

---

1. Introduction

Spatial restrictions make it almost impossible to manipulate a big object in a narrow space by using big-scaled manipulators. For instance, it is impractical to carry an industrial manipulator into our house to move furniture by grasping and lifting. Owing to their small size and flexibility in motion, multiple mobile robots can be adopted to perform tasks in a narrow space [1]. These robots can move in a narrow environment to approach and manipulate objects, but these robots cannot grasp big objects as large-scale industrial manipulators. Therefore, non-prehensile methods [2,3], which involves manipulation without grasping, is practical for such cases. For instance, a preferred way to manipulate a big object is to push it along the floor, or to pivot it with a vertex that makes contact with the floor. In certain cases, the object keeps contacting the environment in such manipulations and the restrictions on the object motion caused by the contacts complicates the kinematics in the manipulation. Furthermore, when multiple mobile robots perform a manipulation task, they themselves form a complex coordinated system [4]. Consequently, non-prehensile manipulations that use multiple mobile robots (as shown in Figure 1) require convoluted manipulation planning than standard manipulation, which comprehensively takes into account each robot's kinematics, the surrounding constraints, and their changes.

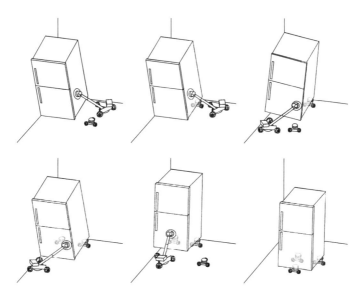

Figure 1. Example of non-prehensile transportation adopting multiple mobile robot: Two types of mobile robots move the refrigerator placed at the corner of a room.

Multiple robot motion planning is often faced with the high dimensional configuration space [5]. Typical planners for such problems are sample-based, such as RRT [6] and PRM [7]. However, manipulation planning problems often encounter particular *multi-modal* structures [8], if the contacts among the robots, the manipulated objects, and the environment change during the process of manipulation. Here, a *mode* refers to a certain set of configurations that hold the same constraints in motion (e.g., the object motion keeping a set of contacts with the environment). Possible configurations under the same constraints form sub-spaces in the configuration space. This requires the planner to be capable not only to probe the sub-spaces of each mode but also to cross among the different sub-spaces. For the application of typical sample-based methods on their original spaces, the expansiveness among the configurations of the different modes is more difficult to achieve than those of the same mode [9]. For instance, we can generate a transitable configuration of the system for a current configuration by sampling the configuration the belongs to the sub-space, while keeping the contact set among the robots, the objects, and the environment unchanged. However, in case the contact set changes, it would be necessary to check the connectivity between the different sub-spaces corresponding to the contact states, which would complicate the problem.

Therefore, typical sample-based methods are usually applied to the modes' sub-spaces after splitting the configuration space into sub-spaces according to the modes. The sample-based methods become realizable by deciding the sequence of modes where connectivity is guaranteed beforehand. Maeda et al. [10] and Miyazawa et al. [11] sampled the manipulation states in a configuration space where a sequence of modes existed and the modes' transitions were prior defined. Some planners have been proposed for creating the modes' roadmap to guide the manipulation sequencing [8,12–14]. In particular, Lee [13] adopted a PRM-based planner [15,16] to split the modes by comprehensively considering the multi-contacts between the object, robots, and the environment to obtain the necessary modes to pass through for a manipulation task. Mode transitions are mainly derived from the compliant transformations of contacts between the object, the environment, and the contacts of the given robots; two robot contacts in the examples [13] were adopted to realize the subsequent mode transitions.

However, when multiple mobile robots manipulated objects by using non-prehensile methods, mode transitions became more complex, because besides considering the mode transitions caused by the changes in the environmental contact, robot contacts also had to be considered. One way of addressing this complexity is by eliminating unfeasible modes in statics from the enormous combinations of contacts among them before considering the transitions among modes. An object should be under sufficient constraints in each mode; otherwise, the robots will fail to manipulate the object (e.g., the object drops down in an unexpected direction because of the lack of constraints). This means that the contacts from the environment and the robots should be able to form full constraints on the objects for non-prehensile manipulations. If we consider the environmental contact alone, the resultant constraints would vary in different contact states. For example, when an object–floor contact state changes from face–face contact into vertex–face contact, the constraints exerted by the floor would reduce. This results in various least requirements of robot constraints under different environmental constraints. Sometimes, we require the object to keep stationary contact with the environment, so that the robot can use friction to move the object (e.g., inclining a box); sometimes, we require the robots to manipulate the object to slide along the contacting part (e.g., sliding a box on a floor). To distinguish these manipulations, the environmental contact should be identified into the fixed contact and the sliding contact. The restraint placed on the object's degrees of freedom by a contact is different between the cases when it is fixed or sliding. Thus, we consider the necessary amount of robot for manipulations, both when relative sliding happened and did not happen. Therefore, a proper consideration of the individual robot's kinematics and the changes in the environmental constraints is essential when splitting the modes.

Furthermore, a robot contact changes when the robot makes or breaks contact with the object either actively or passively (by the object's motion). For example, in Figure 1, a large robot with a manipulator actively makes contact with the object and pushes it over the small robots, and the small robots passively make contact with the object by the action of the large robot. The distinction between the active and passive action appears in many manipulation tasks, and the possible sequence of actions depends on this distinction of actions. Thus, the planner should be able to reason about all such possible mode transitions.

By addressing the problems peculiar to non-prehensile manipulations using the mobile robots described above, we propose a method to generate the modes and transitions between the modes for contact planning. In our method, for a given set of contacts, we identified the modes by analyzing its constraints and least requirements to constrain the object's motion. We determine the mode transitions based on how the robot contact influenced the contact state of a targeted object. The mobile robots were divided into active robots and passive robots according to how their contacts changed in the state transition. Finally, the manipulation actions were determined by sequencing a series of modes. We applied the same concept to the limited cases also [17]. In this paper, we propose a more generalized framework to determine the action sequence including the distinction between the fixed and the sliding contact states.

In the second section, we describe the problem statement, and in the third section, we introduce the generation of the contact state. In the fourth section, the state transition is investigated to sequence the action series. In the fifth section, we describe the applications of our methodology by conducting two simulations of the specified manipulation cases.

2. Problem Statement

In this paper, for simplicity, we have only manipulated objects having convex polyhedrons. Furthermore, we consider that the objects, the environment, and the robots are all rigid. A set of contacts on an object having environmental contact and robotic contact were represented as a contact state and defined as a mode. A contact state is described without the exact position of contact and configuration of the object and the robots.

For a description of the contact state without the exact configurations, we adopt the concept of principal contact (PC) [18] to express the contact states. A PC is a contacting pair between the geometrical primitives (a vertex, an edge, or a face). A PC is denoted by $c = (a, b)$, where a is a geometrical primitive on the object and b is a geometrical primitive on a polyhedron of the environment or a robot. Furthermore, we denote the object–environment PC and object–robot PC by c^{e} and c^{r}, respectively, for clarity.

In our method, we need to distinguish whether relative sliding happening on a contact for determining the constraints required to realize the full constraints of the object. For example, Figure 2 shows the difference of the required constraints in the quasi-static manipulation of an object that lies on a floor with an edge contact when a robot contacts with it as a point–face contact (in the planar case). Manipulation without relative sliding on the point with the help of floor's static friction, as shown in Figure 2c, requires a robot to tilt the object. However, for manipulation with relative sliding between the object and the floor, it is difficult to predict the resultant motion of the object because of dynamic friction, as shown in Figure 2b. To guarantee that the target motion will be realized, an additional robot is required (see Figure 2c), and the number of robots required is different between the two cases.

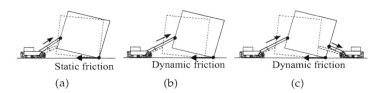

Static friction	Dynamic friction	Dynamic friction
(a)	(b)	(c)

Figure 2. Frictional constraint on object's motion in fixed and sliding cases: (**a**) A robot tilts an object with fixed contact on the floor; (**b**) a robot tilts an object with sliding contact; (**c**) two robots tilts an object with sliding contact.

Therefore, we distinguish whether relative sliding happens or not on a contact. A non-sliding PC is defined as *static* PC, denoted by $c^{\mathrm{e,st}} = (a, b)^{\mathrm{e,st}}$. Correspondingly, a sliding PC is defined as *dynamic* PC, denoted by $c^{\mathrm{e,dn}} = (a, b)^{\mathrm{e,dn}}$.

When multiple robots manipulate objects, not all of them actively make contact with an object. A contact occurs either when the robot actively touches the object or passively touches it when the object is moved. Furthermore, some robots move an object actively and change its state, whereas some robots operate as auxiliaries in the manipulation task. The distinction between the active and passive functions of a contact is important for considering the possible transitions of the states. Accordingly, the contacts are divided into active and passive and are denoted as $c^{\mathrm{r,st,ac}}$ and $c^{\mathrm{r,st,ps}}$, respectively. A robot with active joints can also act as a passive robot. Whether the robot always acts passively or actively or is switchable between passive and active depends on the function of the robot. A contact between an object and a robot usually does not slide. Therefore, for simplicity, we omit the subscript for the static and the dynamic contacts if the contact is between an object and a robot and write as $c^{\mathrm{r,ac}}$ and $c^{\mathrm{r,ps}}$. However, a contact between an object and the environment is passive. Therefore, we omit the subscript for the passive and the active contacts if the contact is between an object and the environment and write as $c^{\mathrm{e,st}}$ and $c^{\mathrm{e,dn}}$.

The set of PCs on an object is called contact formation (CF) [18] and denoted by C. To describe the original CF proposed by Xiao and Zhang [18], which does not concern the sliding of the contact point, we call a CF without distinguishing the static and dynamic as the *primitive* contact states, and denote it as $\widehat{C}$. For example, $\{(a_1, b_1)^{\mathrm{e}}, (a_2, b_2)^{\mathrm{e}}\}$ is the primitive state of $\{(a_1, b_1)^{\mathrm{e,st}}, (a_2, b_2)^{\mathrm{e,st}}\}$ or $\{(a_1, b_1)^{\mathrm{e,st}}, (a_2, b_2)^{\mathrm{e,dn}}\}$. Furthermore, when CFs are the same from the viewpoint of the primitive contact states, we describe them as *isogenous*. For example, $\{(a, b)^{\mathrm{e,st}}\}$ and $\{(a, b)^{\mathrm{e,dn}}\}$ are isogenous, and $\{(a_1, b_1)^{\mathrm{e,st}}, (a_2, b_2)^{\mathrm{e,st}}\}$ and $\{(a_1, b_1)^{\mathrm{e,st}}, (a_2, b_2)^{\mathrm{e,dn}}\}$ are also isogenous.

In this paper, we distinguish the CFs consisting of the contact states between an object and the environment, which we call environmental contact formation (ECF). We also distinguish the CFs consisting of the contact states between the robots and the environment, which we call robot contact formation (RCF); see Figure 3.

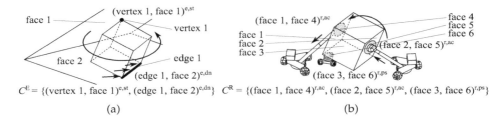

$C^{\mathrm{E}} = \{(\text{vertex 1, face 1})^{\mathrm{e,st}}, (\text{edge 1, face 2})^{\mathrm{e,dn}}\}$ $C^{\mathrm{R}} = \{(\text{face 1, face 4})^{\mathrm{r,ac}}, (\text{face 2, face 5})^{\mathrm{r,ac}}, (\text{face 3, face 6})^{\mathrm{r,ps}}\}$

(a)   (b)

Figure 3. Examples of contact formation (CF): (**a**) environmental contact formation (ECF) when a cuboid contacts with two surfaces of the environment by its vertex and edge; (**b**) robot contact formation (RCF) when a cuboid contacts with two active robots and a passive robot by its faces.

Furthermore, we describe them as C^{E} and C^{R}, respectively. As the result, we denote the mode s in the manipulation as the set of ECF and RCF:

$$s = \{C^{\mathrm{E}}, C^{\mathrm{R}}\}.$$

A multi-contact set on an object was supposed to form the full constraints, so that the robots manipulated the object quasi-statically. Gravity closure [19,20], which is force closure that includes the gravitational force, is introduced later in this paper as the requirement for a mode. Our first goal is to identify all possible modes from the viewpoint of gravity closure.

If one mode can directly transform into another mode without any intermediate ones, it is a possible mode transition. Mode transitions can be described by a graph that comprises nodes and arcs, where the nodes represent individual modes, and the arcs between them represent the transitions; the value of an individual arc represented the cost of the state transition [21]. Our second goal is to generate this mode graph by taking into account some restrictions on robots and the geometrical relationships between an object and the environment. Using the resultant graph, the manipulation action sequences are determined by searching for paths from a given initial mode to a targeted mode before applying a sample-based method to probe the configuration spaces determined by modes.

In the following sections, we have made the following assumptions. The shape and the gravity center of the object and the shape of the environment are given. Furthermore, we have assumed that the type of contact that a robot generates with the object is given, and it does not change in the manipulation. The number of robots is also given. We have considered only the possible RCF on the targeted object, and the contacts between the robots and the environment were ignored; the robots generally make contact with a face (a floor) in the environment.

3. Generation of Modes

In this section, we identified all possible modes from the viewpoint of gravity closure. Given a set of mobile robots and an object lying in a certain environment, we obtained the possible modes using the following steps: (i) The possible ECFs were specially identified. We identified the ECFs before identifying the RCFs because the ECF restrains the robot's accessible area. (ii) For the identified ECF, we identified the possible RCF. (iii) Finally, by combining the ECFs and the RCFs, full constraints could be achieved.

3.1. Generation of ECFs

The geometrical relationship between an object's shape and the shape of its environment determines the possible ECFs. A PRM-based sampling strategy has been proposed to investigate the possible CFs between objects [15,16,22]. In their method, the target object's orientation was incrementally changed with collisions checked to obtain the possible CFs. If a new CF appeared, it was recorded on the list, along with the object's current orientation. In this way, all the possible CFs between the objects could be probed; each CF was recorded with an available object orientation. We adopted this method to investigate the possible ECFs in our planner.

As mentioned in the previous section, we distinguished the contacts into static and dynamic contact based on the resultant CFs obtained by their method. For the given CFs obtained by their method, every PC was split into static and dynamic PCs, and ECFs are given as all combinations of the split static and dynamic PCs. For example, given a CF $C = \{c_1, c_2\}$, the ECFs are $C_1^E = \{c_1^{e,st}, c_2^{e,st}\}$, $C_2^E = \{c_1^{e,st}, c_2^{e,dn}\}$, $C_3^E = \{c_1^{e,dn}, c_2^{e,st}\}$, and $C_4^E = \{c_1^{e,dn}, c_2^{e,dn}\}$.

Using the above method, we created all the possible environmental contact states in a non-prehensile manipulation. However, one thing that needs to be considered in non-prehensile manipulation is that the object does not make contact with the environment when it is loaded onto the mobile robot, as shown in Figure 4a. Thus, this individual state is added with the notation $C^E = \varnothing$.

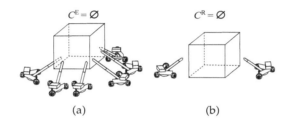

$$C^E = \varnothing \qquad\qquad C^R = \varnothing$$

(a) (b)

Figure 4. Examples of cases where (**a**) ECF is empty and (**b**) RCF is empy.

3.2. Generation of RCFs

RCF was easier to generate than ECF. The geometrical relationships between an object and the robots and the penetrations among them were not checked in this phase of planning because robots could locate flexibly around the targeted object, different from the case of generating the ECFs where geometrical relationships and penetrations should be checked because the geometry of the environment did not change.

When creating the possible RCF, only the robot contact on the object is considered; therefore, all the object's geometrical primitives are assumed to be accessible for robots. PCs between an object and the robots is given a definite label for each of the robots, such as $c_1^{r,ac}, c_2^{r,ac}, c_2^{r,ps}, \ldots c_n^{r,ac}$, where n is the number of robots. As mentioned earlier, whether the robot always acts in passive, in active, or as switchable between passive and active depends on the functions of the robot. Then, all possible PCs between an object and the robots are combined to generate RCFs. For example, when all possible PCs between an accessible object's geometrical primitive and robots are given by $c_1^{r,ac}, c_2^{r,ps}$, the possible RCFs are obtained as $C_1^R = \{c_1^{r,ac}\}$, $C_2^R = \{c_2^{r,ps}\}$ and $C_3^R = \{c_1^{r,ac}, c_2^{r,ps}\}$ by combining the PCs.

In some states, the object lay in a stable pose, and could itself keep the contact state with the environment without any support from the robot, as shown in Figure 4b. In such cases, the RCF is an empty set given as $C^R = \varnothing$, and it is added to possible RCFs.

3.3. ECF-RCF Combination

With ECFs and RCFs created, the contact states of the manipulated object are generated by matching the RCFs with ECFs. In robotic manipulations, generally, force closure is required to achieve full constraints on the targeted object, which means the non-negative combination of primitive wrenches on an object can balance any external load [23]. In the case of non-prehensile manipulation, Maeda et al. [19] and Aiyama et al. [20] proposed gravity closure, which is the force closure formed by robot contacts, environmental contacts, and gravity. External loads applied on an object can be described by wrenches, and they expand the wrench vector space [24]. If sufficient wrench vector bases are provided by the gravitational force and the contacts placed on an object, gravity closure can be achieved. However, the wrench vector bases given by contacts depend on the location and direction of the contacts determined by the configurations of the object and the robots, whereas these configurations are not concerned in the phase to generate the modes and their transitions.

Therefore, we combine an ECF and an RCF if the possible dimension of the wrench space spanned by the ECF and the RCF and the gravity satisfies the condition necessary to realize gravity closure, through which impossible modes are omitted from the viewpoint of statics.

Before calculating the dimensions of the wrench spaces spanned by the ECF and by the RCF for determining their combinations, we eliminated the infeasible combinations for a given ECF by checking whether the object's geometrical primitives were accessible to robots under any of the object's configuration. In this contact planner, we considered only the case when an object's geometrical primitives make absolute contact with the geometrical primitive of the environment; this situation would obviously disable a robot from accessing them. Therefore, for the PCs in ECF, a surface under the PC $(\text{face}, \text{face})^e$, an edge under the PC $(\text{edge}, \text{face})^e$, and a vertex under the PC $(\text{vertex}, \text{face})^e$ are not accessible to robots, as shown in Figure 5. The ECF that contains a certain geometrical primitive of the object that contacts the surface of the environment will not be considered to combine with an RCF that contains the same geometrical primitive of the object that contacts with robots.

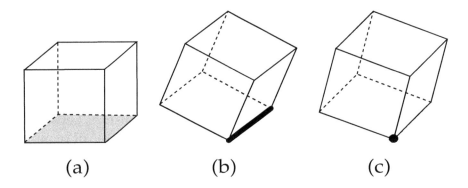

$$\qquad\text{(a)}\qquad\qquad\qquad\text{(b)}\qquad\qquad\qquad\text{(c)}$$

Figure 5. Three types of principal contacts (PCs) that a robot cannot access regardless of the object's configuration: **(a)** $(\text{face}, \text{face})^e$; **(b)** $(\text{edge}, \text{face})^e$; and **(c)** $(\text{vertex}, \text{face})^e$.

With the geometrical feasibility validated, the ECF and RCF individuals are combined according to the condition for gravity closure. Here, we considered the possible dimensions of two kind of wrench vector spaces correspondingly spanned by two kinds of forces: the passive and active forces. The passive force is applied by contacts with the environment and passive robots, and the active force is applied by contacts with the active robots and gravity. The active and passive forces were derived for the given ECFs and RCFs, respectively. Based on this, we combined the ECF and RCF, by taking into account only the derived maximum dimension of the wrench vector bases of the passive and active forces to satisfy the necessary condition of gravity closure. Configurations of the objects and the robots that meet the gravity closure will be decided in our future studies after we decide the mode transitions.

When force is applied on a contact point, the resultant wrench on the object is expressed as

$$w = \begin{bmatrix} f \\ p \times f \end{bmatrix}, \tag{1}$$

where f is the force applied to the contact, and p is the position of contact with respect to the object coordinate frame, as shown in Figure 6a. Certain manipulation studies have taken into account the case when a point contact also can resist a torque, which is called as a soft finger contact. However, we have not considered this type of contact in this paper.

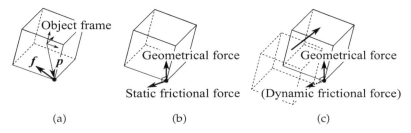

Figure 6. Force applied on a point of contact. (**a**) Definition of force. (**b**) Type of forces applied on a contact when the contact is fixed. (**c**) Types of forces applied on a contact when the contact is sliding.

There are two cases when a contact resists an external force: the geometrical case and the frictional case (Figure 6a,b). The former is the force caused on the object not to penetrate into the environment or the robot. The latter is the frictional force, and there are also two types of frictional forces: static and dynamic. We ignore dynamic friction when we consider the wrench bases to span gravity closure to ensure manipulation, as explained in the section of the problem statement.

For static friction on a point, the frictional cone, which is determined by the linear relationship between the geometrical force and the static frictional force, is usually considered when checking the conditions necessary for force closure. We dealt with static friction as decomposed basis independent of the geometrical force without considering the frictional cone (see Figure 7a) because our aim was to omit the modes that could not realize gravity closure regardless of the object's configuration. Therefore, the wrench space for the jth static or dynamic contact is expressed as

$$W_j = \langle w_{j1}, w_{j2}, w_{j3} \rangle, \quad W_j = \langle w_{j1} \rangle, \tag{2}$$

respectively, where w_{j1} is the basis for the geometrical force, and w_{j2} and w_{j3} are the bases for the static frictional force. The geometric and static frictional forces applied on edge–edge–cross contact can also be represented in the same manner (see Figure 7b). To deal with edge–face contacts and face–face contacts, we consider them as two point contacts on the edge and three non-collinear point contacts on the face (such as the boundary points of the contacting area), as shown in Figure 7c,d. The wrench space spanned by the bases is calculated by summing up the bases for all PCs in the given CF as

$$W = W_1 \cup W_2 \cup \ldots, \cup W_n, \tag{3}$$

where n is the number of bases sets on the object defined in the above manner.

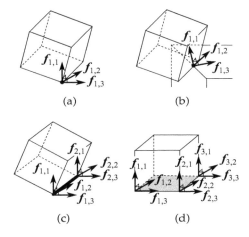

Figure 7. Wrench bases of the force applied on the static contact in the case of (**a**) point–face contact, (**b**) edge–face contact, (**c**) face–face contact, and (**d**) edge–edge–cross contact.

Then, the wrench spaces were separately determined for (a) the passive force applied by the contacts from the environment and the passive robots (denoted by W^{ps}) and (b) the active force applied by gravity and the contacts from the active robots (denoted by W^{ac}); see Figure 8. To achieve gravity closure, the wrench vectors consisting of the above forces must positively span $\mathbb{R}^6$ as follows:

$$\mathrm{pos}(W^{ps} \cup W^{ac}) = \mathbb{R}^6. \tag{4}$$

The reason why we separate the space into W^{ps} and W^{ac} is that the force applied by the environmental contacts and the passive robots cannot realize the force closure by themselves, and the active robots or the gravitational force must act to cause internal force on the object. According to the Carathéodory Theorem, if vectors in a wrench set positively span $\mathbb{R}^6$, the wrench set should contain at least seven vector frames in the six-dimensional space [24]. Thus, at least seven vector frames must exist in the wrench vector space consisting of forces applied by ECFs, RCFs, and gravity to achieve gravity closure. If the total dimension of the three kinds of wrench bases is less than seven, gravity closure would not be realized regardless of the configuration. Therefore, we define the necessary condition for gravity closure when combining ECFs and RCFs to omit the impossible modes as follows. Let $\dim(W^{ps})$ and $\dim(W^{ac})$ be the dimensions of W^{ps} and W^{ac}, respectively, then an ECF and an RCF can be combined for any configuration of the object and the robots satisfying

$$\dim(W^{ps}) + \dim(W^{ac}) \geq 7. \tag{5}$$

To determine the modes satisfying Equation (5), we calculate the maximum dimensions of W^{ps} and W^{ac} as follows:

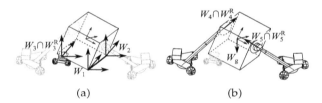

(a)   (b)

Figure 8. Wrench spaces spanned by (**a**) forces applied by the environment–object and the passive robot–object contacts and (**b**) forces applied by the active robot–object contacts and the gravitational force.

Let there be k sets of bases between an object and environment, l sets of bases between an object and passive robots, and m sets of bases between an object and active robots. Let $W_1, \ldots, W_{k+l+m}$ be the wrench space spanned by the contacts, which is derived from Equation (2), and the subscripts correspond with the above order. The wrench spaces simply spanned by the passive and active contacts and gravity are given by

$$W_c^{ps} = W_1 \cup \ldots \cup W_k \cup W_{k+1} \cup \ldots \cup W_{k+l} \tag{6}$$

and

$$W_c^{ac} = W_{k+l+1} \cup \ldots \cup W_{k+l+m} \cup W_g, \tag{7}$$

respectively, where W_g is the wrench space spanned by gravity.

The wrench space spanned by RCFs requires more consideration because not only the contacts between the object and the robots, but also the kinematics of the robots affect the spanned wrench space on the object. The limitation on the available actuators of the robot, the passive joints in the robot, or other kinematics restrictions determine the possible wrench. The possible wrench space for a given robot kinematics and its configuration is also represented by the wrench bases aside from the bases of contact. For example, Figure 9 shows the wrench bases of a robot to push an object and the force bases of a robot to carry an object; these robots have passive joints and can be used in the latter section. The robot to push an object has a basis represented as a force along with the axis of the linear actuator, as shown in Figure 9a. The robot to carry an object has three bases represented as forces passing through the axes of the passive universal joint, as shown in Figure 9b. Let W_j^R represent the wrench space spanned by the robot based on its kinematics. By substituting the wrench spaces spanned by the forces applied by robot–object contacts in Equations (6) and (7) with the wrench spaces spanned by the robots based on their kinematics, the followings equations are obtained

$$W_r^{ps} = W_1 \cup \ldots \cup W_k \cup W_{k+1}^R \cup \ldots \cup W_{k+l}^R \tag{8}$$

and

$$W_r^{ac} = W_{k+l+1}^R \cup \ldots \cup W_{k+l+m}^R \cup W_g. \tag{9}$$

Taking into account the kinematics of the robots, W^{ps} and W^{ac} are obtained by

$$W^{ps} = W_c^{ps} \cap W_r^{ps} = W_1 \cup \ldots \cup W_k \cup [(W_{k+1} \cup \ldots \cup W_{k+l}) \cap (W_{k+1}^R \cup \ldots \cup W_{k+l}^R)] \tag{10}$$

and

$$W^{ac} = W_c^{ac} \cap W_r^{ac} = W_g \cup [(W_{k+l+1} \cup \ldots \cup W_{k+l+m}) \cap (W_{k+l+1}^R \cup \ldots \cup W_{k+l+m}^R)], \tag{11}$$

respectively. By the dimension theorem for union, $\dim(W^{ps})$ and $\dim(W^{ac})$ are given by

$$\dim(W^{ps}) = \dim(W_c^{ps}) + \dim(W_r^{ps}) - \dim(W_c^{ps} \cup W_r^{ps}) \tag{12}$$

and

$$\dim(W^{ac}) = \dim(W_c^{ac}) + \dim(W_r^{ac}) - \dim(W_c^{ac} \cup W_r^{ac}), \tag{13}$$

respectively.

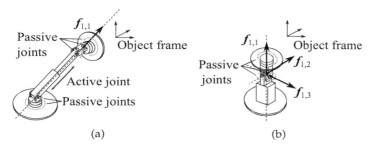

Figure 9. Examples having passive joints and their wrench bases. (**a**) Robot with five passive joints and a linear actuator. (**b**) Robots with three passive joints.

The maximum of $\dim(W^{ps})$ and $\dim(W^{ac})$ are difficult to derive analytically because the exact configuration of each contact state is not concerned in the current planning level. Fortunately, the maximum dimension of a contact state is a constant, except when the object lies in a singular configuration that causes dimensional degeneration. However, such kinds of singular configurations are caused only in specified configurations, such as a point or a line in the wrench space. Therefore, we sampled several configurations of objects and robots to calculate the maximum of $\dim(W^{ps})$ and $\dim(W^{ac})$. If one resultant dimension was smaller than the others, the corresponding configuration was designated as a singularity in the given ECF and RCF. The sampled W^{ps} and W^{ac} that provides the maximum dimension for a given ECF and RCF is considered to be the dimension of the wrench space spanned by W^{ps} and W^{ac}.

Based on the obtained $\dim(W^{ps})$ and $\dim(W^{ac})$, ECFs and RCFs are then combined as modes if the necessary condition (i.e., Equation (5)) could be met. In this way, the modes that would appear when the robots manipulate an object are created. However, this procedure does not guarantee that the generated modes will always satisfy gravity closure, but it reduces the number of modes, as shown in latter section, by omitting the impossible modes from the viewpoint of statics.

4. Generation of Transitions Among Modes

In non-prehensile manipulation planning, the transition among modes is complicated because of the multiple types of contacts and their relationships. Furthermore, to obtain feasible transitions among modes, the possibility of transitions must be well reasoned by taking into account the difference between ECF and RCF in our method.

For the ECF, we consider the transition between two ECFs is determined by geometrical restrictions; therefore, we adopted the goal-contact-relaxation (GCR) graph [15,16,22] to analyze the transition among ECFs. For the RCF, because of the difference between the active robot and the passive robot, further consideration for transition is required. In this section, we propose a method to decide the transitions among modes by taking into account the following requirements for mode transition. By comprehensively considering the transition of ECF and RCF, we can obtain the mode graph, which describes the transition between the modes and represents the manipulation action sequences.

Either ECF or RCF Changes

In non-prehensile manipulations, robots change the contacts between an object and its environment to realize the target motion. Certain manipulation planning requires that ECF and RCF changes do not occur simultaneously. Actually, simultaneous change is highly improbable in the actual manipulation. Consequently, we assume that in a mode transition process, either the RCF changes while maintaining the ECF or the RCF keeps contact to manipulate the object and to change the ECF, as shown in Figure 10a.

Connection of ECFs in GCR Graph

The possible change of contacts between an object and the environment depends on their geometry. The object's CF can transit to its neighboring CF, which connects with it in a GCR graph [15,22]. A GCR graph is a topological graph that represents the transition among the contact states of objects. Nodes that represent the contact states of the object are connected by arcs if the corresponding states can transit to each other by compliant motions. Let $\mathcal{G}$, $V(\mathcal{G})$, and $E(\mathcal{G})$ be a GCR graph, nodes, edges in the graph, respectively. Given two primitive CFs, $\widehat{C}_a \in V(\mathcal{G})$ and $\widehat{C}_b \in V(\mathcal{G})$, $\{\widehat{C}_a, \widehat{C}_b\} \in E(\mathcal{G})$ if it is possible for the two nodes to transit to each other. Though an ECF is further split into the static and dynamic in our method, we adopted a GCR graph to judge the transition between ECFs by checking whether the set of their primitive CFs is an arc in the GCR, as shown in Figure 10b. If the corresponding primitive CFs are the same, they can transit to each other.

Action of Active Robot

The active robot actively exerts contact forces to an object or releases contact from an object. Furthermore, the active robots changed the contact state in a system. In comparison, the passive robots could not exert contact to the object actively. They got contact with an object or released contact from an object when the object moved. Therefore, transitions that caused by the passive contact changes do not occur without the existence of active robots. Similarly, without the existence of active robots, ECFs cannot change except for the transition between isogenous ECFs, as shown in Figure 10c.

Object Motion

As explained above, if a contact is exerted by the environment or a passive robot, the contact changes only under the object's motion. Therefore, if the object is fully constrained by the environment and the passive robots alone, the active robots cannot move the object and the transition is not caused, except for the transition between the isogenous CFs, as shown in Figure 10d.

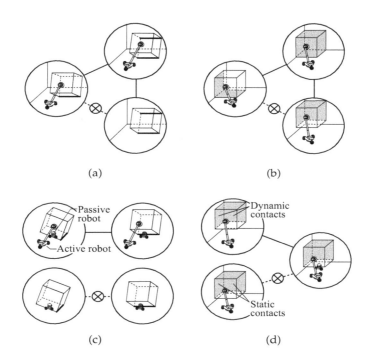

Figure 10. Examples of possible and impossible mode transitions. (**a**) Either ECF or RCF changes. (**b**) ECFs must be the same or connected in a goal-contact-relaxation (GCR) Graph. (**c**) Action of active robots causes a mode transition. (**d**) Object's motion caused change of ECFs and contacts of passive robots.

4.1. Mode Graph

With the above-mentioned requirements for transition, a mode graph, which is a graph representing the possible manipulation sequences of multiple robots, is obtained as follows. Let $\mathcal{M}$, $V(\mathcal{M})$, and $E(\mathcal{M})$ be a mode graph, nodes, and the graph edges, respectively. Let $\mathcal{G}$ be a corresponding GCR graph of ECFs with $V(\mathcal{G})$ and $E(\mathcal{G})$ as its nodes and edges, respectively. Given the two modes, $s_a = \{C_a^E, C_a^R\}, s_b = \{C_b^E, C_b^R\}$, where $s_a \in V(\mathcal{M}), s_b \in V(\mathcal{M}), \{s_a, s_b\} \in E(\mathcal{M})$, the following conditions are satisfied:

1. ECF is changed: $C_a^R = C_b^R$ and $C_a^E \neq C_b^E$:

   (a) ECFs with their primitive CFs connected in $\mathcal{G}$ are transitable under the motion of the non-fully-constrained object caused by the active robots:
   $\{\widehat{C}_a^E, \widehat{C}_b^E\} \in E(\mathcal{G})$, C_a^R and C_b^R include active PC, besides gravity, and $\dim(W^{ps}) < 6$ in either or both s_a and s_b.

   (b) Change in isogenous contacts:
   $\widehat{C}_a^E = \widehat{C}_b^E$

2. RCF is changed: $C_a^R \neq C_b^R$ and $C_a^E = C_b^E$.

   (a) The active robot is added or removed:
   $(C_a^R \backslash C_b^R) \cup (C_b^R \backslash C_a^R)$ comprises only an active PC.

   (b) The passive robot is added or removed under the motion of a non-fully-constrained object caused by the active robots:
   $(C_a^R \backslash C_b^R) \cup (C_b^R \backslash C_a^R)$ that comprise only a passive PC; C_a^R and C_b^R include an active PC, and $\dim(W^{ps}) < 6$ in either or both s_a and s_b.

   (c) Change in isogenous contacts:
   $\widehat{C}_a^R = \widehat{C}_b^R$

4.2. Cost for Transition between Modes

In the generated state transition graph, the nodes were connected by arcs if the corresponding states were able to transit to each other. When we plan the manipulation based on a mode graph, we take into account the cost for transition between the nodes, which is a value defined for each arc according to the targeted manipulation task. For example, as seen in previous section, there are five kinds of connections, including transitions between the isogenous states. They may have different costs. Transitions between the isogenous states have different costs because they are just internal transitions without changes of the contact set on the object.

Given two states, we define the cost function for the transition between them as

$$l(s_a, s_b) = k, \tag{14}$$

where k is the cost for s_a to transform into s_b. In general, the manipulation path is determined by searching for path from the initial to the final states in the graph to lower the total cost. In that case, the objective function is defined as

$$\min \sum_{i=1}^{i=n-1} l(s_i, s_{i+1}), \tag{15}$$

where s_1 is the initial mode, and s_n is the final mode.

5. Simulations

In this section, we showed two examples of mode generation based on the proposed method for the non-prehensile manipulation planning. We showed how the method narrows down the possible modes and their transitions and what manipulation sequences can be chosen from them based on the costs defined on the transitions.

In both examples, a certain number of robots were given, and here the important thing was that we needed to use the given robot to generate valid modes and to search for feasible manipulation paths, where the number of robots was enough to achieve the gravity closure. By searching for the paths with lowest total costs in the state transition graph, we obtained the least amount of manipulation action sequences. Since we only concerned about the transformation of contact sets in the mode transitions, we viewed the cost for each transition as the same. Therefore, we set the cost of transition between the two given modes as $k = 1$ if they contained different primitive states, either when the environmental contact or when any kind of robot contact changed. Otherwise, we set the cost as $k = 0$. Dijkstra's Algorithm was used to search for the shortest path from the initial state to the final state. We inhibited paths where the modes transited back and forth.

In the following examples, we show only the examples for generating modes and determining transitions, and we do not deal with the configuration space of an object–robot system. Therefore, although we show some figures in which the object and the robots are placed in specified configurations, they are only examples of configurations to help understated the resultant modes and transitions.

5.1. Example 1 and Result Discussion

In the first example, we used two types of mobile robots. The first type of robot was a pusher robot that we developed [25–27], as shown in Figure 11b. In this robot, we realized the safety manipulation that avoided the robot from falling; we did this by restricting the force that the robot could apply to the environment by using passive points. The pusher robot has a linear actuator. A face contact between the manipulator and an object acted as an active contact to change the mode of the robot–object system. The pusher robot could move to the target position by using wheels, but those wheels were lifted up and not used while the robot manipulated an object. Because of the special mechanism with passive joints, the kinematics of the robot (shown in Figure 9a) is given as explained in the previous section. See the details of the mechanism in our previous studies [25–27].

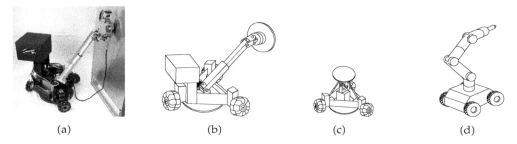

Figure 11. Mobile robots used in examples. (**a**) Pusher robot. (**b**) Schematic of a pusher robot. (**c**) Schematic of a transporter robot. (**d**) Mobile robot with a six-axis manipulator.

Another type of robot is a transporter robot, as shown in Figure 11c. A transporter robot (see Figure 9b) has a structure similar to the pusher robot, but it does not have a linear actuator and has fewer passive joints (as explained in the previous section). The transporter robot can also move to the target position by using wheels but those wheels are not used until it starts transporting the object. As a result, a transporter robot supports an object passively during manipulation, and a face contact between the robot and an object acted as a passive contact. We assume that a pusher robot and a transporter robot act as an active contact and a passive contact, respectively, and these functions are not changed during manipulation for simplicity.

In this manipulation, the task is to load a cuboid up to the transporter robots by using the pusher robots so that the object can be transferred away by the transporter robots. In the initial mode, the cuboid lies against a corner of the two adjacent walls, as shown in Figure 11a. The target final mode is the object held by the three transporter robots, as shown in Figure 11b. We consider the interactions between the cuboid and the environment when loading the cuboid up to the transporter robots by using the proposed method.

As shown in Figure 12, the faces, the edges, and the vertices of the cuboid are denoted as $f_1, f_2, \ldots, f_6, e_1, e_2, \ldots, e_{12}$, and $v_1, v_2, \ldots, v_8$, respectively. The walls are denoted as F_1 and F_2, and the floor is denoted as F_3. For simplicity, we assume that the frictional coefficient of the walls is small enough to deal with the contacts on them as friction-less contacts. Therefore, the ECFs including F_1 and F_2 are always dynamic contacts. We used six pusher robots whose geometrical primitives were denoted as $r_1, r_2, \ldots, r_6$ and three transporter robots whose geometrical primitives were denoted as r_7, r_8, r_9. For simplicity, the transporter robots and the object always made contact at the bottom of the object. Therefore, r_7, r_8, r_9 consisted of RCFs only with f_2. The initial state is given as $s_1 = \{C_1^{\mathrm{E}}, C_1^{\mathrm{R}}\}$, where $C_1^{\mathrm{E}} = \{(f_6, F_1)^{\mathrm{e,dn}}, (f_2, F_3)^{\mathrm{e,st}}, (f_1, F_2)^{\mathrm{e,dn}}\}$, and $C_1^{\mathrm{R}} = \varnothing$. The final state is given as $s_{\mathrm{G}} = \{C_{\mathrm{G}}^{\mathrm{E}}, C_{\mathrm{G}}^{\mathrm{R}}\}$, where $C_{\mathrm{G}}^{\mathrm{E}} = \varnothing$, and $C_{\mathrm{G}}^{\mathrm{R}} = \{(f_2, r_7)^{\mathrm{r,ps}}, (f_2, r_8)^{\mathrm{r,ps}}, (f_2, r_9)^{\mathrm{r,ps}}\}$.

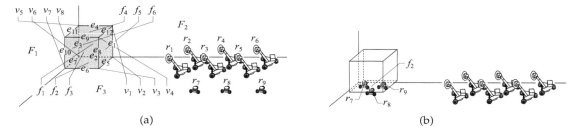

Figure 12. (**a**) Initial mode and (**b**) Target final mode. The targeted object for manipulation is a cuboid. There are two walls and the floor as the environment. We used six pusher robots and three transporter robots.

By applying the proposed method to generate the mode from the viewpoint of statics, 166,553,714 modes were generated, with the type and the corresponding number of robots determined according to the necessary condition to achieve gravity closure. As shown in Table 1, the total number of combinations between the possible ECFs and RCFs was 167,180,587, and we can see that proposed protocol reduced the number of modes. The reduction in the total number of modes was not dramatical because by considering all the possible contact between the robots and the object's geometrical primitives, there existed a huge amount of RCFs. For each ECF, our method eliminated the unavailable robot combinations. However, for most of the ECFs, the environmental constraints enabled most of the robot combinations to provide sufficient constraints to achieve the gravity closure, so the reduction of the candidates of robot combination was not really much. Further, after considering the possible geometrical primitives of the object that made contact with the robots, this reduction became less significant.

The mode transition graph was generated from the above modes by using the rules defined in the previous section. Also, as shown in Table 1, the resultant number of transitions was 8.6×10^9, where we can see that the proposed rules significantly reduced the candidates of transition between modes, which originally would be 1.4×10^{16}. We can see a dramatic reduction in the number of mode transitions here because one contact states can only change to another one under the constraints of geometrical boundary relationships and under the rules that the robot contact changes one by one. With such geometrical boundary relationships and the rules to change the robot contacts considered, a mode has only very few or even no transitable modes, whereas all the other modes will be its transitable modes and a large number of meaningless transitions will be caused if our method is not adopted.

Finally, by searching the mode graph, a total of 27,216 paths which minimized the cost were obtained in the graph.

Table 1. Comparison among the various parameters obtained by Example 1.

	Before Adopting Our Method	After Adopting Our Method
Generated modes	167,180,587	166,553,714
Number of mode transitions	1.4×10^{16}	8.6×10^9

Table 2 shows one of the obtained paths with the lowest cost in the mode graph. As mentioned above, the paths in the modes were decided without planning the exact configuration of robots. Therefore, the configurations of the object and the robots shown in Table 2 are examples of the modes. In the path shown in Table 2, the object was pushed by a pusher robot and rotated about the object's edge e_5 alongside the wall F_1. A transporter robot was then inserted underneath it. The pusher robot then moved the object so that only the object's vertex v_3 was in contact with the floor F_3, and another transporter robot was inserted underneath the object. Finally, with two transporter robots at the bottom, the object lost contact with the floor F_3 through contact with the pusher robot, which allowed the third transporter robot to enter underneath the object. Thus, the object was finally loaded onto three transporter robots.

We can see that the obtained path is reasonable. From the viewpoint of statics and geometry, it was also reasonable to include the paths in which only the combinations of geometrical primitives were different. The resultant paths also contained those paths that were impossible to realize from the viewpoint of the force balance determined by the configurations of the object and the robots. Those paths will be omitted in the later planning phase when deciding the object–robot system's exact configurations based on the transitions of the modes; they will not be dealt with in the current phase where only the transitions of the modes are decided. The important result is that the candidates of the mode transposition were narrowed down, as shown Table 1, and this facilitated the planning to determine the configurations.

Table 2. One of the manipulation paths that minimized the cost.

Image	ECF and RCF in the Mode
	$C_1^E = \{(f_6, F_1)^{e,dn}, (f_1, F_2)^{e,dn}, (f_2, F_3)^{e,st}\}$ $C_1^R = \varnothing$
	$\Downarrow \quad l(s_1, s_2) = 1$
	$C_2^E = \{(f_6, F_1)^{e,dn}, (f_1, F_2)^{e,dn}, (f_2, F_3)^{e,st}\}$ $C_2^R = \{(f_3, r_1)^{r,ac}\}$
	$\Downarrow \quad l(s_2, s_3) = 1$
	$C_3^E = \{(f_1, F_2)^{e,dn}, (e_5, F_3)^{e,st}\}$ $C_3^R = \{(f_3, r_1)^{r,ac}\}$
	$\Downarrow \quad l(s_3, s_4) = 1$
	$C_4^E = \{(f_1, F_2)^{e,dn}, (e_5, F_3)^{e,st}\}$ $C_4^R = \{(f_3, r_1)^{r,ac}, (f_2, r_7)^{r,ps}\}$
	$\Downarrow \quad l(s_4, s_5) = 1$
	$C_5^E = \{(v_3, F_3)^{e,st}\}$ $C_5^R = \{(f_3, r_1)^{r,ac}, (f_2, r_7)^{r,ps}\}$
	$\Downarrow \quad l(s_5, s_6) = 1$
	$C_6^E = \{(v_3, F_3)^{e,st}\}$ $C_6^R = \{(f_3, r_1)^{r,ac}, (f_2, r_7)^{r,ps}, (f_2, r_8)^{r,ps}\}$
	$\Downarrow \quad l(s_6, s_7) = 1$
	$C_7^E = \varnothing$ $C_7^R = \{(f_3, r_1)^{r,ac}, (f_2, r_7)^{r,ps}, (f_2, r_8)^{r,ps}\}$
	$\Downarrow \quad l(s_7, s_8) = 1$
	$C_8^E = \varnothing$ $C_8^R = \{(f_3, r_1)^{r,ac}, (f_2, r_7)^{r,ps}, (f_2, r_8)^{r,ps}, (f_2, r_9)^{r,ps}\}$
	$\Downarrow \quad l(s_8, s_G) = 1$
	$C_G^E = \varnothing$ $C_G^R = \{(f_2, r_7)^{r,ps}, (f_2, r_8)^{r,ps}, (f_2, r_9)^{r,ps}\}$

5.2. Example 2 and Result Discussion

In the second example, a cuboid is loaded onto a step by using mobile robots, as shown in Figure 13. The faces, the edges, and the vertices of the cuboid are denoted as $f_1, f_2, \ldots, f_6, e_1, e_2, \ldots, e_{12}$, and $v_1, v_2, \ldots, v_8$, respectively. The floor is denoted as F_1 and the faces consisting of the step are denoted as F_2 and F_3.

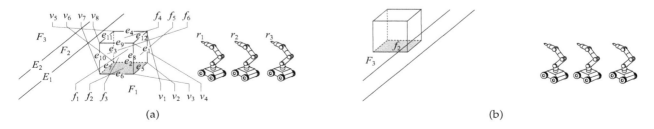

(a)   (b)

Figure 13. (a) Initial mode and (b) target final mode. Targeted object for manipulation is cuboid. There is a step in the environment. We used six pusher robots to bring the cuboid on the step.

A mobile robot with a standard six-axis manipulator is shown in Figure 11d. The robot generates a frictional point of contact with the object. We used three robots and denoted their geometrical primitives as r_1, r_2, and r_3. There were no obstacles around the step; therefore, the manipulation could be performed by either using or not using the environment. By using the proposed method, we generated the modes with the necessary number of robots to achieve gravity closure. Then, in the created mode graph, we took the initial state as $s_1 = \{\{(f_2, F_1)^{e,st}\}, \varnothing\}$ and the final state as $s_G = \{\{(f_2, F_3)^{e,st}\}, \varnothing\}$ and searched for the manipulation paths with the minimum total cost.

Using the procedure given in the previous example, we generated 292,415 modes and 2.9×10^6 transitions. The total number of combinations of the ECFs and RCFs was 299,967, and the total number of combinations of the generated modes was 4.5×10^{10}, as shown in Table 3. For the same reason we analyzed in Example 1, our method also narrowed down the number of possible modes, and especially, dramatically decreased the transitions between them.

Table 3. Comparison among the various parameters obtained by Example 2.

	Before Adopting Our Method	After Adopting Our Method
Generated modes	299,967	292,415
Number of mode transitions	4.5×10^{10}	2.9×10^6

Our search produced 672,352 paths with a minimum total cost. In these obtained paths, the number of robots involved varied. In the two paths shown in Table 4, we found that the number of robots engaged in the two paths were different.

In Path 1, only two robots were needed to perform the manipulation task whereas three robots were needed in Path 2 because the floor was used to exert constraints on the object. Therefore, fewer robots were needed in Path 1. We could choose the path according to some other criteria based on the obtained paths. For example, the manipulation action sequences obtained from Path 1 would be more preferable than those obtained from Path 2 because fewer robots were adopted. Besides, as compared with the previous example, relative sliding occured on the face contact principals in the action sequences of Path 1 between the mode $s_5 = \{\{(e_5, F_1)^{e,dn}\}, \{(f_3, r_1)^{r,ac}, (f_5, r_2)^{r,ac}\}\}$ and the mode $s_6 = \{\{(e_5, F_1)^{e,dn}, (f_2, E_2)^{e,dn}\}, \{(f_3, r_1)^{r,ac}, (f_5, r_2)^{r,ac}\}\}$; therefore, the transition from static contact to dynamic contact appeared between the mode $s_4 = \{\{(e_5, F_1)^{e,st}\}, \{(f_3, r_1)^{r,ac}, (f_5, r_2)^{r,ac}\}\}$ and s_5. A similar transition can also be seen in Path 2.

In the non-prehensile manipulation by multiple mobile robots, there were many possible manipulation sequences each with a difference number of robots. As shown in above example, our method can list up possible patterns of manipulation taking into account their possibility from the viewpoint of statics.

Table 4. Two paths of the manipulation paths which minimized the cost.

Path 1		Path 2	
Image	ECF and RCF in the Mode	Image	ECF and RCF in the Mode
	$C_1^E = \{(f_2, F_1)^{e,st}\}$ $C_1^R = \varnothing$		$C_1^E = \{(f_2, F_1)^{e,st}\}$ $C_1^R = \varnothing$
	$\Downarrow \quad l(s_1, s_2) = 1$		$\Downarrow \quad l(s_1, s_2) = 1$
	$C_2^E = \{(f_2, F_1)^{e,st}\}$ $C_2^R = \{(f_3, r_1)^{r,ac}\}$		$C_2^E = \{(f_2, F_1)^{e,st}\}$ $C_2^R = \{(f_3, r_1)^{r,ac}\}$
	$\Downarrow \quad l(s_2, s_3) = 1$		$\Downarrow \quad l(s_2, s_3) = 1$
	$C_3^E = \{(e_5, F_1)^{e,st}\}$ $C_3^R = \{(f_3, r_1)^{r,ac}\}$		$C_3^E = \{(f_2, F_1)^{e,st}\}$ $C_3^R = \{(f_3, r_1)^{r,ac}, (f_5, r_2)^{r,ac}\}$
	$\Downarrow \quad l(s_3, s_4) = 1$		$\Downarrow \quad l(s_3, s_4) = 1$
	$C_4^E = \{(e_5, F_1)^{e,st}\}$ $C_4^R = \{(f_3, r_1)^{r,ac}, (f_5, r_2)^{r,ac}\}$		$C_4^E = \{(f_2, F_1)^{e,st}\}$ $C_4^R = \{(f_3, r_1)^{r,ac}, (f_5, r_2)^{r,ac}, (f_1, r_3)^{r,ac}\}$
	$\Downarrow \quad l(s_4, s_5) = 0$		$\Downarrow \quad l(s_4, s_5) = 1$
	$C_5^E = \{(e_5, F_1)^{e,dn}\}$ $C_5^R = \{(f_3, r_1)^{r,ac}, (f_5, r_2)^{r,ac}\}$		$C_5^E = \varnothing$ $C_5^R = \{(f_3, r_1)^{r,ac}, (f_5, r_2)^{r,ac}, (f_1, r_3)^{r,ac}\}$
	$\Downarrow \quad l(s_5, s_6) = 1$		$\Downarrow \quad l(s_5, s_6) = 1$
	$C_6^E = \{(e_5, F_1)^{e,dn}, (f_2, E_2)^{e,dn}\}$ $C_6^R = \{(f_3, r_1)^{r,ac}, (f_5, r_2)^{r,ac}\}$		$C_6^E = \{(f_2, F_3)^{e,dn}\}$ $C_6^R = \{(f_3, r_1)^{r,ac}, (f_5, r_2)^{r,ac}, (f_1, r_3)^{r,ac}\}$
	$\Downarrow \quad l(s_6, s_7) = 1$		$\Downarrow \quad l(s_6, s_7) = 0$
	$C_7^E = \{(f_2, E_2)^{e,dn}\}$ $C_7^R = \{(f_3, r_1)^{r,ac}, (f_5, r_2)^{r,ac}\}$		$C_7^E = \{(f_2, F_3)^{e,st}\}$ $C_7^R = \{(f_3, r_1)^{r,ac}, (f_5, r_2)^{r,ac}, (f_1, r_3)^{r,ac}\}$
	$\Downarrow \quad l(s_7, s_8) = 1$		$\Downarrow \quad l(s_7, s_8) = 1$
	$C_8^E = \{(f_2, F_3)^{e,st}\}$ $C_8^R = \{(f_3, r_1)^{r,ac}, (f_5, r_2)^{r,ac}\}$		$C_8^E = \{(f_2, F_3)^{e,st}\}$ $C_8^R = \{(f_3, r_1)^{r,ac}, (f_5, r_2)^{r,ac}\}$
	$\Downarrow \quad l(s_8, s_9) = 1$		$\Downarrow \quad l(s_8, s_9) = 1$
	$C_9^E = \{(f_2, F_3)^{e,st}\}$ $C_9^R = \{(f_3, r_1)^{r,ac}\}$		$C_9^E = \{(f_2, F_3)^{e,st}\}$ $C_9^R = \{(f_3, r_1)^{r,ac}\}$
	$\Downarrow \quad l(s_9, s_G) = 1$		$\Downarrow \quad l(s_9, s_G) = 1$
	$C_G^E = \{(f_2, F_3)^{e,st}\}$ $C_G^R = \varnothing$		$C_G^E = \{(f_2, F_3)^{e,st}\}$ $C_G^R = \varnothing$

6. Conclusions

In this paper, we proposed the modal planning method for multi-contact non-prehensile manipulation using multiple mobile robots. After defining a mode as a set of configurations that hold the same contact state and investigating the transition between modes, a manipulation planner can efficiently probe the configuration space even if the states under varying constraints are difficult to sample in the configuration space. When multiple mobile robots manipulate the object using non-prehensile methods, the modes and their consequent transitions become enormous because of the numerous contacts made by the environment and the robots on the object. For such situations, we proposed a method to narrow down the possible modes and their transitions beforehand by excluding the invalid modes and transitions. In our proposed method, we generated modes that described an object's contact states with the robots and the environment while ignoring their exact configurations, provided each multi-contact set satisfied the necessary condition for gravity closure on the object (along with gravity). Secondly, we investigated the valid transition between the modes by taking into account whether the given robot could actively change an object's contact state under feasible geometrical relationships. Finally, we conducted two simulations on specific manipulation tasks to validate our method and confirmed that the number of modes and transitions had significantly reduced. Also, it was feasible to obtain the sequence of modes obtained by searching the shortest path.

Our method can be adopted to probe the modal spaces in the variant cases of non-prehensile manipulation by mobile robots. If prior sequencing manipulation actions by adopting the generated mode transition graph, the manipulation planner can avoid the heavy computation of searching in a whole large configuration space. Thus, determining the sequence of modes is usually the first hierarchy in manipulation planning, where we do not consider the exact configuration of a system that comprises objects and robots. However, these configurations must be determined to complete manipulation planning by considering certain factors, such as the achievement force closure, the movability of the object, and accessibility for mobile robots. We can obtain those configurations based on the prior determined modal space, by applying sample-based methods to each modal space. This will be significantly more efficient than directly applying sample-based methods to probe a whole configuration space. In our future studies, we propose to further investigate this topic.

Author Contributions: Conceptualization, J.O., S.S. and C.F.; Methodology, C.F., S.S. and J.O.; Software, C.F. and S.S.; Validation, C.F. and S.S.; Data curation, C.F. and S.S.; Formal analysis, C.F., S.S. and J.O.; Investigation, C.F., S.S. and J.O.; Resources, J.O.; Writing–original draft preparation, C.F. and S.S.; Writing–review and editing, C.F., S.S. and J.O.; Visualization, S.S. and C.F.; Supervision, S.S. and J.O.; Project administration, J.O.; Funding acquisition, J.O.

Acknowledgments: Changxiang Fan is financially supported by China Scholarship Council (CSC).

Abbreviations

The following abbreviations are used in this manuscript:

PC   principal contact
CF   contact formation
ECF   environmental contact formation
RCF   robot contact formation
GCR   goal-contact-relaxation

References

1. Stilwell, D.J.; Bay, J.S. Toward the development of a material transport system using swarms of ant-like robots. In Proceedings of the IEEE International Conference on Robotics and Automation, Atlanta, GA, USA, 2–6 May 1993; pp. 766–771.
2. Mason, M.T. *Mechanics of Robotic Manipulation*; MIT Press: Cambridge, MA, USA, 2001.
3. Ruggiero, F.; Lippiello, V.; Siciliano, B. Nonprehensile Dynamic Manipulation: A Survey. *IEEE Robot. Autom. Lett.* **2018**, *3*, 1711–1718. [CrossRef]
4. Arai, T.; Ota, J. Let us work together-Task planning of multiple mobile robots. In Proceedings of the 1996 IEEE/RSJ International Conference on Intelligent Robots and Systems' 96, Osaka, Japan, 8 November 1996; Volume 1, pp. 298–303.
5. Latombe, J.C. *Robot Motion Planning*; Springer US: Boston, MA, USA, 1991; Volume 124.
6. LaValle, S.M. *Rapidly-Exploring Random Trees: A New Tool for Path Planning*; Technical Report TR98-11; Computer Science Department, Iowa State University, IA, USA, 1998.
7. Kavraki, L.; Svestka, P.; Latombe, J.; Overmars, M.H. Probabilistic Roadmaps for Path Planning in High-Dimensional Configuration Spaces. *IEEE Trans. Robot. Autom.* **1996**, *12*, 566–588. [CrossRef]

8. Hauser, K.; Latombe, J.C. Multi-modal motion planning in non-expansive spaces. *Int. J. Robot. Res.* **2010**, *29*, 897–915. [CrossRef]

9. Hsu, D.; Kavraki, L.E.; Latombe, J.C.; Motwani, R.; Sorkin, S. On finding narrow passages with probabilistic roadmap planners. In *Proceedings of the Workshop on the Algorithmic Foundations of Robotics, Robotics: The Algorithmic Perspective*; A. K. Peters, Ltd.: Natick, MA, USA, 1998; pp. 141–154.

10. Maeda, Y.; Arai, T. Planning of graspless manipulation by a multifingered robot hand. *Adv. Robot.* **2005**, *19*, 501–521. [CrossRef]

11. Miyazawa, K.; Maeda, Y.; Arai, T. Planning of graspless manipulation based on rapidly-exploring random trees. In Proceedings of the 6th IEEE International Symposium on Assembly and Task Planning: From Nano to Macro Assembly and Manufacturing, Montreal, QC, Canada, 19–21 July 2005; pp. 7–12.

12. Barry, J.; Kaelbling, L.P.; Lozano-Pérez, T. A hierarchical approach to manipulation with diverse actions. In Proceedings of the 2013 IEEE International Conference on Robotics and Automation, Karlsruhe, Germany, 6–10 May 2013; pp. 1799–1806.

13. Lee, G.; Lozano-Pérez, T.; Kaelbling, L.P. Hierarchical planning for multi-contact non-prehensile manipulation. In Proceedings of the 2015 IEEE/RSJ International Conference on Intelligent Robots and Systems (IROS), Hamburg, Germany, 28 September–2 October 2015; pp. 264–271.

14. Schmitt, P.S.; Neubauer, W.; Feiten, W.; Wurm, K.M.; Wichert, G.V.; Burgard, W. Optimal, sampling-based manipulation planning. In Proceedings of the 2017 IEEE International Conference on Robotics and Automation (ICRA), Singapore, 29 May–3 June 2017; pp. 3426–3432.

15. Xiao, J. Goal-contact relaxation graphs for contact-based fine motion planning. In Proceedings of the 1997 IEEE International Symposium on Assembly and Task Planning, Marina del Rey, CA, USA, 7–9 August 1997; pp. 25–30.

16. Xiao, J.; Ji, X. Automatic generation of high-level contact state space. *Int. J. Robot. Res.* **2001**, *20*, 584–606. [CrossRef]

17. Fan, C.; Shirafuji, S.; Ota, J. Least Action Sequence Determination in the Planning of Non-Prehensile Manipulation with Multiple Mobile Robots. In *the 15th International Conference on Intelligent Autonomous Systems (IAS–15)*; Springer: Cham, Switzerland, 2018; pp. 174–185.

18. Xiao, J.; Zhang, L. A General Strategy to Determine Geometrically Valid Contact Formations from Possible Contact Primitives. In Proceedings of the IEEE International Conference on Robotics And Automation, Atlanta, GA, USA, 2–6 May 1993; Volume 1, p. 2728.

19. Maeda, Y.; Aiyama, Y.; Arai, T.; Ozawa, T. Analysis of object-stability and internal force in robotic contact tasks. In Proceedings of the IEEE/RSJ International Conference on Intelligent Robots and Systems, Osaka, Japan, 8 November 1996; Volume 2, pp. 751–756.

20. Aiyama, Y.; Arai, T.; Ota, J. Dexterous assembly manipulation of a compact array of objects. *CIRP Ann.* **1998**, *47*, 13–16. [CrossRef]

21. LaValle, S.M. *Planning Algorithms*; Cambridge University Press: Cambridge, UK, 2006.

22. Ji, X.; Xiao, J. Planning motions compliant to complex contact states. *Int. J. Robot. Res.* **2001**, *20*, 446–465. [CrossRef]

23. Ponce, J.; Faverjon, B. On computing three-finger force-closure grasps of polygonal objects. *IEEE Trans. Robot. Autom.* **1995**, *11*, 868–881. [CrossRef]

24. Murray, R.M.; Li, Z.; Sastry, S.S. *A Mathematical Introduction to Robotic Manipulation*; CRC Press: Boca Raton, FL, USA, 1994.

25. Shirafuji, S.; Terada, Y.; Ota, J. Mechanism Allowing a Mobile Robot to Apply a Large Force to the Environment. In *The 14th International Conference on Intelligent Autonomous Systems (IAS–14)*; Springer: Cham, Switzerland, 2016; pp. 795–808.

26. Shirafuji, S.; Terada, Y.; Ito, T.; Ota, J. Mechanism allowing large-force application by a mobile robot, and development of ARODA. *Robot. Auton. Syst.* **2018**, *110*, 92–101. [CrossRef]

27. Ito, T.; Shirafuji, S.; Ota, J. Development of a Mobile Robot Capable of Tilting Heavy Objects and its Safe Placement with Respect to Target Objects. In Proceedings of the 2018 IEEE International Conference on Roboics and Biomimetics (ROBIO2018), Kuala Lumpur, Malaysia, 12–15 December 2018; pp. 716–722.

11

Topological Map Construction based on Region Dynamic Growing and Map Representation Method

Fei Wang *, Yuqiang Liu, Ling Xiao, Chengdong Wu and Hao Chu

Faculty of Robot Science and Engineering, Northeastern University, Shenyang 110819, China; yuqiang0616@163.com (Y.L.); 1870719@stu.neu.edu.cn (L.X.); wuchengdong@mail.neu.edu.cn (C.W.); chuhao@mail.neu.edu.cn (H.C.)
* Correspondence: wangfei@mail.neu.edu.cn.

Featured Application: It is suitable for robots in the human-machine collaboration category, especially service robots.

Abstract: In the human–machine interactive scene of the service robot, obstacle information and destination information are both required, and both kinds of information need to be saved and used at the same time. In order to solve this problem, this paper proposes a topological map construction pipeline based on regional dynamic growth and a map representation method based on the conical space model. Based on the metric map, the construction pipeline can initialize the region growth point on the trajectory of the mobile robot. Next, the topological region is divided by the region dynamic growth algorithm, the map structure is simplified by the minimum spanning tree, and the similar region is merged by the region merging algorithm. After that, the parameter TM (topological information in the map) and the parameter OM (occupied information in the map) are used to represent the topological information and the occupied information. Finally, a topological map represented by the colored picture is saved by converting to color information. It is highlighted that the topological map construction pipeline is not limited by the structure of the environment, and can be automatically adjusted according to the actual environment structure. What's more, the topological map representation method can save two kinds of map information at the same time, which simplifies the map representation structure. The experimental results show that the map construction method is flexible, and that resources such as calculation and storage are less consumed. The map representation method is convenient to use and improves the efficiency of the map in preservation.

Keywords: human–machine interactive navigation; mobile robot; topological map; regional growth

1. Introduction

Human–machine interactive navigation refers to the process through which machines and operator cooperate with each other to control the movement of devices and realize interactive navigation [1]. This type of navigation has been widely used in indoor service robots [2,3], self-driving cars [4,5], automatic guided vehicles (AGV) [6,7], and so on [8,9]. Among them, the most important part between the operator and machines is the map. The map needs to combine continuous spatial environment information with human abstract intentions, and discover the relationships between intentions and real spatial areas; finally, these connections are abstracted into a sequence of events (that is, a topological map) [10].

Many researchers began to study topological maps very early. Nodes are jointly represented by the sector feature of a laser and a proportional invariant feature of vision, which do not depend on any artificial landmark, and the global location of robots in the process of map creation [11]. However, the laser sector feature is required in the intersection area among different channels, and the visual feature is also required to be proportional invariant. Therefore, the application effect will be affected in such large-scale and featureless scenarios as a living room. The final topological map is sparse, and cannot achieve accurate navigation obstacle avoidance [12]. A self-organizing method of hierarchical clustering (Map-TreeMaps) is proposed in [13]. Each unit of the map represents the structured data of the tree, while the treemap method provides a global view of the local hierarchy. This method enhances the generality of the construction of topological maps and solves the problem that some environmental geometric structures are constrained. Similar to the Chow–Liu tree model in [14], as long as the number of searching layers is big enough, various spatial structures can be detected and clustered separately. In addition, the segmentation results can be easily optimized by thresholding the weights of local subgraphs. Both methods abandon the details of the local subgraph, which makes it difficult to achieve accurate navigation obstacle avoidance. An auxiliary graph is used to solve the problem of local subgraph association in [15], which improves the efficiency of segmentation and doesn't solve the problem of subgraph details.

Moreover, the above clustering segmentation cannot deal with large-scale and open space, which is not human-friendly for human–machine interactive navigation. For example, in the application of service robots, large living rooms and long corridors cannot be regarded as a region, but should be divided into several regions according to the actual situation [16]. A novel and efficient method for updating Voronoi diagrams was proposed in [17], which only updates those units that are actually affected by the environment, and finally lower the number of visits and computing time. In addition, a skeleton-based Voronoi diagram method is also proposed, which is particularly effective for noise removal. A simultaneous location and mapping algorithm (VorSLAM) based on Voronoi map representation is proposed in [18]. One of the basic features of this algorithm is that the features correspond to the local map one by one, and each feature is associated with a local map defined on the feature. This not only retains the details of the local map, it also alleviates the problem of large scene segmentation by Voronoi partitioning. However, the Voronoi graph is based on the principle of distance or special structure to divide the space, so that the area obtained is "basically the same size", and the special structure also limits the applicability of the algorithm.

From the research history of topological map construction, a generalized Voronoi map [19] and spectral clustering [20] are the two main methods for topological map construction at present [21]. For example, a lightweight method is proposed to create maps by combining metric maps and topological information in [22]. By combining the information of two maps, the robot can realize autonomous navigation and obstacle avoidance in a large area. In paper [23], spectral clustering and an extended Voronoi graph are used to construct a topological graph from a metric graph. The specific idea is to use spectral clustering to segment the metric graph and get the center of the cluster. After determining the first vertex, other vertices are established by an extended Voronoi graph, and vertices are divided into connection points and tail nodes. Although the combination of two maps achieves navigation avoidance, the difficulty of map preservation is increased; the combination of two segmentation methods improves the efficiency and scope of application of topological segmentation, and the redundancy and uncontrollability of segmentation results is coming, which easily leads to the accumulation of topological vertices in some areas.

To solve the above problems, this paper proposes a novel pipeline of constructing a topological map and an efficient way of expressing a topological map. The main contributions of this paper are as follows:

- A complete system of building, saving, and loading topographic maps is proposed, which can make the topographic maps readily applied.
- A topological segmentation method based on region dynamic growth is proposed, which makes the region segmentation no longer limited by the geometrical structure of the environment, and also more in line with the actual needs of human–machine interactive navigation.

- A representation method of a topological map based on the conical space model is proposed, which makes the map retain not only the information of the topological relationship, but also the information regarding the obstacle occupied.

Finally, several comparative experiments are carried out in the Gazebo simulator provided by ROS (Robot Operating System, ROS) and the author's lab. The experimental results verify the effectiveness of the proposed system and method. The overall block diagram of the topological map building system is shown in Figure 1.

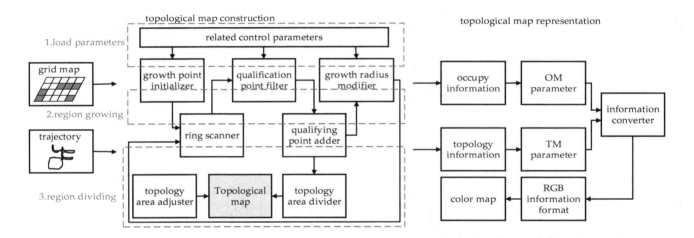

Figure 1. The architecture of the topological map construction system.

2. Topological Map Construction Based on Regional Dynamic Growth

The process of building a topological map based on region dynamic growth is shown in the left part of Figure 1. The system first randomly selects the initial growth point on the trajectory of the robot, and then grows dynamically according to the control parameters. When the regional characteristics meet the requirements, the growth will stop. This method is inspired by the spherical expansion of voxel clusters in paper [24]. Then, after each region has grown, it will be segmented from the meter map, and the color identification of the topological region will be established. The center of gravity of the topological region will be used as the center of the region, and the neighborhood of the region will be scanned. Finally, after all the topological areas have been established, the region adjustment part will delete, merge, and grow the unqualified areas, and finally complete the construction of the whole environment's topological map.

2.1. Regional Growth Process

Based on the metric map and the trajectory of the robot, the map building system will randomly initialize the region growing point on the trajectory (which is in the same coordinate system as the metric map). It is worth mentioning that this point is not the final center of the region (also called the topological vertex). Then, the circular growth of the region is achieved by using a circular scanner with the same degree of growth in all directions on the plane, rather than depending on the geometric structure or special characteristics of the environment.

There are several important parameters in the regional dynamic growth algorithm, which are described below:

- r_a: The points' addition ratio in the circular scan, which refers to the ratio of qualified points in each circular scan. It can prevent malformation, making the growth area approximately circular or elliptical rather than elongated.
- r_o: The obstacle ratio in the circular scan, which refers to the ratio of unqualified points to the total qualified points in the convex hull region composed of qualified points in the added points

and areas in each annular scanning. It can prevent the region from growing into a concave region, and ensure the convexity of the topological region.

■ r_p: The points pass ratio of the topological region, which refers to the ratio of all the points in the topological region to all the points in the circular region (ideal region). It can reflect the contrast between the growth area and the ideal area in the current state. It is used to modify the growth radius in real time, so its role is to prevent growth deficiency.

■ R_w: The control weight of dynamic modification of the growth radius, which refers to the extent of radius modification in each region.

The concept of the qualified point is mentioned above. If one of them is not satisfied, it is called an unqualified point. The qualified point needs to satisfy all the following conditions:

1. The metric map area that corresponded to the point must be in a free space, and cannot be an occupied space or an unknown space.

2. The topological map area corresponding to the point must be a non-topological identifier area, which can only be a free space or a local area identifier.

3. Conditions 1 and 2 must be satisfied for all points through which the ray emitted from the region vertex passes.

As the location of the robot's trajectory must be free and can basically traverse the whole environment, it is better to sample the growing points from the robot's trajectory. Then, the main steps of the region dynamic growth algorithm are as follows (Figure 2, the symbols in the Figure 2d will be described later):

(1) Determine the growth point and scan all the adjacent points around the current region vertex in a circular way.

(2) Rays from the growth point to each point were calculated using the Bresenham algorithm [25], and all of the qualified points were screened.

(3) Calculate the convex hull after each qualified point is added using the Graham scanning algorithm [26], and retain the point where the obstacle ratio satisfies the requirements.

(4) Add qualifying points to the topological area. At the same time, the points pass ratio and the points add ratio in the current state are calculated. If the points' addition ratio is less than the threshold, the growth of the current region is exited, indicating that the region growth is completed ahead of time. If the addition rate is normal, the points pass ratio will be used to modify the maximum growth radius. The modified formula is shown in Equation (1):

$$R_{max}^{t+1} = R_{max}^t + R_w \cdot r_p \cdot R_{max}^t \tag{1}$$

where R_{max}^t is the maximum growth radius at the current time t, and R_{max}^{t+1} is the maximum growth radius at the time $t + 1$.

After that, the regional dynamic growth algorithm will grow in other areas in the same way until the topological area covers most of the environment.

The center of each region will be updated after the end of each region growth. This paper considers that every topological region is an irregular convex polygon composed of convex hull points in the region, and the polygon is approximately circular or elliptic, so the center of gravity of the polygon can be used as a new center.

As shown in (d) of Figure 2, the polygon is considered to be composed of multiple triangles of area $S(S_0, S_1, S_2, \dots)$, where p_c represents a vertex and $p_0p_1, p_1p_2, p_2p_3, \dots$ represent the other two points in triangles. Then, if the coordinates of three vertices of the triangle $\triangle p_1p_2p_3$ are known: $p_1(x_1, y_1), p_2(x_2, y_2), p_3(x_3, y_3)$, the common center of gravity coordinates can be obtained, as shown in Equation (2).

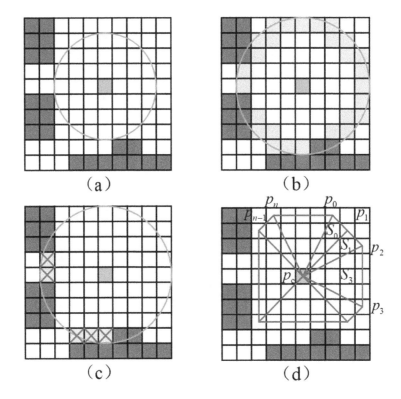

(a)   (b)

(c)   (d)

Figure 2. Main steps of the regional dynamic growth algorithm. (**a**) Growth state at a certain moment. (**b**) The next moment is grown in a circular expansion manner to obtain adjacent candidate points. (**c**) Remove the unqualified points in the adjacent candidate points. (**d**) Add qualified candidate points to the topological region.

However, Equation (2) is not suitable for multi-triangle calculation, so the triangle area calculation method is used in this paper. Firstly, the area of the triangle is calculated by vector cross-multiplication, as shown in Equation (3). In addition, it is not necessary to consider whether the traversal order of the three points is clockwise or counter-clockwise in the calculation process, because it can be cancelled in the subsequent calculations.

$$\begin{cases} x_g = \frac{x_1+x_2+x_3}{3} \\ y_g = \frac{y_1+y_2+y_3}{3} \end{cases} \tag{2}$$

$$S = \frac{(x_2 - x_1) * (y_3 - y_1) - (x_3 - x_1) * (y_2 - y_1)}{2} \tag{3}$$

Assuming that an irregular convex polygon is composed of n triangles, in which each triangle has an area of S_i and a center of gravity of $G_i(x_i, y_i)$, the integral can be transformed into an accumulation sum by using the formula for calculating the center of gravity of a planar thin plate, as shown in Equation (4):

$$\begin{cases} x_g = \frac{\iint_D x dS}{S} = \frac{\sum\limits_{i=1}^{n} x_i S_i}{\sum\limits_{i=1}^{n} S_i} \\ y_g = \frac{\iint_D y dS}{S} = \frac{\sum\limits_{i=1}^{n} y_i S_i}{\sum\limits_{i=1}^{n} S_i} \end{cases} \tag{4}$$

Then, in any irregular convex n polygon $(p_0, p_1, p_2, \ldots, p_n)$ with $p_i(x_i, y_i)(i = 0, 1, 2, 3, \ldots, n)$ as a vertex, it is divided into n triangles composed of a center point $p_c(x_c, y_c)$ as a vertex and any two points p_i. Then, the coordinates of the center of gravity of the polygon are calculated as shown in the following Equation (5):

$$\begin{cases} x' = \dfrac{\sum\limits_{i=1}^{n}(x_{\mathrm{c}}+x_i+x_{i-1})S_i}{3\sum\limits_{i=1}^{n}S_i} \\[4mm] y' = \dfrac{\sum\limits_{i=1}^{n}(y_{\mathrm{c}}+y_i+y_{i-1})S_i}{3\sum\limits_{i=1}^{n}S_i} \end{cases} \tag{5}$$

Finally, the principal semi-axial length and the short semi-axis length of the region are calculated by principal component analysis. The pseudo code of the regional dynamic growth algorithm is shown in Algorithm 1.

Algorithm 1: Regional dynamic growth

Objectives: The growing point is regarded as a temporary topological vertex, and then all the neighboring points are scanned annularly to select eligible points and add them to the current topological region.

Input: Growth point $p_{\mathrm{c}}(x,y)$, maximum growth radius $R_{\max}$.

Output: Central point p'_c of the topological region, radius R'_{new} of the topological region.

1:    initial growth radius r
2:    initialization of eligible point set $P_{\mathrm{p}}=\{\}$
3:    initialization of unqualified point set $P_{\mathrm{f}}=\{\}$
4:    **for** $(r=1;r<R_{\max};r++)$
5:        $P_{\mathrm{t}}=\{\}$
6:        $P_{\mathrm{t}} \leftarrow$ ring scanner (p_c, r)
7:        **for each** p_{i} **in** P_{t}
8:            **if** $(p_{\mathrm{i}}$ is failure$)$
9:                $P_{\mathrm{f}} \leftarrow$ fail point filter (p_{i})
10:           **continue**
11:        $P_{\mathrm{r}} \leftarrow \{\}$
12:        $P_{\mathrm{r}} \leftarrow$ calculate the point at which ray $(p_c \rightarrow p_i)$ passes
13:        **if** $(P_{\mathrm{r}}$ is failure$)$
14:            $P_{\mathrm{f}} \leftarrow$ fail point filter(p_{i})
15:           **continue**
16:        $P_{\mathrm{h}} \leftarrow$ calculate the current convex hull (P_{p}, p_i)
17:        calculate the number of unqualified points containing P_{f} in convex hull P_{h}
18:        **if** (obstacle ratio r_a doesn't satisfies the requirements)
19:            $P_{\mathrm{f}} \leftarrow$ fail point filter (p_{i})
20:        $P_{\mathrm{p}} \leftarrow P_{\mathrm{t}}$, and add a topological area identifier
21:        **if** (addition rate r_a satisfies the requirements)
22:            $R'_{\max} \leftarrow$ modify the growth radius $(R_{\max}, r_{\mathrm{p}}, R_{\mathrm{w}})$
23:            $R_{\max} \leftarrow R'_{\max}$
24:        **else break**
25:    $P'_{\mathrm{c}} \leftarrow$ recalculate the regional center (P_c, P_{p})
26:    $R'_{\mathrm{new}} \leftarrow$ recalculate the radius information of the area (P_{p})
27:    **return** $R'_{\max}, P'$

2.2. Regional Adjustment Process

After the growth of the previous section, the whole environment will be divided into regions of different sizes because of the dynamic growth. The small areas not only waste the vertex resources of the topological map, but also cannot be used for the interactive navigation of the actual scene. Therefore, it is necessary to merge small areas and optimize the topological map.

In order to improve the coverage of topographic maps, the region can be regenerated. During the secondary growth process, if the radius of the current region is less than the threshold value and no new points are added, or the maximum radius of secondary growth is meeted, the growth process will quit. The method is similar to regional growth in Section 2.1, which is not discussed here again.

In addition, the Kruskal algorithm is used to adjust the topological map before region merging, and the minimum spanning tree form of the topological map is obtained, which further simplifies the map structure. The next step is to merge some areas. The region merging algorithm first counts the topological vertices that meet the merge requirements. It needs to satisfy the following two conditions:

1.   The radius of the main vertex area is less than the threshold.
2.   The total area of the merged area is less than the threshold.
3.   The obstacle ratio of the merged area is less than the threshold.

Algorithm 2: Region Merging

Objectives: To merge small areas, delete the original topological vertices and generate new topological vertices, and change the color identification of the topological area.

Input: Topological region vertex sequence set P_m to be merged.

Output: The status of this subarea merge: true means the merge was successful; false means no merge occurred.

1:    initialize the set of topological vertex ordinals $V_{com} = \{\}$ that need to be merged
2:    initialize the point set $P_{com} = \{\}$ of the points contained in the merged region
3:    initialize the point set $P_f = \{\}$ of the disqualified points around the merged area
4:    initialize the first vertex information ($p_{first} \leftarrow P_m^0$) of the merge process
5:    calculate the weighted center coordinate $p_c \leftarrow P_m$
6:    calculate the average scan radius $r_c \leftarrow P_m$
7:    **for** r **in** $[0, r_c]$
8:        $P \leftarrow$ loop traversal of all the points contained within the scan radius.
9:        **for** P_i **in** P
10:           **if** (P_i belongs to the topological area)
11:               $P_{com} \leftarrow P_i$
12:           **else** $P_f \leftarrow P_i$
13:    **if** (P_{com} is empty)
14:        **return false**
15:    $P_h \leftarrow$ calculate the current convex hull (P_{com})
16:    calculate the number of unqualified points containing P_f in convex hull P_h
17:    **if** (obstacle ratio r_a doesn't satisfies the requirements)
18:        **return false**
19:    delete all of the vertex information in P_m
20:    Merge the vertex regions in P_m and add the region identifier at the same time
21:    $P_c' \leftarrow$ recalculate the center of the area
22:    $r_n \leftarrow$ recalculate the radius of the merged area
23:    Add a new topological vertex (r_n, P_c', P_{com})
24:    **return true**

After the vertex numbers that need to be merged are obtained, the region merge can be performed. The main steps are as follows:

(1)   The points in the corresponding topological region of all the vertices are extracted, and the qualified points (identical with the vertex color marker) and the failure points (different from the vertex color) are counted.

(2)   The convex hulls of qualified points are calculated, the unqualified points of convex hulls are counted from the failure points, and the obstacle ratio is calculated. If the obstacle ratio is greater than the threshold, then exit.

(3)   If the obstacle ratio is less than the threshold, the region merging is carried out, and the original vertex is deleted, and a new vertex is established.

The pseudo code of the algorithm is shown in Algorithm 2.

3. Topological Map Representation

The process of map preservation is shown in the block diagram on the right side of Figure 1, from which it can be seen that the representation method of the topological map is a process of combining the metric map with the topological map. The method is inspired by the ROS package [27], which expands from a gray value to an RGB color, so that the occupied information and the topological information is saved efficiently while saving and reusing the topological map. In this paper, the parameter TM and the parameter OM are used to represent the topological information and the occupied information, respectively; then, the information converter is used to convert the parameters into the RGB value of the color picture, so that the obtained RGB value can not only be used to display and identify the topology area in real time, but can also contribute to save the map information as a color image.

3.1. Online Representation of the Map

The metric map divides the space into a finite number of grid cells $M = \{m_i | i = 0, 1, 2, \ldots n\}$, each m_i corresponding to an occupied variable. If the grid cell is completely occupied by "1" and is not occupied as "0", then $p(m_i = 1)$ or $p(m_i)$ indicates the possibility that the grid cell is occupied [28]. Therefore, the smallest unit of the metric is the grid cell, and a grid cell is represented by only one variable.

This only represents the information-occupied obstacles; although it can achieve accurate navigation and obstacle avoidance, it cannot achieve a higher level of control. For example, the user tells the robot "I want to go to the kitchen!", and the robot does not know how to perform unless the robot knows the specific location of the kitchen [29]. In order to solve this problem, a scheme combining the metric map and the topological map has been proposed.

A topological map is an environment representation method based on an adjacent graph, which can be represented by an adjacency list structure. Similar to the concept of graphs, topological maps have two basic elements: nodes (which are called vertices in this paper) and edges. Nodes represent different locations that can be distinguished in the environment or various states of distinguishable robots; edges represent relationships between nodes, such as distance or motion control commands. Such representations have low storage requirements, and also support efficient path planning, especially for large-scale unknown environments or outdoor environments.

In this paper, the vertices mainly include the position of the vertices, the vertex number, the adjacent vertices, and so on. The edges mainly include starting and ending vertices, weights, etc. In addition, metric maps and topological maps exist in the form of a "hierarchical map" in the framework proposed in this paper. The two maps interact with each other, learn from each other, and complement each other, providing different map services for the system.

3.2. Topological Map Preservation

The preservation of topological map is the process of integrating topological information and occupied information to form RGB image information. In order to describe the two kinds of map information conveniently, a conical space model is proposed, which is controlled by the parameter TM, the parameter OM, and integer one. The height of the cone is determined by the integer one, and the height is always one. The radius of the bottom surface of the cone is determined by the parameter OM, and the value of the polar coordinates of the bottom surface is determined by the parameter TM, which is inspired by the HSV color space theory [30]. An example process is shown in Figure 3 below.

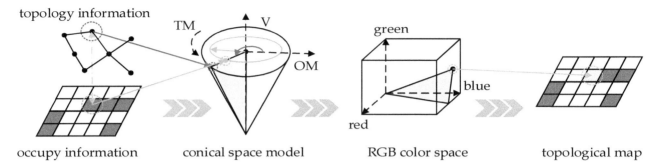

Figure 3. Topological map preservation process.

In the figure, the red line represents the conversion process of the topological information, the green line represents the conversion process of the occupied information, and the blue line represents the synthesis process between the two pieces of information. It should be noted that the relationship between the grid cell and topological vertex is "many-to-one", which means that a topological vertex corresponds to multiple grid elements. In other words, multiple grid cells form a region, and a region is represented by a topological vertex.

In the process of preservation, the information of the topological vertex is extracted from the topological map firstly; then, the occupied information of the grid cell corresponding to the topological vertex is found in the metric map, and the two pieces of information are converted into the parameters TM and OM. In the conical space model, a cross-section can be obtained by these two parameters. The cross-section represents the result of the synthesis of two kinds of information. However, such information is still unusable, and needs to be converted again. The blue triangle cross-section is shown in Figure 3. Therefore, the information converter is proposed to convert the cross-section information into RGB information, which is often used in the field of vision. The RGB value is used to represent a grid cell, and then each grid cell is operated in the same way. Finally, the color map containing the topological information and occupancy information is saved. In the following, the TM and OM parameters are explained in detail:

TM is measured by the angle, and the range of values is [0, 360]. It is calculated counterclockwise from red. This value is used to represent the topological neighborhood, which can represent at most 360 topological neighborhoods. In order to show a higher degree of discrimination, values can be taken at intervals, such as 20 color values for the identification of the topological region, and 18 kinds of topological region identifiers can be used.

OM represents the degree of proximity to the spectral color, and the range of values is [0, 1]. This value is used to represent whether the obstacle is occupied at a point in the map. The larger the value, the greater the likelihood that there is an obstacle at that point; the smaller the value, the greater the likelihood that the point may be free.

Then, in the beginning of topological map preservation, the occupied information and topological information need to be converted. The conversion formulas are as shown in Equation (6):

$$
\begin{cases}
TM = a + b \times P_i \\
OM = k \times p(m_i)
\end{cases}
\tag{6}
$$

where a is the start value, b is the interval value, k is the certain coefficient, and $a = 0, b = 20, k = 1$ is taken in the paper.

After that, some intermediate variables should be computed from these two parameters, and the formulas are shown in Equation (7), where $V = 1$ is taken in the paper. Then, with the help of intermediate variables, the RGB information can be obtained by Equation (8):

$$\begin{cases} h = \left\lfloor \frac{TM}{60} \right\rfloor \quad (\mathrm{mod}\,6) \\ f = \frac{TM}{60} - h \\ p = V \times (1 - OM) \\ q = V \times (1 - f \times OM) \\ g = V \times (1 - (1 - f) \times OM) \end{cases} \tag{7}$$

In addition, the yaml file and the picture in pgm format are used to save the metric map traditionally, where the picture in pgm picture is a grayscale picture, and only the gray level data can be saved. Therefore, the picture in ppm picture is used to save the topological map, because the picture in ppm format can save RGB color data. Moreover, the ppm image format is divided into the ASCII encoding format (file descriptor is P3) and the binary-encoding format (file descriptor is P6), because the binary encoding format consumes less memory than the ASCII encoding format, so the binary encoding format is used.

$$(R, G, B) = \begin{cases} (255, g, p) & \text{if } h = 0 \\ (q, 255, p) & \text{if } h = 1 \\ (p, 255, g) & \text{if } h = 2 \\ (p, q, 255) & \text{if } h = 3 \\ (g, p, 255) & \text{if } h = 4 \\ (255, p, q) & \text{if } h = 5 \end{cases} \tag{8}$$

Except for the ppm image, a yaml file that contains the pixel coordinates of topological vertices, adjacent vertices, and other information is also saved, and the details saved in this yaml file will be explained in the next section. In order to simplify the calculation, in the following experiments, the area indicated by red is occupied, the area indicated by green is free and the area indicated by blue is unknown.

3.3. Topological Map Reading

The reading of the topological map means that the map file is parsed first, and then the obstacle occupied information and topological information are restored. The reading process takes two steps: reading the yaml file and the ppm image.

Here, to explain the contents of the yaml file, the file contains the following main parts:

◆   Image file path: Refers to the saved path of the ppm image file.

◆   Resolution: Refers to the resolution of the map, and is used to represent the scale of a pixel in the real world, with a unit (meters/pixel).

◆   Origin: Refers to the two-dimensional (2D) pose of the lower left pixel in the map, as (x, y, yaw), with yaw as the counter-clockwise rotation (yaw = 0 means no rotation). The yaw is ignored in this paper.

◆   Free thresh: Pixels with an occupancy probability less than this threshold are considered completely free.

◆   Occupied thresh: Pixels with an occupancy probability greater than this threshold are considered completely occupied.

◆   Mode: The way the file is saved, which can be one of three values: trinary, scale, or raw. Trinary is the default.

◆   Number of topological vertices.

◆   Topological information: A series of lists that contain the pixel coordinates of topological vertices, adjacent vertices, and other information (such as semantic information; this is empty in the paper).

In the map representation method proposed in this paper, the metric map is a two-dimensional map, so the map coordinates, pixel coordinates, initial points, and other relationships can be shown in Figure 4 below. In the image, different color regions represent different topological regions, and dark color regions are more likely to be occupied.

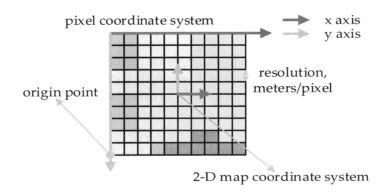

Figure 4. Relations between coordinate systems in the picture in ppm format.

Traditionally, the origin of the pixel coordinate system is in the upper left corner of the picture, while the origin of the map coordinate system is in the center of the picture, and the pixel coordinates of the center can be calculated by the origin point. Then, with the help of the resolution of the map, all the pixels can be restored to grid cells.

In the first place, the pixel coordinates of the pixel points in the picture need to be converted into map coordinates by the map coordinates of the lower left pixel, and the specific conversion formula is as shown in Equation (9):

$$\begin{cases} x_{map} = x_{orign} + x_{pixel} * res \\ y_{map} = y_{orign} + h_{map} * res - y_{pixel} * res \end{cases} \tag{9}$$

where (x_{pixel}, y_{pixel}) is the pixel coordinate, (x_{map}, y_{map}) is the map coordinate, res is the map resolution, and h_{map} is the height of the map.

At the same time, the RGB image information in the ppm image file needs to be converted into two parameters in the conical space model, and the conversion formula is shown in Equation (10).

$$\begin{aligned} &V \leftarrow \max(R, G, B) \\ &OM \leftarrow \begin{cases} \frac{V - \min(R,G,B)}{V} & \text{if } V \neq 0 \\ 0 & \text{otherwise} \end{cases} \\ &TM \leftarrow \begin{cases} \frac{60(G-B)}{V - \min(R,G,B)} & \text{if } V = R \\ \frac{120 + 60(B-R)}{V - \min(R,G,B)} & \text{if } V = G \\ \frac{240 + 60(R-G)}{V - \min(R,G,B)} & \text{if } V = B \end{cases} \end{aligned} \tag{10}$$

where if $TM < 0$, then $TM = TM + 360$. After the above calculation, the final range of the three values is: $0 \leq V \leq 1, 0 \leq OM \leq 1, 0 \leq TM \leq 360$.

Afterwards, the topological information can be obtained from the parameter TM, and the topological map is established. In addition, the occupied information of the point is read from the parameter OM, and the metric map is established. In this way, the topological information and occupied information can be completely recovered.

4. Experiment and Result

Finally, several experiments will be carried out to illustrate the effectiveness of the proposed algorithm and the rationality of the topological map representation method. Relevant experiments were completed in the simulation environment, and the metric map was obtained by using the gmapping algorithm [31].

In the Gazebo simulation environment, four environments were built representing an indoor home, office, pillar, and open space. The experimental environment is shown in Figure 5. In the experiment, the turtlebot2 robot equipped with a 2D laser sensor was used as an experimental platform. All of the experiments were performed on a notebook with a memory 8G, i7 processor, and a GTX1050 graphics card.

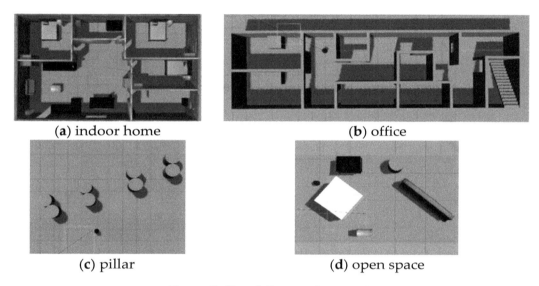

(a) indoor home   (b) office

(c) pillar   (d) open space

Figure 5. Simulation environment.

The first experiment was carried out in the indoor home environment, and each part of the whole system was tested. The experimental results are shown in Figure 6, which includes the construction of the metric map, the construction of the topological map, the adjustment of the topological map, and the preservation of the topological map. These continuous processes truly implement the process of topological mapping from build to save.

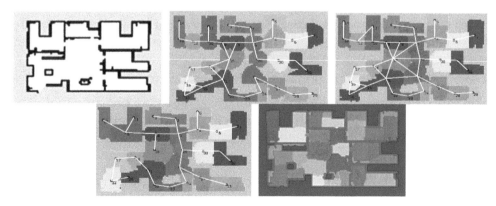

Figure 6. Topological map construction process.

In the following, relevant experiments were carried out in the office environment, pillar environment, and open space. The results are shown in Figure 7, which shows the metric map and topological map constructed in different environments. Among them, the growth radius is between 1.2–3 m, the map resolution is 0.20 m, and the control weight is one.

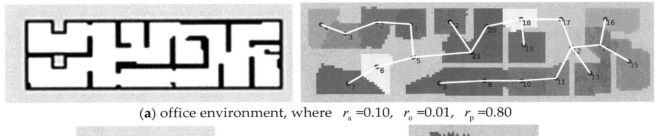

(**a**) office environment, where r_a =0.10, r_o =0.01, r_p =0.80

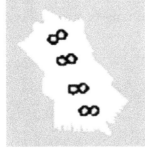 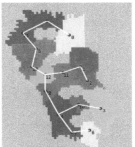

(**b**) pillar environment, where r_a =0.12, r_o =0.05, r_p =0.70

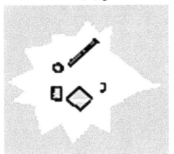 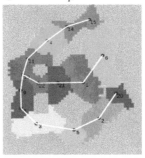

(**c**) open space, where r_a =0.13, r_o =0.06, r_p =0.65

Figure 7. Occupied map (**left**) and its corresponding topological map (**right**).

The office environment consists of many small spaces and long corridors. It can be seen that the map construction algorithm performs poorly at the door and the area is somewhat deformed. However, in the latter two examples, the map construction algorithm is more perfect: the region is more similar to the ellipse, and the dynamic growth and merging process makes the small region fully covered.

Next, in the indoor environment, this paper chooses a special location (next to the dining table in the living room) to build the map, and discusses the impact of these three parameters on the map construction under the circumstances of changing the obstacle ratio, the point addition ratio, and the points pass ratio. Here, the growth radius is set to 1.5 m, the map resolution is 0.2 m, and the control value R_w is set to one. The details of regional growth are shown in Figure 8.

The obstacle ratio is to prevent the region from growing into a concave region, and ensure the convexity of the topological region. As can be seen in group (**a**) of Figure 7, with the increase of the obstacle ratio, the area will grow toward a small space (such as a door) next to it, and a few obstacle points will be added continuously.

The points addition ratio prevents malformation, making the growth area approximately circular or elliptical rather than elongated. It can be seen in group (**a**) of Figure 8 that when the point addition ratio is too small or too large, it is easy to cause growth malformation or insufficient growth.

The points pass ratio can reflect the contrast between the growth area and the ideal area in the current state. It is used to modify the growth radius in real time, so its role is to prevent growth deficiency. When the growth area is far from the ideal area, it can make the region grow seriously. This effect is well reflected in the (**c**) group, but it also destroys the convex nature of the area. Therefore, the above three parameters need to be adjusted according to the actual situation, and the topological map will represent the environment better.

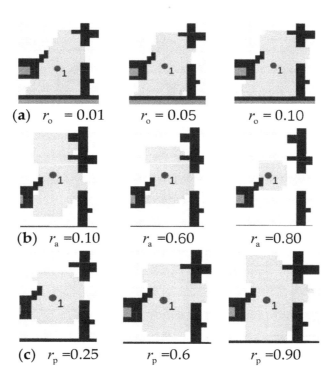

Figure 8. The influence of three parameters on the construction of the topological map. (a) Impact of changes in the obstacle ratio on the region. (b) Impact of changes in the point addition ratio on the region. (c) Impact of changes in the points pass ratio on the region.

In addition, map resolution is another important factor affecting the construction of topological maps. In the indoor home environment, a large number of experiments were carried out for different resolutions in order to obtain an optimal topological map. Figure 9 shows the effect of map resolution on topological map connectivity and occupancy.

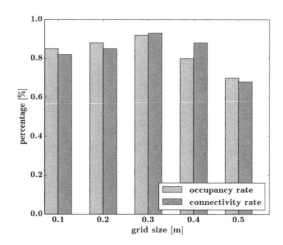

Figure 9. The impact of map resolution on topological map construction.

Finally, topological maps built in the office environment, pillar environment, and open space are saved, as shown in Figure 10. Experiments show that the topological map representation proposed in this paper can save the occupied information and topological information well.

Figure 10. Topological maps saved in three environments.

What's more, the system proposed in this paper was integrated into the turtlebot2 equipped with the hokuyo laser sensor and NVidia TK1, and the topological map construction was successfully completed in the author's laboratory, refer to the Supplementary Materials. This experiment proves the practicality of the pipeline proposed in this paper on mobile platforms such as service robots. The configuration and map of the turtlebot2 experimental system is shown in Figure 11.

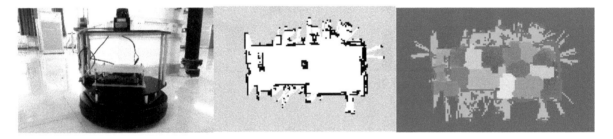

Figure 11. Setup of the turtlebot2 experiments and maps.

Not only is the topological map construction method not limited by the environment geometry, but the topological map preservation method can also simultaneously save the occupied information and topological information to achieve the map conditions required for accurate navigation. In order to better illustrate that, Table 1 summarizes some of the information about the topological map creation in various environments. As can be seen from the table, the topological map can cover the entire environment, and the more complex the environment, the more topological vertices the map construction method will use to describe it. Although the topological map storage requirements are larger than the metric maps, more information will be saved to provide the same precision navigation environment.

Table 1. Topological map information for each scene.

	Topological Map Information		Storage Requirements (kB)	
	Number of Vertexes	**Occupied Ratio**	**Metric Map**	**Topological Map**
Indoor (171.224 m^2)	31	0.922444	36.0183	109.46
Office (115.389 m^2)	22	0.985809	30.0172	91.27
Pillar (129.821 m^2)	15	0.984625	20.0172	72.061
Open space (171.261 m^2)	15	0.936123	16.0172	48.0853
Laboratory (277.999 m^2)	16	0.883049	30.0181	90.0903

5. Conclusions

In this paper, a new framework for creating high-precision topological maps and efficient map representations for diverse environments is presented. In various structured environments, the regional dynamic growth algorithm proposed in this paper can divide the free space into multiple convex regions, each forming a topological region to represent the environment. This greatly reduces the use requirements compared to the most advanced methods, and extends the range of use. The topological map representation method proposed in this paper uses the parameter TM and the parameter OM of the conical space model to represent the occupation information and topological information of the map efficiently, which satisfies the conditions of high-precision navigation.

Through a large number of experiments, the topological map construction method proposed in this paper can construct a topological map that conforms to a specific environment without being limited by the environment geometry. In addition, the topological map can be made to better describe the entire environment by modifying the corresponding control parameters. It also verifies that the topological map preservation method proposed in this paper can save the occupied information and topological information at the same time by occupying a small amount of storage space. The universality of the construction method and the efficiency of the preservation method provide conditions for human–machine interactive navigation, so that the topological map can be truly applied in real life.

For the future research, since this paper only studies two-dimensional topological maps, three-dimensional topological maps will be the direction of future work. In addition, the semantic information of the actual scene can be added to the topological node to improve the robot's understanding of the actual environment.

6. Patents

A patent named "Human–machine interactive navigation system and method based on brain–computer interface" is pending.

Author Contributions: This study was completed by the co-authors. F.W., C.W., and H.C. conceived and led the research. The major experiments and analyses were undertaken by Y.L. and L.X.; F.W. supervised and guided this study. Y.L. and L.X. wrote the paper. All the authors have read and approved the final manuscript. Conceptualization, F.W.; Formal analysis, Y.L.; Investigation, Y.L.; Methodology, Y.L.; Project administration, F.W., C.W. and H.C.; Resources, F.W.; Software, Y.L.; Validation, L.X.; Writing—original draft, Y.L.; Writing—review & editing, Y.L. and L.X.

References

1. Mavridis, N. A review of verbal and non-verbal human–robot interactive communication. *Robot. Auton. Syst.* **2015**, *63*, 22–35. [CrossRef]
2. Luo, R.C.; Shih, W. Autonomous Mobile Robot Intrinsic Navigation Based on Visual Topological Map. In Proceedings of the 2018 IEEE International Symposium on Industrial Electronics (ISIE), Cairns, Australia, 13–15 June 2018; pp. 541–546.
3. Chung, M.J.Y.; Pronobis, A.; Cakmak, M.; Fox, D.; Rao, R.P.N. Autonomous question answering with mobile robots in human-populated environments. In Proceedings of the 2016 IEEE/RSJ International Conference on Intelligent Robots and Systems (IROS), Daejeon, Korea, 9–14 October 2016; pp. 823–830.
4. Jo, K.; Kim, C.; Sunwoo, M. Simultaneous localization and map change update for the high definition map-based autonomous driving car. *Sensors* **2018**, *18*, 3145. [CrossRef] [PubMed]
5. Lin, H.Y.; Yao, C.W.; Cheng, K.S. Topological map construction and scene recognition for vehicle localization. *Auton. Robot.* **2018**, *42*, 65–81. [CrossRef]
6. Qing, G.; Zheng, Z.; Yue, X. Path-planning of automated guided vehicle based on improved Dijkstra algorithm. In Proceedings of the 2017 Chinese Control and Decision Conference (CCDC), Chongqing, China, 28–30 May 2017; pp. 7138–7143.
7. Papoutsidakis, M.; Kalovrektis, K.; Drosos, C.; Stamoulis, G. Design of an Autonomous Robotic Vehicle for Area Mapping and Remote Monitoring. *Int. J. Comput. Appl.* **2017**, *167*, 36–41. [CrossRef]
8. Chandrasekaran, B.; Conrad, J.M. Human-robot collaboration: A survey. In Proceedings of the 2015 Southeast Conference, Fort Lauderdale, FL, USA, 9–12 April 2015; pp. 1–8.

9.   Alitappeh, R.J.; Pereira, G.A.S.; Araújo, A.R.; Pimenta, L.C.A. Multi-robot deployment using topological maps. *J. Intell. Robot. Syst.* **2017**, *86*, 641–661. [CrossRef]

10. Johnson, C. Topological Mapping and Navigation in Real-World Environments. Master's Thesis, University of Michigan, Ann Arbor, MI, USA, 12 December 2017.

11. Li, X.; Qiu, H. An effective laser-based approach to build topological map of unknown environment. In Proceedings of the 2015 IEEE International Conference on Robotics and Biomimetics (ROBIO), Zhuhai, China, 6–9 December 2015; pp. 200–205.

12. Konolige, K.; Marder-Eppstein, E.; Marthi, B. Navigation in hybrid metric-topological maps. In Proceedings of the 2011 IEEE International Conference on Robotics and Automation (ICRA), Shanghai, China, 9–13 May 2011; pp. 3041–3047.

13. Azzag, H.; Lebbah, M.; Arfaoui, A. Map-TreeMaps: A New Approach for Hierarchical and Topological Clustering. In Proceedings of the 2010 International Conference on Machine Learning and Applications (ICMLA), Washington, DC, USA, 12–14 December 2010; pp. 873–878.

14. Liu, M.; Colas, F.; Siegwart, R. Regional topological segmentation based on mutual information graphs. In Proceedings of the 2011 IEEE International Conference on Robotics and Automation (ICRA), Shanghai, China, 9–13 May 2011; pp. 3269–3274.

15. Bandera, A.; Sandoval, F. Spectral clustering for feature-based metric maps partitioning in a hybrid mapping framework. In Proceedings of the 2009 IEEE International Conference on Robotics and Automation (ICRA), Kobe, Japan, 12–17 May 2009; pp. 1868–1874.

16. Yu, W.; Amigoni, F. Standard for Robot Map Data Representation for Navigation. In Proceedings of the 2014 IEEE/RSJ International Conference on Intelligent Robots and Systems (IROS), Chicago, IL, USA, 14–18 September 2014; pp. 3–4.

17. Lau, B.; Sprunk, C.; Burgard, W. Improved updating of Euclidean distance maps and Voronoi diagrams. In Proceedings of the 2010 IEEE/RSJ International Conference on Intelligent Robots and Systems (IROS), Taipei, Taiwan, 18–22 October 2010; pp. 281–286.

18. Guo, S.; Ma, S.; Li, B.; Wang, M.; Wang, Y. Simultaneous Location and Mapping Through a Voronoi-diagram-based Map Representation. *Acta Autom. Sin.* **2011**, *37*, 1095–1104.

19. Ramaithitima, R.; Whitzer, M.; Bhattacharya, S.; Kumar, V. Automated creation of topological maps in unknown environments using a swarm of resource-constrained robots. *IEEE Robot. Autom. Lett.* **2016**, *1*, 746–753. [CrossRef]

20. Kaleci, B.; Senler, C.M.; Parlaktuna, O.; Gürel, U. Constructing Topological Map from Metric Map Using Spectral Clustering. In Proceedings of the 2016 International Conference on TOOLS with Artificial Intelligence (ICTAI), San Jose, CA, USA, 6–8 November 2016; pp. 139–145.

21. Liu, M.; Colas, F.; Pomerleau, F.; Siegwart, R. A Markov semi-supervised clustering approach and its application in topological map extraction. In Proceedings of the 2012 International Conference on Intelligent Robots and Systems (2012), Vilamoura, Algarve, Portugal, 7–12 October 2012; pp. 4743–4748.

22. Ravankar, A.A.; Ravankar, A.; Emaru, T.; Kobayashi, Y. A hybrid topological mapping and navigation method for large area robot mapping. In Proceedings of the 2017 Society of Instrument and Control Engineers of Japan (SICE), Kanazawa, Japan, 19–22 September 2017; pp. 1104–1107.

23. Kaleci, B.; Parlaktuna, O.; Gurel, U. A comparative study for topological map construction methods from metric map. In Proceedings of the 2018 Signal Processing and Communications Applications Conference (SIU), Izmir, Turkey, 2–5 May 2018; pp. 1–4.

24. Blochliger, F.; Fehr, M.; Dymczyk, M.; Schneider, T.; Siegwart, R. Topomap: Topological mapping and navigation based on visual slam maps. In Proceedings of the 2018 IEEE International Conference on Robotics and Automation (ICRA), Brisbane, QLD, Australia, 21–25 May 2018; pp. 1–9.

25. The Bresenham Line-Drawing Algorithm. Available online: https://www.cs.helsinki.fi/group/goa/mallinnus/lines/bresenh.html (accessed on 9 December 2018).

26. Graham, R.L. An efficient algorithm for determining the convex hull of a finite planar set. *Inf. Process. Lett.* **1972**, *1*, 132–133. [CrossRef]

27. map_server. Available online: http://wiki.ros.org/map_server (accessed on 9 December 2018).

28. Colleens, T.; Colleens, J.J.; Ryan, D. Occupancy grid mapping: An empirical evaluation. In Proceedings of the 2007 Mediterranean Conference on Control & Automation, Athens, Greece, 27–29 June 2007; pp. 1–6.

29. Saunders, J.; Syrdal, D.S.; Koay, K.L.; Burke, N.; Dautenhahn, K. "Teach Me–Show Me"—End-User Personalization of a Smart Home and Companion Robot. *IEEE Trans. Hum.-Mach. Syst.* **2016**, *46*, 27–40. [CrossRef]
30. Hanbury, A. Circular statistics applied to colour images. In Proceedings of the 2003 Computer Vision Winter Workshop, Valtice, Czech Republic, 3–6 February 2003; pp. 53–71.
31. Grisettiyz, G.; Stachniss, C.; Burgard, W. Improving Grid-based SLAM with Rao-Blackwellized Particle Filters by Adaptive Proposals and Selective Resampling. In Proceedings of the 2005 IEEE International Conference on Robotics and Automation, Barcelona, Spain, 18–22 April 2005; pp. 2432–2437.

12

Multi-Robot Trajectory Planning and Position/Force Coordination Control in Complex Welding Tasks

Yahui Gan [1,2,*], Jinjun Duan [1,2], Ming Chen [1,2] and Xianzhong Dai [1,2]

[1]   School of Automation, Southeast University, Nanjing 210096, China; duan_jinjun@yeah.net (J.D.);
    chen_ming@yeah.net (M.C.); xzdai@seu.edu.cn (X.D.)
[2]   Key Lab of Measurement and Control of Complex Systems of Engineering, Ministry of Education,
    Nanjing 210096, China
*   Correspondence: ganyahui@yeah.net.

Abstract: In this paper, the trajectory planning and position/force coordination control of multi-robot systems during the welding process are discussed. Trajectory planning is the basis of the position/ force cooperative control, an object-oriented hierarchical planning control strategy is adopted firstly, which has the ability to solve the problem of complex coordinate transformation, welding process requirement and constraints, etc. Furthermore, a new symmetrical internal and external adaptive variable impedance control is proposed for position/force tracking of multi-robot cooperative manipulators. Based on this control approach, the multi-robot cooperative manipulator is able to track a dynamic desired force and compensate for the unknown trajectory deviations, which result from external disturbances and calibration errors. In the end, the developed control scheme is experimentally tested on a multi-robot setup which is composed of three ESTUN industrial manipulators by welding a pipe-contact-pipe object. The simulations and experimental results are strongly proved that the proposed approach can finish the welding task smoothly and achieve a good position/force tracking performance.

Keywords: trajectory planning; position/force cooperative control; hierarchical planning; object-oriented; symmetrical adaptive variable impedance

1. Introduction

With the complication and diversification of industrial production tasks, multi-robot cooperative systems have demonstrated stronger operational capabilities, more flexible system structures, and stronger collaboration capabilities. Therefore, multi-robot collaboration has become an important challenge for robot control.

Multi-robot collaboration has been applied in many fields such as handing, assembly, welding, etc. The application of this paper is focused on the welding field. In arc welding, the traditional welding workstation consisting of "welding robot + positioner" is not able to meet the current demand for small-volume, customized, flexible, and automated production. Multi-robot cooperative welding adopts a universal and high-degree of freedom handing robot instead of a low-degree of freedom positioner, which can effectively improve the flexibility of welding tasks and welding automation.

Arc welding is a complex system that contains both pose constraints and wrench constraints, the use of multi-robot systems for arc welding has many advantages, but it also brings more complex control problems. Two of the most critical issues are the trajectory planning in the multi-robot collaboration process and the coordinated control of position/force among multi-robot.

For the trajectory planning of multi-robot systems, the current research issues include motion constraints, control methods for cooperative motion, and implementation of multi-robot cooperative systems.

(1) For the motion constraint problem of multi-robot systems, the main task is to derive the end-effector of robot pose, velocity and acceleration constraints. For example, the idea is first pointed out in [1] that when two or more robots grab a common object, the robot end-effector is subject to kinematic constraints, and gives the speed constraint relationship of the multi-robot end-effector when the multi-robot is holding an object and rotates around its center. The pose, speed and acceleration constraints of the end-effector in the case of two robots grasp the common object, operate a pair of pliers and grab an object with a ball joint are deduced in [2,3]. The concept of absolute motion when the two robots cooperatively clamp the workpiece is proposed in [4]. The multi-robot cooperated trajectory planning approach based on the closed kinematic chain model is proposed in [5]; (2) For the control methods of collaborative movements, a method of establishing a constraint model with differential algebraic equations which using feedback linearization is addressed in [6]. The common control methods of general differential algebraic systems is summarized base on the above idea in [7]; (3) The realization for multi-robot collaboration system, the problem of trajectory planning and programming of general multi-robot cooperative system are analyzed. Such as ABB MultiMove function, KUKA RoboTeam function, Yaskawa independence/collaboration function, FANUC cooperative action function, etc.

Although the above research has given a certain impetus to the kinematics constraints and trajectory planning, the current multi-robot coordination tasks are relatively simple and most of them are focus on the coordination of dual robots. The above trajectory planning method is not scalable and feasible for multi-robot collaboration to accomplish specific tasks in a more complex and unstructured environment.

Multi-robot trajectory planning is the basis for completing the welding tasks, there are not only pose constraints, but also the wrench constraints in the process of welding displacement. Therefore, multi-robot position/force coordination control is another key issue in the cooperative welding process. Research methods for multi-robot position/force coordination control include master/slave control, hybrid motion/force control, synchronization control and impedance control. (1) Master/slave control. The control idea is to define one of the robotic arms as the master arm and the other as the salve arm. The master and the slave arm should've meet the certain constraints. The master arm is controlled by the position control mode, and the slave arm follows the motion trend of the master arm detected by the force/torque sensor mounted at the wrist of the slave arm. The master/slave force control approach for the coordination of two arms which carries a common object cooperatively was proposed in [8,9], and the necessity of force control for cooperative multiple robots is also pointed out. However, the slave arm needs to have a fast following response [10], otherwise it will lead to system instability. Force/torque sensor and position controller are difficult to achieve high-speed response in actual control systems. Therefore, the master-slave control strategy is only suitable for low-speed applications; (2) Hybrid motion/force control. The basic idea is that two arms work equally and coordinated by the centralized control. The position control is used in the free space, and the force control is used in the constrained space, such as [11–14]. A difficulty in implementing the hybrid control law in rigid environment is knowing the form of the constraints active at any time. And it also needs to sacrifice some performance by choosing low feedback gains, which makes the motion controller "soft" and the force controller more tolerant of force error; (3) Synchronization control. The basic idea is to track the desired trajectory which is generated by the desired force based on the dynamic model of the manipulators. The control problem is formulated in terms of suitably defined errors accounting for the motion synchronization between the manipulators involved in the cooperative task. The concept of motion synchronization was used in [15,16]. An adaptive control strategy was adopted to track the desired trajectory, ensuring the synchronization position errors converged to zero. In addition, intelligent control strategies were also used in the coordination control of nonlinear cooperative manipulator systems [17–19]. However, the synchronization control is based on the dynamic model. Although the synchronization error at the end effector of the manipulator was considered, due to the difficulties in dynamic modeling, over-complexity control model, strong coupling and nonlinearity, it has not been applied in most actual control systems; (4) Impedance control. The basic idea is to achieve the adjustment of the position based on the force error. Impedance control is a stable and effective method widely used in many fields including coordination. The coordination strategy of the object based on impedance control was studied in [20–24]. The force acting on the object was decomposed into the external force that contributes to the object's motion and the internal force by the end-effector of both arms. Following the guidelines in above references, the external impedance and the internal impedance were combined in a unique control framework [25]. Compared with the first three control strategies, impedance control overcomes the shortcomings of the above control methods, and can effectively control the internal and external force.

In contrast, impedance control has been widely used in multi-robot position/force coordination control. However, a closed-chain system is often subject to external dynamic disturbances and calibration errors in the actual industrial system. These factors can cause time-varying trajectory deviations at the end-effector, and time-varying trajectory deviations can cause unknowns and dynamically changing external forces and internal forces between multi-robot and the operated object. Most of the current impedance control methods that used in the above studies are constant, and the presented control strategies are not feasible for the unknown and dynamically changing trajectory errors. To the best of our knowledge, no research has been reported to solve the trajectory deviation problem during the multi-robot coordination with actual industrial robotic systems.

This paper uses multi-robot systems to complete the pipe-contact-pipe welding task as the research object. A study on the trajectory planning and position/force coordination control in the welding process is conducted.

The remaining of this paper is organized as follows. The system overview of multi-robot cooperative welding system is introduced in Section 2. The object-oriented Multi-robot trajectory planning based on "hierarchical scheme" is proposed in Section 3. The Coordination Strategy based on Symmetrical Adaptive Variable Impedance Control is given in Section 4. A series of experiments are carried out in Section 5, followed by conclusions in Section 6.

2. Problem Description

2.1. System Description

A multi-robot cooperative welding system refers to the collaboration of many sets of industrial robots on a welding task. In general, three or more robots are necessary. The cooperative system includes at least one welding robot, two or more robots are required which are used to splice the different pieces together. To discuss the key issue in welding process, the multi-robot systems cooperate welding a pipe-connect-pipe are given as an example in this paper. The typical multi-robot cooperation welding system schematic diagram is shown in Figure 1.

Figure 1 shows that the multi-robot welding system that includes three industrial robots, one welding robot, and two transfer robots. When the controller receives the specified task, two transfer robots grasp different welding workpieces, respectively in the initial welding position of relocating and splicing, at the same time, the welding robot begins to welding. After the two transfer robots cooperate with the workpiece, they will change the pose in the whole welding process, and then complete welding with the welding robot.

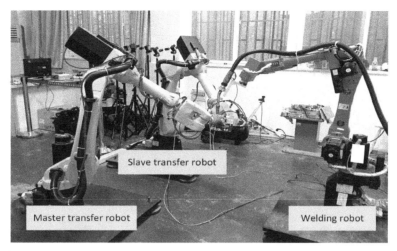

Figure 1. The diagram of multi-robot cooperation welding system.

2.2. Problem Description

The multi-robot cooperative welding system is shown in Figure 1. To solve the constraint pose and wrench at the same time in welding process, it is difficulty to plan the cooperative motion trajectory and achieve the desired position/force cooperative effect during the transportation.

In order to satisfy the welding process requirements in the continuous welding process, the transfer robots must continuously change the posture of the clamping workpiece to ensure that the welding spot is always in the preset welding position. The welding robot must constantly adjust the position of the welding torch to ensure the welding torch meets the welding requirements. It is required that both the transfer robots and welding robot should satisfy a certain position and posture constraints during the whole welding process.

With the process of welding, the workpiece held by transfer robot is gradually welded, then the transfer robots and the workpiece will be formed as a closed-chain system. In the actual control system, the external disturbance and calibration errors are often existing. These disturbances and calibration errors will cause the time-varying trajectory deviations in the robot end-effector during cooperation, and the dynamic time-varying trajectory deviations can cause a huge internal forces between the robot and the workpiece. Improper control may lead a damage to the workpiece or robot.

From the above analysis, the two key issues in the control of multi-robot cooperative welding system are:

- How to achieve the coordinated motion of multi-robot, which is the problem of planing the multi-robot trajectory;
- How to effectively control the internal force between the robot and the workpiece, which is the problem of position/force cooperation.

3. Task-Oriented Multi-Robot Trajectory Planning

3.1. Problem Formulation

The schematic diagram of the multi-robot cooperating to complete the welding task is shown in Figure 2.

Where $robot\_1$ and $robot\_2$ are used for transfer robots, $robot\_3$ is used for a welding robot. The transfer robots grab the workpiece R_a and R_b, respectively. The welding robot is equipped with a welding torch at the end-effector, and the workpiece R_a and R_b are saddle-shaped after splicing.

The relevant coordinate system shown in Figure 2 is defined as follows. T_w represents the world coordinated system, T_o represents the reference coordinate system of the object. T_{bi} denotes the base coordinate of the i-th robot, T_i denotes the coordinate system of the i-th robot end-effector. T_{weld} is the weld coordinate system of the workpiece, and T_{object} is the coordinate system of the workpiece.

To facilitate the coordinate conversion, it is usually assumed that the world coordinate system coincides with the base coordinate system of *robot_1*.

The requirements for the multi-robot cooperation to complete the welding task of the tube include:

- The welding torch should meet the requirements of the ship-type welding posture;
- The configuration of each robot is reachable, and the configuration has no singularity in the welding process;
- The motion planning results of each robot are in a common collaborative space;
- There is no collision between the robot itself and another robots in the welding process.

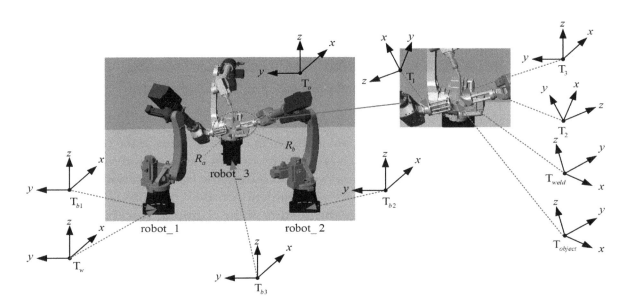

Figure 2. The system coordinates of multi-robot pipe-connect-pipe welding task.

3.2. Trajectory Planning Strategy

As shown in Figure 2, the coordinate systems involved in the multi-robot cooperative welding process are numerous, and the transformation between coordinate systems is complicated. At the same time, the welding requirements as described above must also be met in order to finish the welding task smoothly. We can conduct that the initial welding position affects whether the entire welding task can proceed smoothly or not. Based on the above constraints and requirements, a multi-robot trajectory planning based on "hierarchical scheme", which considered the optimal initial welding position, is proposed in this paper.

The basic idea of the above scheme is to first determine the optimal initial welding position, that is to determine the position of the reference coordinate system of the object in the world coordinate system. Then according to the welding task, the trajectory of the object in its reference coordinate system is planned. Finally, the robot end-effector trajectory is planned through the constraint relationship between the robot and the object.

The following steps are used to obtain the trajectory of multi-robot.

The first step is to determine the layout of the optimal welding position in the initial state, it is same to determine the position p (x,y,z) which T_o is related to the world coordinate T_w.

In combination with the requirements of the multi-robot cooperative welding process, the following indicators are considered to affect the layout selection of initial welding position.

1. The dual-arm task-based directional manipulability measure (DATBDMM) is mainly aimed at the transfer robots.
2. Flexible measure (FM) is mainly for the welding robot.
3. Global joint exercise (GJE) is used for the transfer robots and the welding robot.

According to the above performance indicators, the mathematical model for establishing the optimal initial welding position layout is shown in Equation (1).

$$
\begin{cases}
\text{DATBDMM}_{cv} = \max \left(\boldsymbol{p}^{\mathrm{T}} \boldsymbol{J}_{vc} \boldsymbol{p} \right)^{-1} \\[2mm]
\text{FM} = \max \dfrac{\sum\limits_{m-1}^{a} \sum\limits_{n-1}^{b} D_P(\alpha, \beta)}{\sum\limits_{m-1}^{a} \sum\limits_{n-1}^{b} 1} \\[4mm]
\text{GJE} = \min \sum\limits_{j=1}^{3} \sum\limits_{i=1}^{n} {}^{j}w_i \left({}^{j}\theta_i(k) - {}^{j}\theta_i(k-1) \right)^2
\end{cases}
\tag{1}
$$

where $\boldsymbol{p} \in \boldsymbol{R}^{3 \times 1}$ represents the velocity unit vector at the point of mass center of the workpiece. $\boldsymbol{J}_{vc} = \left(\boldsymbol{J}_{1cv} \boldsymbol{J}_{1cv}{}^{\mathrm{T}} \right)^{-1} + \left(\boldsymbol{J}_{2cv} \boldsymbol{J}_{2cv}{}^{\mathrm{T}} \right)^{-1}$, $\boldsymbol{J}_{icv}$ represents the speed Jacobian matrix of the robot at the center of operated object. α, β indicate the rotation angle of welding torch around the y-axis and z-axis, respectively. After rotation, it can meet the welding requirements (shipping welding requirements). $D_P(\alpha, \beta)$ denotes a pose reachable function. If there is an inverse solution, it is denoted as $D_P(\alpha, \beta) = 1$. Otherwise, it is denoted as $D_P(\alpha, \beta) = 0$. ${}^{j}w_i$ represents the influence factor of each joint. ${}^{j}\theta_i(k) - {}^{j}\theta_i(k-1)$ indicates the amount of joint change at a certain moment.

The optimal initial welding position $\boldsymbol{p}$ (x,y,z) is determined according to the above performance index, and the mathematical model of the optimal initial welding position for multi-robot cooperative welding is shown in Equation (2), then the optimal solution can be solved.

$$
\max f(x, y, z) = \frac{k_{dm} \cdot \text{DATBDMM}_{cv} + k_{fm} \cdot \text{FM}}{k_{gje} \cdot \text{GJE}},
\tag{2}
$$

where f represents a multi-objective optimal function, k_{dm}, k_{fm}, k_{gje} represent the influence factors corresponding to the degree of dual-directional operation of the task, the flexibility, and the amount of joint change, respectively. According to the importance of each parameter and our preliminary experience, the value of k_{dm} is set to 0.5, k_{fm} is set to 0.3 and k_{gje} is set to 0.2.

In the second step, the trajectory of the operated object in the reference coordinate system is planned according to the welding task. The workpiece coordinate system which formed by the pipe connection is shown in Figure 3, the weld formed by the pipe is a saddle-shaped curve.

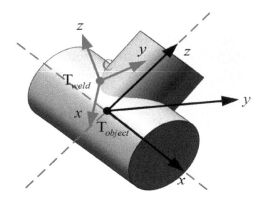

Figure 3. The schematic diagram of the object coordinate system and welding coordinate system.

The principle of establishing the reference coordinate system of the operated object is, the intersection point of the pipe center-line is set as the origin of the coordinate system, the z-axis is along the center-line of the upper tube and the x-axis is along the center-line of the lower tube. In Figure 3, C is a point on the saddle-shaped curve, and the coordinate in the reference coordinate system of the operated object is $C(x, y, z)$. It can use the parametric equation to represent the equation of the saddle-shaped curve in the reference coordinate system as shown in Equation (3).

$$
\begin{cases}
x = r \cdot \cos \theta \\
y = r \cdot \sin \theta \\
z = \sqrt{R^2 - r^2 \cdot \sin^2 \theta}
\end{cases}
,
\tag{3}
$$

where r is the radius of the upper tube R_b, R is the radius of the lower tube R_a, and θ is the rotation angle parameter.

The solder point C can be adjusted to the ship-type welding posture by two-step, the weld coordinate system rotates α around the z_{object}, and rotates β around the x_{object}. Where $\beta \in (0, 2\pi)$, α is the angle between z_{weld} and the reference coordinate system z_o of the operated object. The formula is shown in Equation (4).

$$
\alpha = \cos^{-1} \frac{\sqrt{R^2 - r^2 \cdot \sin^2 \theta}}{R \cdot \sqrt{2 + 2 \cdot \frac{r}{R} \cdot \sin^2 \theta}}.
\tag{4}
$$

Based on the above analysis, the trajectory of the operated object in its reference coordinate system can be obtained as shown in Equation (5).

$$
\begin{cases}
x = 0 \\
y = 0 \\
z = 0 \\
R = f(2\pi \cdot i \cdot t / T) \\
P = 0 \\
Y = 2\pi \cdot i \cdot t / T
\end{cases}
,
\tag{5}
$$

where i denotes the i-th communication cycle, t denotes the total time which is required for the welding task, and T denotes the interpolation period. $f(\cdot)$ is a variable attitude function, it can be expressed as Equation (6).

$$
f(\theta) = \cos^{-1} \frac{\sqrt{R^2 - r^2 \sin^2 \theta}}{R \sqrt{2 + 2 \frac{r}{R} \sin^2 \theta}}.
\tag{6}
$$

In the third step, the trajectory of each robot is planned according to the constraint relationship between the robot's end-effector and the operated object. The movement of the operated object m (assumed coincident with the T_{object}) in T_o can be described as $T_m^o(t)$. $T_m^o(t)$ is obtained according to the welding task (Equation (5)), and it can be expressed as Equation (7).

$$
T_m^o(t) = \begin{bmatrix} R_m^o(t) & p_m^o(t) \\ 0 & 1 \end{bmatrix},
\tag{7}
$$

where $R_m^o(t)$ represents the (3×3) rotation matrix of the centroid with respect to the reference coordinate system of the operated object, and $p_m^o(t)$ represents the (3×1) position matrix of the centroid with respect to the reference coordinate system of the operated object.

According to the closed-chain constraint formed by the transfer robot and the operated object, the pose constraint relationship of the following Equation (8) can be obtained.

$$
\begin{cases}
T_{bi}^w \cdot T_i^{bi}(q_i) \cdot T_m^i = T_m^w \\
T_o^w \cdot T_m^o(t) = T_m^w
\end{cases}
,
\tag{8}
$$

where $\boldsymbol{T}_{bi}^w$ is homogeneous transform representing the robot base frame $\boldsymbol{T}_{bi}$ with respect to the world frame $\boldsymbol{T}_w$. $\boldsymbol{T}_i^{bi}(\boldsymbol{q}_i)$ is homogeneous transform representing the end-effector frame of robot $\boldsymbol{T}_i$ with respect to its base frame $\boldsymbol{T}_{bi}$. $\boldsymbol{T}_m^i$ is homogeneous transform representing the mass frame of object $\boldsymbol{T}_m$ with respect to the end-effector frame of robot $\boldsymbol{T}_i$. $\boldsymbol{T}_m^w$ is homogeneous transform representing the mass frame of object $\boldsymbol{T}_m$ with respect to the world frame $\boldsymbol{T}_w$. $\boldsymbol{T}_o^w$ is homogeneous transform representing the object frame $\boldsymbol{T}_o$ with respect to the world frame $\boldsymbol{T}_w$.

From Equation (8), the kinematics of the i-th manipulator can be obtained as

$$
\begin{aligned}
\boldsymbol{T}_i^{bi}(\boldsymbol{q}_i) &= (\boldsymbol{T}_{bi}^w)^{-1} \cdot \boldsymbol{T}_o^w \cdot \boldsymbol{T}_m^o(t) \cdot (\boldsymbol{T}_m^i)^{-1} \\
&= \boldsymbol{T}_w^{bi} \cdot \boldsymbol{T}_o^w \cdot \boldsymbol{T}_m^o(t) \cdot \boldsymbol{T}_i^m,
\end{aligned}
\tag{9}
$$

where $\boldsymbol{T}_w^{bi}$, $\boldsymbol{T}_o^w$ and $\boldsymbol{T}_i^m$ are constant matrix.

In the entire cooperative welding task, the transfer robots coordinate the workpiece to meet the requirements of the ship-type welding. The welding robot doesn't need to adjust the posture during the whole welding process, only the position of the welding torch in the operated object coordinate system needs to adjust. The position transformation matrix p_3^o of the tip of the welding torch which relative to the reference coordinate system of the operated object can be expressed as Equation (10).

$$
p_3^o = \begin{pmatrix}
r \cdot \cos\alpha \cdot \cos\beta \cdot \cos\theta - r \cdot \cos\alpha \cdot \cos\beta \cdot \sin\theta + \sin\alpha \cdot \sqrt{R^2 - r^2 \cdot \sin^2\theta} \\
r \cdot \sin\beta \cdot \cos\theta + r \cdot \cos\alpha \cdot \cos\theta \\
-r \cdot \sin\alpha \cdot \cos\beta \cdot \cos\theta + r \cdot \sin\alpha \cdot \sin\beta \cdot \sin\theta + \cos\alpha \cdot \sqrt{R^2 - r^2 \cdot \sin^2\theta}
\end{pmatrix}.
\tag{10}
$$

According to the conversion relationship between the welding robot and the operated object, and the coordinate transform between the welding robot and the world coordinate system, Equation (11) can be obtained.

$$
\begin{cases}
\boldsymbol{T}_{b3}^w \cdot \boldsymbol{T}_3^{b3}(\boldsymbol{q}_i) = \boldsymbol{T}_3^w \\
\boldsymbol{T}_o^w \cdot \boldsymbol{T}_3^o(t) = \boldsymbol{T}_3^w
\end{cases}
\tag{11}
$$

According to the Equation (11), the motion of the welding robot relative to its own coordinate system can be obtained as in Equation (12).

$$
\boldsymbol{T}_3^{b3}(\boldsymbol{q}_i) = \boldsymbol{T}_3^{b3} \cdot \boldsymbol{T}_w^w \cdot \boldsymbol{T}_3^o(t).
\tag{12}
$$

According to Equations (9) and (11), the trajectory of the transfer robot and the welding robot in their own coordinate system can be obtained. Based on the solution formula of inverse kinematics, the trajectory of each robot's joint angles can be obtained.

4. Symmetrical Adaptive Variable Impedance Control for Multi-Robot Coordination

Multi-robot trajectory planning is the foundation for the cooperative welding process, because there is no generalized wrench constraint between the welding robot and the transfer robot, so the multi-robot position/force coordination control problems can be converted to the coordination between the transfer robots.

4.1. Problem Formulation

When the transfer robots are operating with a common object, the system becomes a closed-chain system. The closed-chain system which is constitute by the robots and the operated object is shown in Figure 4.

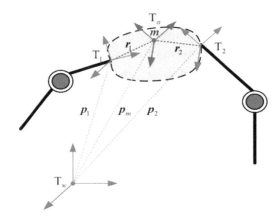

Figure 4. The diagram of the closed-chain system.

In Figure 4, $p_i(i = 1, 2)$, p_m denote the position vector of the robot end-effector with respect to the world coordinate system and the position vector of the operated object centroid with respect to the world coordinate system, respectively. $r_i(i = 1, 2)$ represents the position vector of the center of mass which is relative to the coordinate system of the robot. The movement of the operated object can be described by the center of mass m. And the Newton–Euler equation is shown as Equation (13).

$$\begin{cases} \boldsymbol{f}_m = \boldsymbol{M}\ddot{\boldsymbol{c}} - \boldsymbol{M}\boldsymbol{g} \\ \boldsymbol{n}_m = \boldsymbol{I}\dot{\omega} + \omega \times \boldsymbol{I}\omega \end{cases}, \tag{13}$$

where M and I are the mass and inertia matrices of the object, c and ω are the position and the angular velocity vector of the object, g is the acceleration of gravity, respectively.

If the movement of the operated object is known (as shown in Equation (5)). According to Equation (13), the wrench which the transfer robots need to exert on the center of operated object can be obtained. The wrench diagram of the force exerting on the operated object is shown in Figure 5.

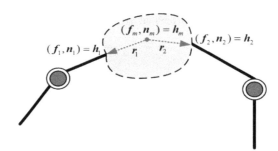

Figure 5. The wrench diagram of force exerting on the operated object.

The force formula of the transfer robots exerting on the operated object is shown in Equation (14).

$$\begin{cases} \boldsymbol{f}_m = \sum\limits_{i=1}^{2} \boldsymbol{f}_i \\ \boldsymbol{n}_m = \sum\limits_{i=1}^{2} \boldsymbol{n}_i + \sum\limits_{i=1}^{2} \boldsymbol{r}_i \times \boldsymbol{f}_i \end{cases} \tag{14}$$

where f_i and n_i represent the force and moment of the i-th robot end-effector exerting on the operated object, respectively.

According to the concept of "virtual chain", Equation (13) can be expressed in the form of Equation (15).

$$h_m = Wh \tag{15}$$

where $h_m = \begin{bmatrix} f_m \\ n_m \end{bmatrix}$, $h = [f_1^T, n_1^T, f_2^T, n_2^T]^T$, $W = \begin{bmatrix} I_3 & 0_3 & I_3 & 0_3 \\ S(r_1) & I_3 & S(r_2) & I_3 \end{bmatrix}$, $S(r_i) = \begin{bmatrix} 0 & -r_{iz} & r_{iy} \\ r_{iz} & 0 & -r_{ix} \\ -r_{iy} & r_{ix} & 0 \end{bmatrix}$, $r_i = [r_{ix} r_{iy} r_{iz}]^T$.

In Equation (15), W denotes the grip matrix and h denotes the wrench matrix which is exerting on the contact point of the operated object. If the wrench is known, the wrench which is exerting on the centroid point of the operated object can be obtained according to the Equation (15). But the actual situation is the inverse problem of Equation (15), it can be attributed to the load distribution problem.

From Equation (15), we can see that once the wrench h_m at the center of the operated object is known, the wrench h which the transfer robots need to exert on the contact point can be obtained by solving the pseudo-inverse matrix. Since W is a row full rank matrix, theoretically there are infinitely many solutions for h. According to the conclusion in [20], the general form of the following equation is obtained.

$$h = W^\dagger h_m + (I - W^\dagger W)\varepsilon, \tag{16}$$

where $W^\dagger = AW^T(WAW^T)^{-1}$, A is a positive definite matrix and ε is an arbitrary vector.

According to the conclusion in [25], Equation (16) can be converted to the form of Equation (17).

$$h = W^\dagger h_m + Vh_i, \tag{17}$$

where h_i indicates the internal force at the center of mass of the operated object, which can be set according to the actual requirements. According to [26,27], $W^\dagger$ and V can be selected as shown in Equations (18) and (19), respectively.

$$W^\dagger = \frac{1}{2}\begin{bmatrix} I & 0 \\ -S(r_1) & I \\ I & 0 \\ S(r_2) & I \end{bmatrix}, \tag{18}$$

$$V = \begin{bmatrix} I & 0 \\ -S(r_1) & I \\ -I & 0 \\ S(r_2) & -I \end{bmatrix}. \tag{19}$$

In general, the external forces h_m and the internal forces h_i are given quantity. According to Equation (17), the wrench which needs to exert on the single robot can be obtained.

Although the resultant force of the transfer robots need to exert on the contact point of the object can be operated by the load distribution according to Equation (17). However, when there is the trajectory deviation existing which caused by external distribution or calibration error. The trajectory deviation will affect the motion of the operated object, but also affect the internal force between the transfer robots. If only the resultant force is simply tracked, not only the target trajectory can't be tracked, but also failing to track the desired internal force.

4.2. Symmetrical Adaptive Variable Impedance Control for Coordination Control

A symmetrical adaptive variable impedance position/force coordination strategy is proposed to solve the influence of the unknown trajectory deviation which is caused by the external disturbance forces. The basic idea is to first decompose the resultant forces into the internal and external force which are exerting on the contact points of the operated object. And ideally, the desired internal and external force can be obtained. Consider the disturbance force which is caused by the trajectory deviation is dynamically changing, so the adaptive variable impedance is proposed to track the desired position and force.

The first step is to decompose the internal and external force. The desired resultant force of the transfer robots need to exert on the contact point of the operated object is given as shown Equation (17). In order to achieve tracking the desired external force and internal force of the operated object by the transfer robots, Equation (17) can be expressed as a form of force exerting on the contact point of the operated object with transfer robots, it is shown as Equation (20).

$$h = h_E + h_I. \tag{20}$$

where h_E and h_I represent the external force and internal force exerting on the contact point of the operated object with the transfer robots, respectively. They are meet the following equation.

$$\begin{cases} h_E = W^\dagger h_m \\ h_I = V h_i \end{cases}. \tag{21}$$

According to the Equation (21), the proposed internal and external force can be known.

In actual control, the wrench of the transfer robots exerting on the contact point of the operated object can be detected by a six-dimensional force/torque sensor which is installed at the end-effector of the robot, and the wrench can be denoted by h_r. It is further possible to decompose h_r into the external forces and internal forces as shown in Equation (22).

$$\begin{cases} h_{Er} = W^\dagger W h_r \\ h_{Ir} = (I - W^\dagger W) h_r \end{cases}, \tag{22}$$

where h_{Er} and h_{Ir} represent the actual values of the external force and the internal force which are resolved based on the measured values by the six-dimensional force/torque sensor.

In the second step, position/force coordination control is based on symmetrical adaptive variable impedance. The schematic diagram of multi-robot position/force coordination control based on a symmetrical adaptive variable impedance is shown in Figure 6.

From Figure 6, a symmetric internal and external impedance coordination strategy is used for the transfer robots, and the position control strategy is used for the welding robot to just follow the object's motion. For the transfer robots, the inner impedance controller is composed of transfer robots and the operated object, and the outer impedance controller is composed of the operated object and the environment (external disturbance or external force caused by trajectory deviations). In the actual control, a symmetrical coordination method is adopted to convert the internal and the external force which is exerted on the contact point of the operated object at the end-effector of the single robot. The purpose of the inner impedance controller is to track the desired internal force and modify the movement trajectory of the end of the transfer robot based on the force deviation. The purpose of the outer impedance controller is to track the desired external force and modify the movement trajectory of the operated object according to the force deviation. Then the transfer robots and welding robot update the respective movement trajectories according to the corrected trajectory of the operated object.

Considering that the desired internal forces, external forces, and the uncertain wrench which is caused by external disturbances are time-varying functions, an adaptive variable impedance control strategy is adopted. Therefore, the system is divided into an adaptive variable impedance inner loop controller and an adaptive variable impedance outer loop controller. The diagram of symmetrical internal and external adaptive impedance control is shown in Figure 7.

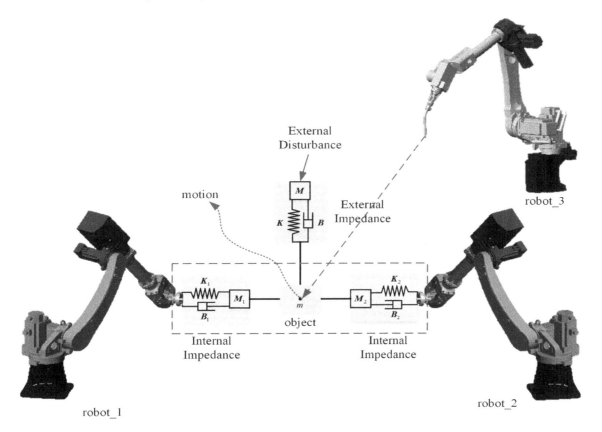

Figure 6. The schematic diagram of multi-robot position/force coordination control.

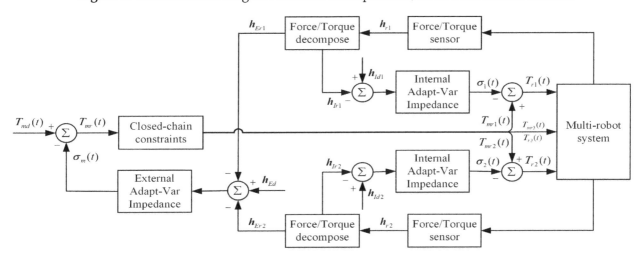

Figure 7. The diagram of symmetrical internal and external adaptive impedance control.

As shown in Figure 7, the desired movement trajectory of the operated object which inputs by the system is $T_{md}(t)$. In the ideal conditions, it exists $T_{md}(t) = T_{mr}(t)$. According to the closed-chain constraint conditions, the trajectories $(T_{mr1}(t), T_{mr2}(t), T_{mr3}(t))$ of multi-robot can be obtained, where $T_{mr1}(t) = T_{r1}(t)$, $T_{mr2}(t) = T_{r2}(t)$, $T_{mr3}(t) = T_{r3}(t)$. When uncertainty forces caused by the trajectory deviations exist, they can be obtained by the two six-dimensional force/torque sensor which are denoted by h_{r1} and h_{r2}. According to Equation (22), the measurement of the resultant forces can be decomposed into external forces(h_{Er1}, h_{Er2}) and internal forces(h_{Ir1}, h_{Ir2}), respectively.

When there is a force deviation between the measured external force and the desired external force h_{Ed}, the system will obtains an error trajectory $\sigma_m(t)$ through the adaptive variable impedance external loop controller, then the corrected trajectory $T_{mr}(t)$ of the operated object is obtained, and the trajectory of each robot is further corrected. When there is a force deviation between the measured internal force and the desired internal force (h_{Id1} and h_{Id2}), the system obtains error trajectories ($\sigma_1(t)$ and $\sigma_2(t)$) through the adaptive variable impedance inner loop controller, and then transfer robots's correction trajectories ($T_{r1}(t)$ and $T_{r1}(t)$) can be obtained.

The adaptive variable external impedance law is proposed as Equation (23), as shown in Figure 7.

$$
\begin{cases}
M[\ddot{T}_{mr}(t) - \ddot{T}_{md}(t)] + [B + \Delta B(t)][\dot{T}_{mr}(t) - \dot{T}_{md}(t)] = h_{Er1}(t) + h_{Er2}(t) - h_{Ed}(t) \\[2mm]
\Delta B(t) = \dfrac{B}{\dot{T}_{mr}(t) - \dot{T}_{md}(t)}\Phi(t) \\[2mm]
\Phi(t) = \Phi(t - \lambda) + \sigma \dfrac{h_{Ed}(t - \lambda) - h_{Er1}(t - \lambda) - h_{Er2}(t - \lambda)}{B}
\end{cases}
\tag{23}
$$

$$
\begin{cases}
M_I[\ddot{T}_{ri}(t) - \ddot{T}_{mri}(t)] + [B_I + \Delta B_I(t)][\dot{T}_{ri}(t) - \dot{T}_{mri}(t)] = h_{Iri}(t) - h_{Idi}(t) \\[2mm]
\Delta B_I(t) = \dfrac{B_I}{\dot{T}_{ri}(t) - \dot{T}_{mri}(t)}\Phi(t) \\[2mm]
\Phi(t) = \Phi(t - \lambda) + \sigma \dfrac{h_{Idi}(t) - h_{Iri}(t)}{B_I}
\end{cases}
\tag{24}
$$

where M is the desired inertia matrix, B is the damping matrix, λ is the sampling period of the controller and σ is the update rate.

The adaptive variable internal impedance law showns in Equation (7) is expressed as Equation (24).

Where M_I denotes the desired inertia matrix of the internal wrench, B_I denotes the damping matrix of the internal wrench, and i denotes i-th robot.

In our previous work [28], the adaptive variable impedance control has been proven stable and convergent and been used to track the dynamic force with unknown trajectory deviations.

5. Simulations and Experiments

5.1. Simulation

This section mainly verifies the feasibility of position/force coordination for multi-robot based on the symmetrical adaptive variable impedance control. Matlab SimMechanics was used to simulate the multi-robot cooperative welding. In order to verify the effectiveness of the algorithm, the simulation is close to the actual physical experiment. During the experiment, it is assumed that there is a certain expected pressure and external disturbances, the purpose is to test the tracking effect of external and internal forces.

The object is composed into two rigid pipes, R_a and R_b, respectively. The radius $r_a = 0.051$ m, $r_b = 0.0445$ m, the length $l_a = 0.12$ m, $l_b = 0.08$ m, respectively. The thickness of the pipe is $h = 0.003$ m, and the density is 1000 kg/m^3. The center of the 1st robot's end-effector and the center of pipe R_a coincide the center of the 2nd robot's end-effector and the centerline of pipe R_b is coincide.

The optimal initial welding position [0.5 m -0.75 m 0.404 m] can be obtained by solving Equation (2) through genetic algorithm, the optimization process of genetic algorithm is shown in Figure 8. According to the requirements of the welding task, the variable pose trajectory of the operated object is shown as Equations (5) and (6). Furthermore, the motion of each robot with respect to its respective base coordinate system can be solved according to Equations (9) and (12). Assuming that the desired internal wrench is [0 N 0 N 15 N 0 N·m 0 N·m 0 N·m], the desired internal force is among z-axis. The external disturbance is operated among x and y axis are shown in Figure 9.

The schematic diagram of the system simulation process for multi-robot completing pipe welding is shown in Figure 10.

The desired motion trajectory of the operated object and the actual motion trajectory after the external disturbance in x-axis and y-axis are shown in Figure 11, and the internal force tracking effect of the transfer robots is shown in Figure 12.

Combining Figures 11 and 12, it shows that after the external disturbance exerting on the x and y axis, the closed-chain system is flexible. From Figure 12, it shows that during the coordinated welding process, the tracking of the desired internal force can be achieved, and the internal force is maintained within the allowable tolerance range throughout the entire process.

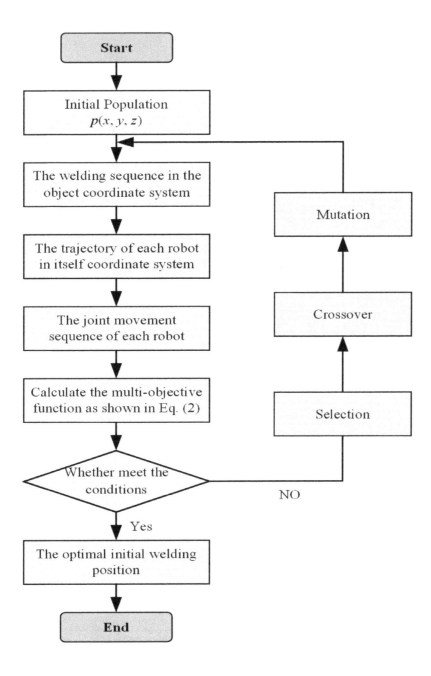

Figure 8. The optimization process of genetic algorithm.

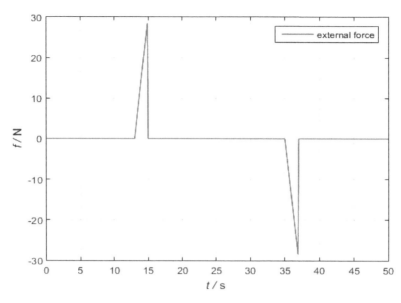

Figure 9. The external disturbance exerting on the operated object.

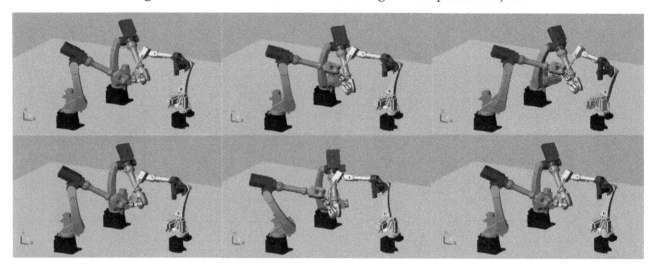

Figure 10. The schematic diagram of the system simulation process for multi-robot systems.

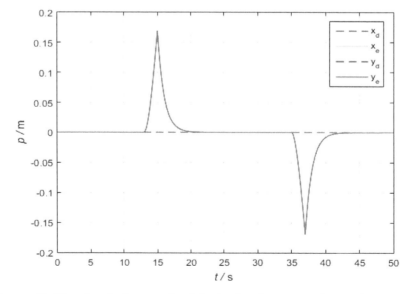

Figure 11. The simulation result of desired trajectory and actual trajectory.

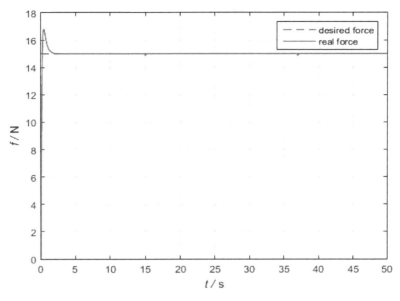

Figure 12. The simulation result of the internal force tracking effect.

5.2. Experimental Studies

To demonstrate the performance of the proposed algorithm, experiments were conducted using the test-bed as shown in Figure 13. And the logical scheme of the control software is shown in Figure 14.

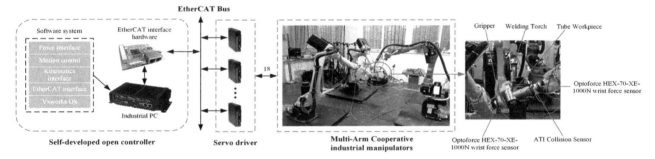

Figure 13. Hardware architecture of the test-bed.

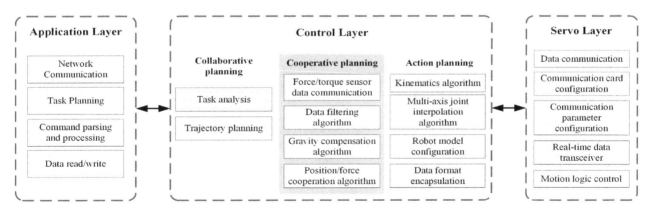

Figure 14. The logical scheme of the control software.

The test platform consisted of aself-developed open controller, servo drivers, two ESTUN ER16 industrial manipulators, one ESTUN ER4 industrial manipulators, two force/torque sensors, a collision sensor, and two pairs of gripper. The self-developed open controller used an industrial PC with a configuration of Intel Celeron @1.2 GHz, 512 MB of RAM, VxWorks RTOS. The servo drivers used ESTUN ProNet series, the bandwidth of the servo driver is from 125 Hz to 1000 Hz. Hischer CIFX communication card is used to EtherCAT communication between industrial PC and servo drivers. The force/torque sensors use Optoforce HEX-70-XE-1000N, the collision sensor uses ATI SR-61. The self-developed open controller was used for task coordination, position/force control, motion planning, forward/inverse kinematics, 18-axis cycle synchronization interpolation and human-machine interface. The force/torque sensor was mounted at the wrist of each manipulator. An ATI collision sensor was installed between the end-effector and the gripper of the 1-*st* robot was added for the protection of the whole system. The force/torque sensor provided the UDP protocol with the fastest frequency of 1 kHz, so the force sensor and the controller communicate through UDP. Consider the controller computing power, the bandwidth of the servo drivers and communication frequency of the force sensor at the same time, the communication cycles of the controller and the servo drivers, the controller and the force sensor are both set to 5 ms. Two force sensors were initialized by gravity compensation.

The physics experiment was consistent with the simulation. The transfer robots separately grasp a part of the workpiece to be welded, which was spliced at the initial welding point. The welding robot started arcing at the initial welding point. Then the transfer robots coordinated the workpiece to be displace, and the welding robot completed the welding of the weld seam. The key frames of the whole welding task is shown in Figure 15.

Figure 15. The key frames of the whole welding task.

During the welding process, due to the presence of unknown factors such as mechanical calibration error, base coordinate calibration error, and external disturbance, the transfer robots produced an uncertain wrench to the welding workpiece. The internal wrench exerting on the end-effector without force control is shown in Figure 16.

From the above results, we can see that the internal wrench without the force control is large. In order to control the internal forces in a proper range, a symmetrical adaptive variable impedance control is proposed in this paper. To certify the performance of the proposed algorithm, the traditional constant impedance control and the proposed algorithm are compared as shown in Figure 17.

Figure 17 shows that the force control effect has been significantly improved. The desired trajectory of transfer robots' end-effector and the center of workpiece are shown in Figure 18.

From Figure 17, we can conclude that the variable adaptive impedance control can achieve a better effect than the traditional constant impedance control. By using the proposed algorithm, the trajectory deviations of the transfer robots are shown in Figure 19, the trajectory deviations of the welding robot is shown in Figure 20.

Through Figures 17–20, it shows that the internal force of the transfer robots exerting on the workpiece to be welded is within the controllable range and does not cause damage to the welding system. The welding result has been shown in Figure 21. It indicates that a uniform smooth weld seam without cracks has been got by our method. And the welding quality is much better compared with novice welders.

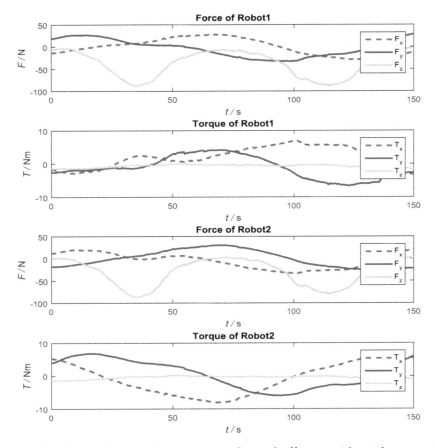

Figure 16. The internal wrench exerting on the end-effector without force control.

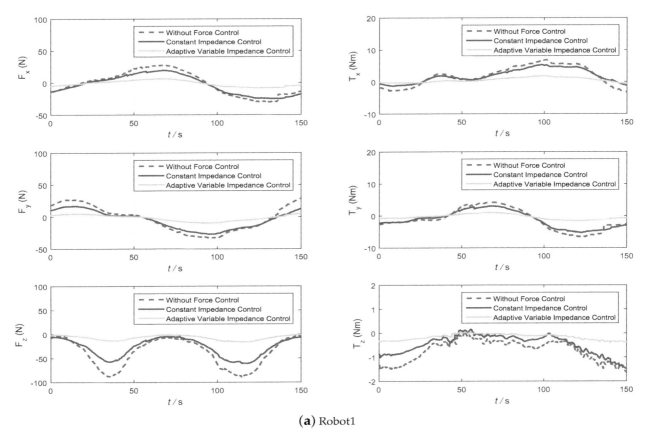

(**a**) Robot1

Figure 17. *Cont.*

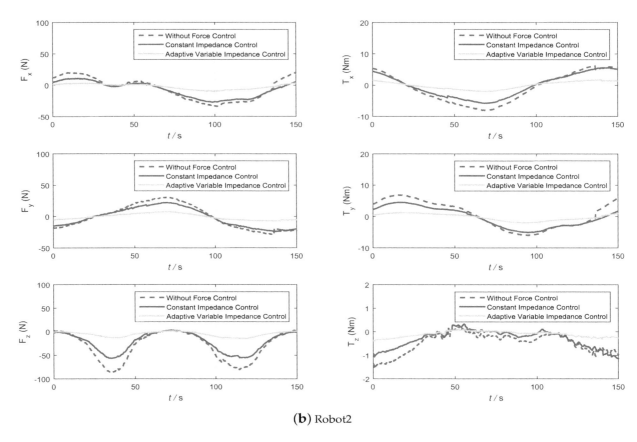

(**b**) Robot2

Figure 17. Comparison of two algorithm results.

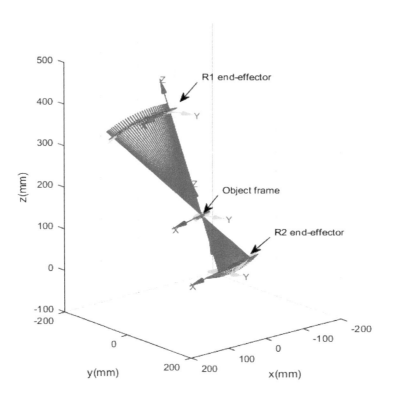

Figure 18. The desired trajectory of transfer robots' end-effector and the center of workpiece.

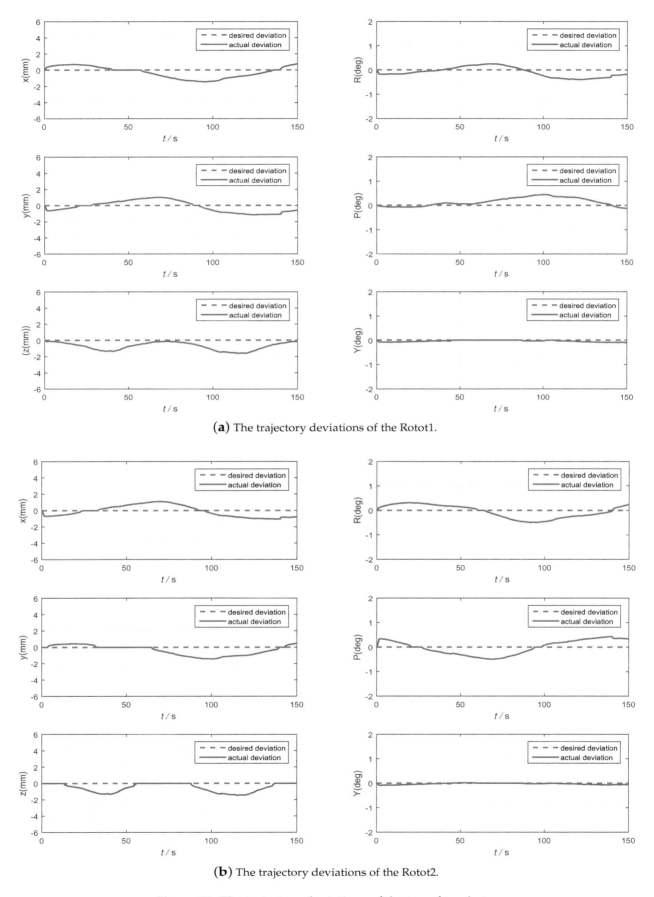

(**a**) The trajectory deviations of the Rotot1.

(**b**) The trajectory deviations of the Rotot2.

Figure 19. The trajectory deviations of the transfer robots.

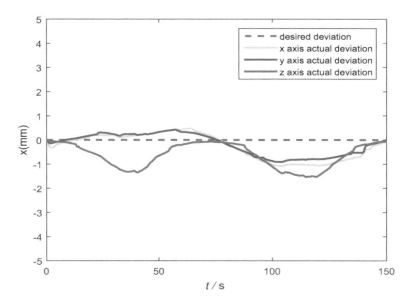

Figure 20. The trajectory deviations of the welding robot.

Figure 21. Pipe-connect-pipe arc welding result.

In the simulations and experiments, the selection method of the design parameters of the control system (M, B, λ and σ) can be referenced in [28]. The following parameters can obtain the good performance in the above simulations and experiments, the inertia coefficient $M = diag\{1,1,1,0,0,0\}$, $M_I = diag\{1,1,1,0,0,0\}$, the initial damping coefficient $B = diag\{65,65,65,0,0,0\}$, $B_I = diag\{90,90,90,0,0,0\}$, $\lambda = 0.005s$ and $\sigma = 0.01$.

6. Conclusions

In the actual multi-robot cooperative welding system, the trajectory planning and position/force coordination control are the two most critical issues. The trajectory planning is the basis of position/force coordination control. The biggest difference between multi-robot cooperative welding and other types of multi-robot cooperative tasks is that they have more constraints and the coordinate system is more cumbersome and complicated. During the welding of the workpiece, an unknown changing wrench can be generated between the transfer robots and the workpiece due to the external disturbance and calibration errors.

In the face of problems such as complex changes of coordinate system and welding constraints in the process of multi-robot cooperative welding, a planning strategy for "hierarchical planning" is proposed in this paper. Firstly, the optimal initial welding position is determined according to the optimization index, and then the trajectory of the operated object in its coordinate system is planned. Finally, the trajectory of each robot relative to its base coordinate system is obtained.

In actual control systems, due to the calibration errors and external disturbances, the trajectory of robot end-effector is often deviated. In response to this problem, a symmetric adaptive internal and external variable impedance control strategy is used to track the internal and external forces exerting on the operated object. Through simulation and physical tests, it is concluded that the proposed control strategy is applicable to multi-robot cooperative welding systems and can effectively solve the two problems in the process of cooperative welding.

Author Contributions: Methodology, Y.G., J.D.; validation and data curation, M.C., J.D.; writing—original draft preparation, Y.G., J.D.; writing—review and editing, Y.G.; visualization, M.C.; supervision, Y.G.; project administration, X.D.; funding acquisition, Y.G., X.D.

References

1. Mason, M.T. Compliance and Force Control for Computer Controlled Manipulators. *IEEE Trans. Syst. Man Cybern.* **1981**, *11*, 418–432. [CrossRef]
2. Zheng, Y.F.; Luh, J.Y.S. Control of two coordinated robots in motion. In Proceedings of the 1985 24th IEEE Conference on Decision and Control, Fort Lauderdale, FL, USA, 11–13 December 1985; pp. 1761–1766.
3. Luh, J.Y.S.; Zheng, Y.F. Constrained Relations between Two Coordinated Industrial Robots for Motion Control. *Int. J. Robot. Res.* **1987**, *6*, 60–70. [CrossRef]
4. Nagai, K.; Iwasa, S.; Watanabe, K.; Hanafusa, H. Cooperative control of dual-arm robots for reasonable motion distribution. In Proceedings of the 1995 IEEE/RSJ International Conference on Intelligent Robots and Systems. Human Robot Interaction and Cooperative Robots, Pittsburgh, PA, USA, 5–9 August 1995; p. 54.
5. Zhou, B.; Xu, L.; Meng, Z.; Dai, X. Kinematic cooperated welding trajectory planning for master-slave multi-robot systems. In Proceedings of the 2016 35th Control Conference, Chengdu, China, 27–29 July 2016; pp. 6369–6374.
6. Mcclamroch, N.H.; Wang, D. Feedback stabilization and tracking of constrained robots. *IEEE Trans. Autom. Control* **1988**, *33*, 419–426. [CrossRef]
7. Krishnan, H.; Mcclamroch, N.H. *Tracking in Nonlinear Differential-Algebraic Control Systems with Applications to Constrained Robot Systems*; Pergamon Press, Inc.: Oxford, UK, 1994.
8. Nakano, E. Cooperational Control of the Anthropomorphous Manipulator "MELARM". In Proceedings of the 4th International Symposium on Industrial Robots, Tokyo, Japan, 19–21 November 1974; pp. 251–260.
9. Barbieri, L.; Bruno, F.; Gallo, A.; Muzzupappa, M.; Russo, M.L. Design, prototyping and testing of a modular small-sized underwater robotic arm controlled through a Master-Slave approach. *Ocean Eng.* **2018**, *158*, 253–262. [CrossRef]
10. Scaradozzi, D.; Sorbi, L.; Zingaretti, S.; Biagiola, M.; Omerdic, E. Development and integration of a novel IP66 Force Feedback Joystick for offshore operations. In Proceedings of the 22nd Mediterranean Conference on Control and Automation, Palermo, Italy, 16–19 June 2014; pp. 664–669.
11. Hayati, S. Hybrid position/Force control of multi-arm cooperating robots. In Proceedings of the 1986 IEEE International Conference on Robotics and Automation, San Francisco, CA, USA, 7–10 April 1986; pp. 82–89.
12. Uchiyama, M.; Iwasawa, N.; Hakomori, K. Hybrid position/Force control for coordination of a two-arm robo. In Proceedings of the 1987 IEEE International Conference on Robotics and Automation, Raleigh, NC, USA, 31 March–3 April 1987; pp. 1242–1247.
13. Uchiyama, M.; Dauchez, P. A symmetric hybrid position/force control scheme for the coordination of two robots. In Proceedings of the 1988 IEEE International Conference on Robotics and Automation, Philadelphia, PA, USA, 24–29 April 1988; Volume 1, pp. 350–356.
14. Masaru, U.; Pierre, D. Symmetric kinematic formulation and non-master/slave coordinated control of two-arm robots. *Adv. Robot.* **1992**, *7*, 361–383.
15. Sun, D.; Mills, J.K. Adaptive synchronized control for coordination of multirobot assembly tasks. *IEEE Trans. Robot. Autom.* **2002**, *18*, 498–510.
16. Rodriguez-Angeles, A.; Nijmeijer, H. Mutual synchronization of robots via estimated state feedback: A cooperative approach. *IEEE Trans. Control Syst. Technol.* **2004**, *12*, 542–554. [CrossRef]

17. Lian, K.Y.; Chiu, C.S.; Liu, P. Semi-decentralized adaptive fuzzy control for cooperative multirobot systems with H-inf motion/internal force tracking performance. *IEEE Trans. Syst. Man Cybern. Part B Cybern.* **2002**, *32*, 269–280. [CrossRef] [PubMed]

18. Gueaieb, W.; Karray, F.; Al-Sharhan, S. A robust adaptive fuzzy position/force control scheme for cooperative manipulators. *IEEE Trans. Control Syst. Technol.* **2003**, *11*, 516–528. [CrossRef]

19. Gueaieb, W.; Karray, F.; Al-Sharhan, S. A Robust Hybrid Intelligent Position/Force Control Scheme for Cooperative Manipulators. *IEEE/ASME Trans. Mechatron.* **2007**, *12*, 109–125. [CrossRef]

20. Walker, I.D.; Freeman, R.A.; Marcus, S.I. Analysis of Motion and Internal Loading of Objects Grasped by Multiple Cooperating Manipulators. *Int. J. Robot. Res.* **1991**, *10*, 396–409. [CrossRef]

21. Bonitz, R.G.; Hsia, T.C. Force decomposition in cooperating manipulators using the theory of metric spaces and generalized inverses. In Proceedings of the 1994 IEEE International Conference on Robotics and Automation, San Diego, CA, USA, 8–13 May 1994; Volume 2, pp. 1521–1527.

22. Leidner, D.; Dietrich, A.; Schmidt, F.; Borst, C.; Albu-Schäffer, A. Object-centered hybrid reasoning for whole-body mobile manipulation. In Proceedings of the IEEE International Conference on Robotics and Automation, Hong Kong, China, 31 May–7 June 2014; pp. 1828–1835.

23. Dietrich, A.; Wimbock, T.; Albu-Schaffer, A. Dynamic whole-body mobile manipulation with a torque controlled humanoid robot via impedance control laws. In Proceedings of the IEEE/RSJ International Conference on Intelligent Robots and Systems, San Francisco, CA, USA, 25–30 September 2011; pp. 3199–3206.

24. Ott, C.; Hirzinger, G. Comparison of object-level grasp controllers for dynamic dexterous manipulation. *Int. J. Robot. Res.* **2012**, *31*, 3–23.

25. Caccavale, F.; Chiacchio, P.; Marino, A.; Villani, L. Six-DOF Impedance Control of Dual-Arm Cooperative Manipulators. *IEEE/ASME Trans. Mechatron.* **2008**, *13*, 576–586. [CrossRef]

26. Chiacchio, P.; Chiaverini, S.; Sciavicco, L.; Siciliano, B. Global task space manipulability ellipsoids for multiple-arm systems. *IEEE Trans. Robot. Autom.* **1991**, *7*, 678–685. [CrossRef]

27. Erhart, S.; Hirche, S. Internal Force Analysis and Load Distribution for Cooperative Multi-Robot Manipulation. *IEEE Trans. Robot.* **2017**, *31*, 1238–1243. [CrossRef]

28. Duan, J.; Gan, Y.; Chen, M.; Dai, X. Adaptive variable impedance control for dynamic contact force tracking in uncertain environment. *Robot. Auton. Syst.* **2018**, *102*, 54–65. [CrossRef]

Permissions

All chapters in this book were first published in MDPI; hereby published with permission under the Creative Commons Attribution License or equivalent. Every chapter published in this book has been scrutinized by our experts. Their significance has been extensively debated. The topics covered herein carry significant findings which will fuel the growth of the discipline. They may even be implemented as practical applications or may be referred to as a beginning point for another development.

The contributors of this book come from diverse backgrounds, making this book a truly international effort. This book will bring forth new frontiers with its revolutionizing research information and detailed analysis of the nascent developments around the world.

We would like to thank all the contributing authors for lending their expertise to make the book truly unique. They have played a crucial role in the development of this book. Without their invaluable contributions this book wouldn't have been possible. They have made vital efforts to compile up to date information on the varied aspects of this subject to make this book a valuable addition to the collection of many professionals and students.

This book was conceptualized with the vision of imparting up-to-date information and advanced data in this field. To ensure the same, a matchless editorial board was set up. Every individual on the board went through rigorous rounds of assessment to prove their worth. After which they invested a large part of their time researching and compiling the most relevant data for our readers.

The editorial board has been involved in producing this book since its inception. They have spent rigorous hours researching and exploring the diverse topics which have resulted in the successful publishing of this book. They have passed on their knowledge of decades through this book. To expedite this challenging task, the publisher supported the team at every step. A small team of assistant editors was also appointed to further simplify the editing procedure and attain best results for the readers.

Apart from the editorial board, the designing team has also invested a significant amount of their time in understanding the subject and creating the most relevant covers. They scrutinized every image to scout for the most suitable representation of the subject and create an appropriate cover for the book.

The publishing team has been an ardent support to the editorial, designing and production team. Their endless efforts to recruit the best for this project, has resulted in the accomplishment of this book. They are a veteran in the field of academics and their pool of knowledge is as vast as their experience in printing. Their expertise and guidance has proved useful at every step. Their uncompromising quality standards have made this book an exceptional effort. Their encouragement from time to time has been an inspiration for everyone.

The publisher and the editorial board hope that this book will prove to be a valuable piece of knowledge for researchers, students, practitioners and scholars across the globe.

List of Contributors

Qing Wu, Zeyu Chen, Lei Wang, Hao Lin, Zijing Jiang and Dechao Chen
School of Computer Science and Technology, Hangzhou Dianzi University, Hangzhou 310018, China

Shuai Li
Department of Computing, The Hong Kong Polytechnic University, Hung Hom, Kowloon, Hong Kong 999077, China

Xuyou Li, Shitong Du, Guangchun Li and Haoyu Li
College of Automation, Harbin Engineering University, Harbin 150001, China

Wojciech Kowalczyk
Institute of Automation and Robotics, Poznán University of Technology (PUT), Piotrowo 3A, 60-965 Poznán, Poland

Wojciech Giernacki
Institute of Control, Robotics and Information Engineering, Electrical Department, Poznan University of Technology, Piotrowo 3a Street, 60-965 Poznan, Poland

Oscar Alonso-Ramirez, Antonio Marin-Hernandez, Homero V. Rios-Figueroa and Ericka J. Rechy-Ramirez
Artificial Intelligence Research Center, Universidad Veracruzana, Sebastian Camacho No. 5, Xalapa 91000, Mexico

Michel Devy
CNRS-LAAS, Université Toulouse, 7 avenue du Colonel Roche, F-31077 Toulouse CEDEX, France

Saul E. Pomares-Hernandez
Department of Electronics, National Institute of Astrophysics, Optics and Electronics, Luis Enrique Erro No. 1, Puebla 72840, Mexico

Adrian Burlacu and Marius Kloetzer
Department of Automatic Control and Applied Informatics, "Gheorghe Asachi" Technical University of Iasi, 700050 Iasi, Romania

Cristian Mahulea
Department of Computer Science and Systems Engineering, University of Zaragoza, 50018 Zaragoza, Spain

Qiao Cheng, Xiangke Wang, Jian Yang and Lincheng Shen
College of Intelligence Science and Technology, National University of Defense Technology, Changsha 410073, China

Baifan Chen, Dian Yuan, Chunfa Liu and Qian Wu
School of Automation, Central South University, Changsha 410083, China

Guobin Wang, Yubin Lan, Haixia Qi, Pengchao Chen, Fan Ouyang and Yuxing Han
National Center for International Collaboration Research on Precision Agricultural Aviation Pesticides Spraying Technology (NPAAC), South China Agricultural University, Guangzhou 510642, China

Huizhu Yuan
State Key Laboratory for Biology of Plant Disease and Insect Pests, Institute of Plant Protection, Chinese Academy of Agricultural Sciences, Beijing 100193, China

Changxiang Fan
Department of Precision Engineering, Graduate School of Engineering, The University of Tokyo, 7-3-1 Hongo, Bunkyo-ku, Tokyo 113-8656, Japan

Shouhei Shirafuji and Jun Ota
Research into Artifacts, Center for Engineering, The University of Tokyo, 5-1-5 Kashiwanoha, Kashiwa-shi, Chiba 277-8568, Japan

Fei Wang, Yuqiang Liu, Ling Xiao, Chengdong Wu and Hao Chu
Faculty of Robot Science and Engineering, Northeastern University, Shenyang 110819, China

Yahui Gan, Jinjun Duan, Ming Chen and Xianzhong Dai
School of Automation, Southeast University, Nanjing 210096, China
Key Lab of Measurement and Control of Complex Systems of Engineering, Ministry of Education, Nanjing 210096, China

Index

Printed in the USA
CPSIA information can be obtained
at www.ICGtesting.com
JSHW051411091023
49903JS00006B/379